Principles of Physical Optics

Principles of Physical Optics

Albie Reid

Willford Press
www.willfordpress.com

Published by Willford Press,
118-35 Queens Blvd., Suite 400,
Forest Hills, NY 11375, USA

ISBN: 978-1-68285-737-3

Cataloging-in-Publication Data

Principles of Physical Optics / Albie Reid.
p. cm.
Includes bibliographical references and index.
ISBN 978-1-68285-737-3
1. Physical optics. 2. Optics. 3. Physics. I. Reid, Albie.
QC395.2 .P75 2019
535.2--dc23

For information on all Willford Press publications
visit our website at www.willfordpress.com

Willford Press

Printed in China.

Contents

Preface

Physical optics is a branch of optics that studies optical phenomena, such as polarization, interference, diffraction, dispersion, scattering, etc. All these phenomena cannot be explained by using the approximations of geometric optics. In physical optics, the wave property of light is considered. Light waves are electromagnetic waves. The interaction of light with materials involves the modeling of both electric and magnetic fields of the light wave. When the propagation of light in systems cannot be solved analytically, numerical modeling techniques are used. These include the boundary element method, finite element method and the transmission-line matrix method. This book is a valuable compilation of topics, ranging from the basic to the most complex theories and principles in the field of physical optics. It aims to shed light on some of the unexplored aspects of this field. It aims to serve as a resource guide for students and experts alike and contribute to the growth of the discipline.

A detailed account of the significant topics covered in this book is provided below:

Chapter 1- Physical optics is a branch of optics, which is concerned with the study of optical phenomena. This chapter will introduce briefly the optical phenomena under its scope of study, such as interference, diffraction and polarization as well as the methods of Block wave- MoM method, transfer-matrix method, Gerchberg-Saxton algorithm, etc.

Chapter 2- Geometrical optics is a branch of optics, which describes the propagation of light as rays. Some of the optical phenomena addressed under geometrical optics include reflection, refraction and total internal reflection. This chapter has been carefully written to provide an easy understanding of such phenomena. It also includes topics such as specular reflection, diffused reflection, signal reflection, reflection phase change and refractive index, besides others.

Chapter 3- Wave interference is the phenomenon in which two waves superimpose to generate a wave with a lower, greater or the same amplitude. The aim of this chapter is to explore the fundamental concepts of constructive wave interference, destructive wave interference, Young's interference experiment and thin-film interference.

Chapter 4- When a wave encounters a slit or an obstacle, the bending of light waves occur around the corners of the obstacle or aperture and penetrates into the geometrical shadow behind the obstacle. This phenomenon is called diffraction. This chapter explores the different forms of diffraction such as Fresnel diffraction, Fraunhofer diffraction and particle diffraction as well as the key concepts of diffraction grating, diffraction-limited system, Bragg law and diffraction.

Chapter 5- Polarization is a property characteristic of transverse waves that helps to specify the orientation of oscillations. It is a complex topic that requires understanding of Jones calculus, polarizers, Brewster's angle and birefringence, which have been extensively dealt within this chapter.

I would like to make a special mention of my publisher who considered me worthy of this opportunity and also supported me throughout the process. I would also like to thank the editing team at the back-end who extended their help whenever required.

Albie Reid

1

Fundamental Principles of Physical Optics

Physical optics is a branch of optics, which is concerned with the study of optical phenomena. This chapter will introduce briefly the optical phenomena under its scope of study, such as interference, diffraction and polarization as well as the methods of Block wave- MoM method, transfer-matrix method, Gerchberg-Saxton algorithm, etc.

Physical optics is the study of the physical properties and behavior of electromagnetic waves and their interactions with matter. The optical range of wavelengths may be taken as the range from about 1 nanometer (4×10^{-8} in.) to about 1 millimeter (0.04 in.). More narrowly, physical optics deals with the relationship between the atomic structure of a system and the absorption, emission, and propagation of light inside the system. The chief founder of this branch of science was Michael Faraday, who in 1845 provided the first clue to the electromagnetic nature of light by showing that the optical properties of glass could be altered by a magnetic field. Branches of physical optics include wave optics and Fourier optics. Fourier optics utilizes the mathematical technique of the Fourier transform to analyze the propagation of light and the formation of images.

PO is often used to compute the back scatter.

Review of Kirchoff-Huygens Principle

The position vectors of the source and the observer are expressed as rho' and ρ, respectively. The Green function G(ρ, ρ') for the scalar Helmholtz equation is defined in the following equation.

$$\nabla^2 G(\rho, \rho') + k^2 G(\rho, \rho') = -\delta(\rho - \rho').$$

By using $G(\rho, \rho')$ the solution $\phi(\rho)$ for the general inhomogeneous Helmholtz equation

$$\nabla^2 \phi(\rho) + k^2 \phi(\rho) = -\rho(\rho)$$

is expressed by using the Green function as

$$\phi(\rho) = \int_S G(\rho, \rho')\rho(\rho')\, dS' + \oint_{\partial S} \{G(\rho, \rho')\nabla'\phi(\rho') - \nabla' G(\rho, \rho')\phi(\rho')\}\, dn',$$

where S the domain under consideration, ∂S the boundary of S composed of a set of closed contours, and dn is the outward unit normal vector of ∂S. The solution of the Helmholtz equation is uniquely determined when (i) the source distribution and (ii) the wave function on the boundary.

The wave function on the boundary is, however, usually unknown. Therefore, it is the boundary element method that solves the boundary integral equation with respect to this unknown wave function.

Contrary in the physical optics, the wave function on the boundary is approximated by that for the infinite plane boundary which is a canonical solution.

Infinite Plane Boundary

The behavior of the wave function on the infinite plane boundary is first described as a canonical problem.

As shown in Fig., infinite plane boundary is considered at. $y = 0$

The 2D semi-infinite space problem with a unit point source at $(0, y')$ in the domain $y \geq 0$ is considered

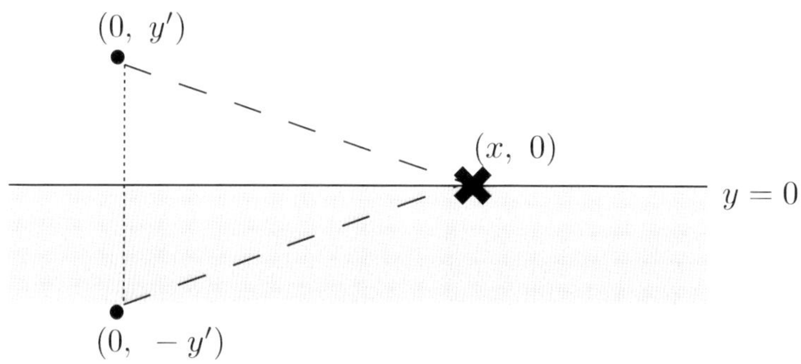

Figure: Treatment of infinite plane boundary

Free Space

In the free space, the wave function $\phi(\rho)$ generated from the unit point source at $(0, y')$ is expressed as

$$\phi(\rho) = G(\rho, \rho')$$

$$= \frac{1}{4j} = H_0^{(2)}(k\rho_1),$$

$$\rho_1 = \sqrt{x^2 + (y - y')^2}.$$

The derivative of the wave function along y, *i.e.* $\frac{\partial}{\partial y}\phi(\rho)$ is expressed as

$$\frac{\partial}{\partial y}\phi'(\rho) = \frac{jk}{4} H_1^{(2)}(k\rho_1)\frac{\partial \rho_1}{\partial y}$$

$$= \frac{jk}{4} H_1^{(2)}(k\rho_1)\ \frac{y - y'}{\rho_1}.$$

It is noted that

$$\frac{d}{dx} H_0^{(2)}(x) = -H_1^{(2)}(x)$$

is used.

Semi-infinite Space

Next, the infinite plane boundary is considered. To satisfy the boundary condition, the image method is applied. The boundary is removed first, and then an image source at $(0, -y')$ is considered so as to satisfy the boundary condition at y = 0. In the following discussion, the wave function due to the real source is denoted by $\phi_d(\rho)$and that due to the image source is denoted by $\phi_r(\rho)$,

Dirichlet Condition

To satisfy the Dirichlet condition

$$\phi(\rho) = 0, \quad y = 0,$$

the amplitude of the image source shall be −1. In case,

$$\begin{aligned}\phi(\rho) &= \phi_d(\rho) + \phi_r(\rho) \\ &= \frac{1}{4j} H_0^{(2)}(k\rho_1) - \frac{1}{4j} H_0^{(2)}(k\rho_2),\end{aligned}$$

where

$$\rho_2 = \sqrt{x^2 + (y+y')^2}.$$

It is clear that $\rho_1 = \rho_2 \overset{\text{def}}{=} \rho_0$ at, $y = 0$, then $\phi(\rho)\big|_{y=0} = 0$ is obtained. In case, the normal derivative $\frac{\partial}{\partial y}\phi(\rho)$ is expressed as

$$\begin{aligned}\frac{\partial}{\partial y}\phi(\rho)\bigg|_{y=0} &= \frac{\partial}{\partial y}\phi_d(\rho)\bigg|_{y=0} + \frac{\partial}{\partial y}\phi_r(\rho)\bigg|_{y=0} \\ &= \frac{jk}{4} H_1^{(2)}(k\rho_1)\frac{-y'}{\rho_0} - \frac{jk}{4} H_1^{(2)}(k\rho_2)\frac{y'}{\rho_0} \\ &= 2\frac{\partial}{\partial y}\phi_d(\rho)\bigg|_{y=0}\end{aligned}$$

where

$$\rho_0 = \sqrt{x^2 + y'^2}.$$

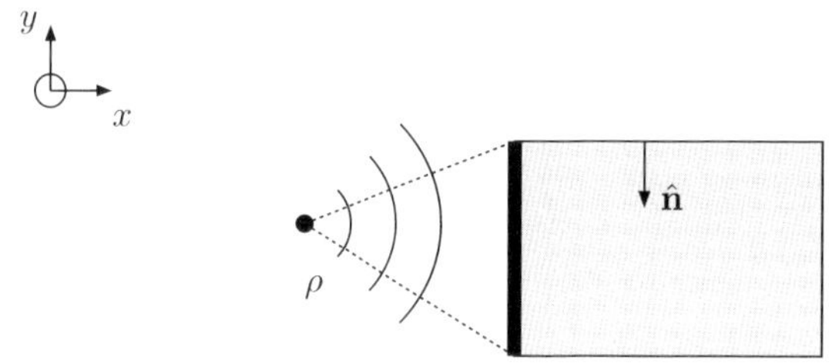

Figure : Physical Optics Approximation Model
An equivalent source is assumed only along the bold line

Neumann Condition

To satisfy the Neumann condition

$$\frac{\partial}{\partial y}\phi(\rho) = 0, \quad y = 0,$$

the amplitude of the image source shall be $+1$. In case,

$$\frac{\partial}{\partial y}\phi(\rho) = \frac{\partial}{\partial y}\phi_d(\rho) + \frac{\partial}{\partial y}\phi_r(\rho)$$

$$\frac{jk}{4}H_1^{(2)}(k\rho_1)\frac{-y'}{\rho_0} + \frac{jk}{4}H_1^{(2)}(k\rho_2)\frac{y'}{\rho_0}.$$

It is clear that $\rho_1 = \rho_2 \overset{\text{def}}{=} \rho_0$ at $y = 0$, then $\frac{\partial}{\partial y}\phi(\rho)\Big|_{y=0} = 0$ is obtained. In case, the wave function $\phi(\rho)$ is expressed as

$$\phi(\rho)\Big|_{y=0} = \phi_d(\rho)\Big|_{y=0} + \phi_r(\rho)\Big|_{y=0}$$

$$= \frac{1}{4j}H_0^{(2)}(k\rho_0) + \frac{1}{4j}H_0^{(2)}(k\rho_0)$$

$$= 2\phi_d(\rho)\Big|_{y=0}$$

Physical Optics Approximation

Although the equivalent source in the contour integral of Eq. $\phi(\rho) = \int_S G(\rho,\ \rho')\rho(\rho')dS' + \oint_{\partial S}\{G(\rho,\ \rho')\nabla'\phi(\rho') - \nabla'G(\rho,\ \rho')\phi(\rho')\}dn'$ is unknown, the same relation as is described in the previous section is applied to approximate the equivalent source by using the incident wave as

Dirichlet condition

$$\phi(\rho') \to 0,$$

$$\frac{\partial \phi(\rho')}{\partial n'} \rightarrow \begin{cases} 2\dfrac{\partial \phi^{inc}(\rho')}{\partial n'}, & \textit{visible point from the source, } 0, \textit{ otherwise.} \\ 0, & \end{cases}$$

Neumann condition

$$\phi(\rho') \rightarrow \begin{cases} 2\phi^{inc}(\rho'), & \textit{visible point from the source, otherwise,} \\ 0, & \end{cases}$$

$$\frac{\partial \phi(\rho')}{\partial n'} \rightarrow 0.$$

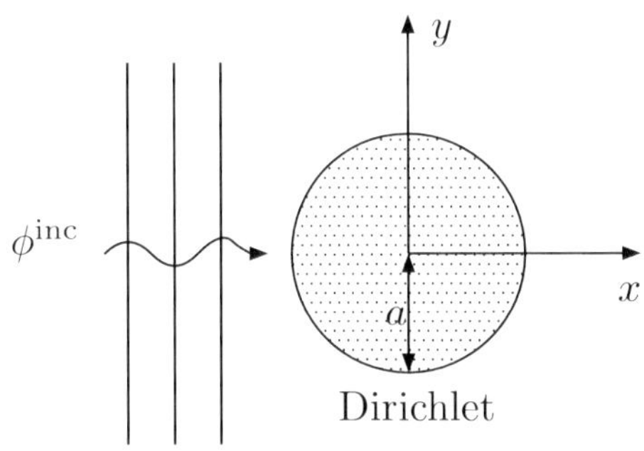

Figure: Plane wave scattering from infinite circular cylinder

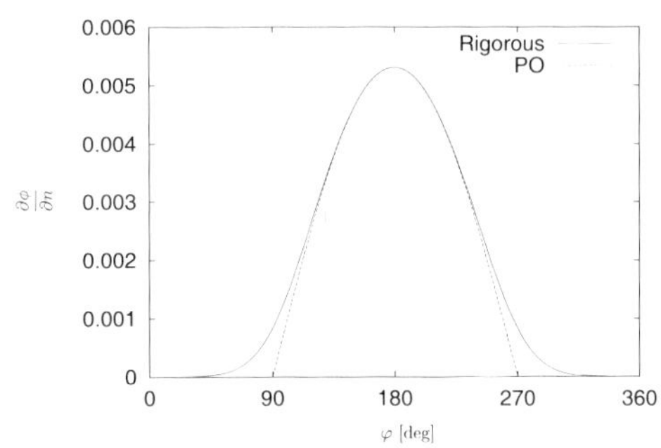

Figure: Equivalent secondary source on the scatterer (amplitude)

(a = 5λ, d = 100λ)

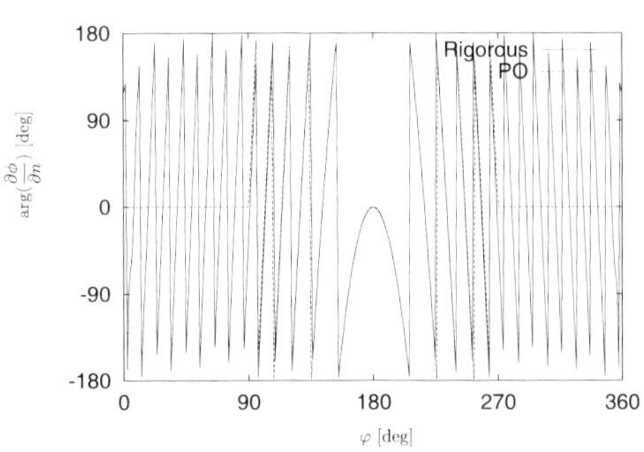

Figure: Equivalent secondary source on the scatterer (phase)

(a = 5λ, d = 100λ)

This approximation is called the physical optics approximation (PO).

As PO uses the geometrical optics incident wave to compute the scattered wave, it is classified as a high frequency technique. High frequency techniques are advantageous when the scatterer is large relative to the wavelength.

Properties of PO

- Different from geometrical optics approximation, the wave function is continuous at the reflection and the shadow boundaries. The same computation is applicable independent of the position of the observer.
- It is time consuming to integrate the equivalent source which exists on the visible region of the boundary. However, the contribution from reflection and diffraction points can be extracted by using the stationary phase method.
- The error is larger outside the specular reflection region, especially in the shadow region. This is due to the inaccuracy of the equivalent source at the diffracting edges.

Numerical Example

As a 2D example, the scattering wave from the infinitely long circular cylinder with the Dirichlet condition for the +x-propagating plane wave incidence is considered. The analysis model is shown in Fig. Since the problem of Fig. can be solved analytically by using eigenfunction expansion method, the accuracy of the PO approximation can be evaluated by comparison.

Figures compares respectively the amplitude and the phase of the rigorous and PO equivalent secondary sources on the boundary. The rigorous solution has nonzero amplitude in the shadowed region, but the PO approximation is zero in the region.

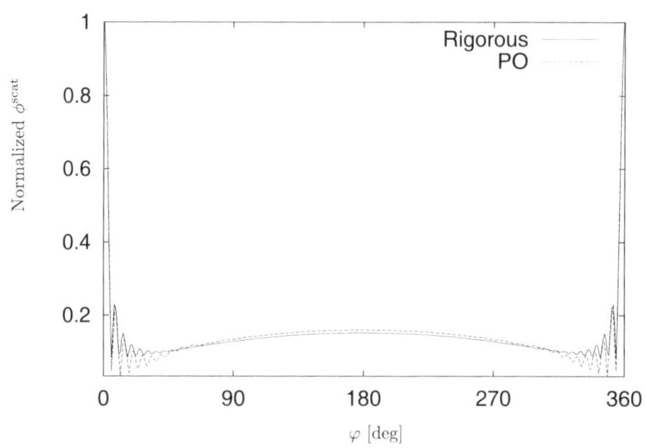

Figure: Directive pattern of the scattered wave in the far field region

$$(a = 5\lambda, d = 100\lambda)$$

Figure shows the far field directive pattern of the scattered wave which is normalized by the maximal. In the back scattering region $\varphi = 90^\circ \sim 270^\circ$ and the GO boundaries $\varphi = 0^\circ, 360^\circ$, PO approximation gives the accurate results.

Interference

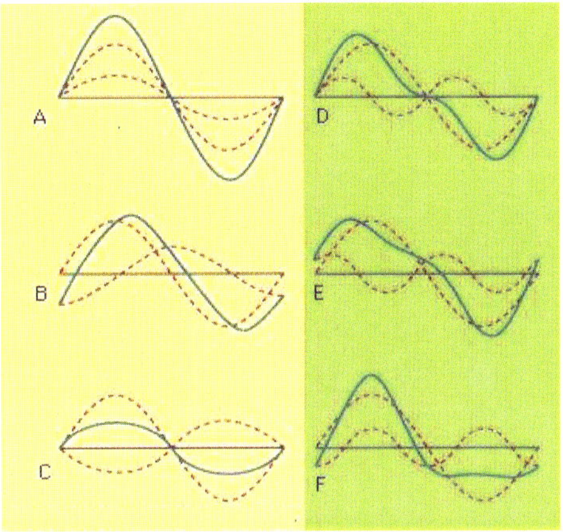

Interference, in physics, the net effect of the combination of two or more wave trains moving on intersecting or coincident paths. The effect is that of the addition of the amplitudes of the individual waves at each point affected by more than one wave.

If two of the components are of the same frequency and phase (i.e., they vibrate at the same rate and are maximum at the same time), the wave amplitudes are reinforced, producing constructive interference; but, if the two waves are out of phase by 1/2 period (i.e., one is minimum when the other is maximum), the result is destructive interference, producing complete annulment if they are of equal amplitude. The solid line in Figures represents the resultant of two waves (dotted lines) of slightly different amplitude but of the same wavelength. The two component waves are in phase in Figure but out of phase by 1/4 period and 1/2 period in Figures.

When two stones are dropped into a pool of water, waves spread out from each source, and interference occurs where they overlap. Constructive interference results where the crest of one coincides with the crest of the other. Two wave trains of light from a double slit produce interference, an effect that is visible on a screen as a pattern of alternating dark and light bands caused by intensification and extinction at points at which the waves are in phase and out of phase, respectively.

Interference also occurs between two wave trains moving in the same direction but having different wavelengths or frequencies. The resultant effect is a complex wave. A pulsating frequency, called a beat, results when the wavelengths are slightly different. Figures show complex waves (solid lines) composed of two component interfering waves (dotted lines), the ratio of their wavelengths being 1:2 and of their amplitudes 1:3.

Interference between waves traveling in opposite directions produces standing waves.

Diffraction

Diffraction, the spreading of waves around obstacles. Diffraction takes place with sound; with electromagnetic radiation, such as light, X-rays, and gamma rays; and with very small moving particles such as atoms, neutrons, and electrons, which show wavelike properties. One consequence of diffraction is that sharp shadows are not produced. The phenomenon is the result of interference(i.e., when waves are superimposed, they may reinforce or cancel each other out) and is most pronounced when the wavelength of the radiation is comparable to the linear dimensions of the obstacle. When sound of various wavelengths or frequencies is emitted from a loudspeaker, the loudspeaker itself acts as an obstacle and casts a shadow to its rear so that only the longer bass notes are diffracted there. When a beam of light falls on the edge of an object, it will not continue in a straight line but will be slightly bent by the contact, causing a blur at the edge of the shadow of the object; the amount of bending will be proportional to the wavelength. When a stream of fast particles impinges on the atoms of a crystal, their paths are bent into a regular pattern, which can be recorded by directing the diffracted beam onto a photographic film.

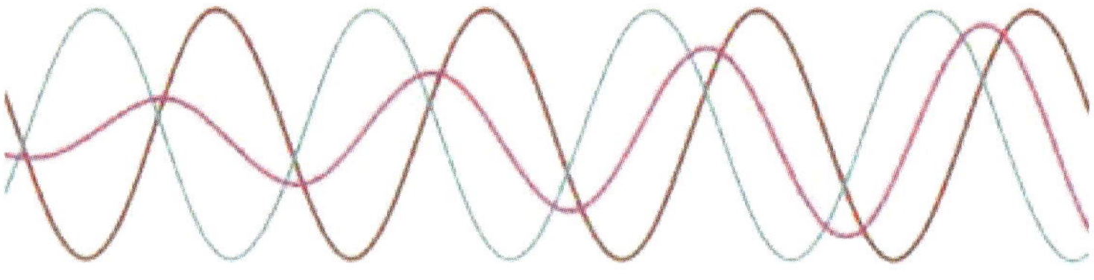

Explanation of diffraction.

Polarization

Polarization, property of certain electromagnetic radiations in which the direction and magnitude of the vibrating electric field are related in a specified way.

Light waves are transverse: that is, the vibrating electric vector associated with each wave is perpendicular to the direction of propagation. A beam of unpolarized light consists of waves moving in the same direction with their electric vectors pointed in random orientations about the axis of propagation. Plane polarized light consists of waves in which the direction of vibration is the same for all waves. In circular polarization the electric vector rotates about the direction of propagation as the wave progresses. Light may be polarized by reflection or by passing it through filters, such as certain crystals, that transmit vibration in one plane but not in others.

Beam Propagation Method

Beam propagation method is a mathematical procedure used to study the evolution of electromagnetic fields in arbitrary inhomogenious medium. BPM yields the response of a given device to an external optical sigal, similar to in an experiment.

The Beam Propagation Method (BPM)

The beam propagation method (BPM) is a powerful technique that can be used to solve optical circuits where $\Delta << 1$. In this case, the modes are governed by the scalar Helmholtz equation.

$$\left[\nabla^2 + \frac{\omega^2}{c^2} n^2(\omega, x, y)\right] E = 0$$

where

$$\nabla^2 = \frac{\partial^2}{\partial x^2} + \frac{\partial^2}{\partial y^2} + \frac{\partial^2}{\partial z^2}$$

h is the height of the waveguide and E is the transverse component of the electric field as a function of the circular frequency ω and the dimensional coordinates x, y and z, whereas the refractive index is a function of ω, x and y

Describing E as

$$E(\omega, x, y, z) = \xi(\omega, x, y, z) e^{-jkz}$$

and substituting it into equation yields

$$\frac{\partial^2 \xi}{\partial x^2} + \frac{\partial^2 \xi}{\partial y^2} + \frac{\partial^2 \xi}{\partial z^2} - 2jk\frac{\partial \xi}{\partial z} - k^2\xi + \frac{h^2}{c^2} n^2 \xi = 0$$

$$\therefore -\frac{\partial^2 \xi}{\partial z^2} + 2jk_0 \frac{\partial \xi}{\partial z} = \nabla_\perp^2 \xi + k_0^2 \left\{ \left[\frac{n(x,y)}{n_0}\right]^2 - 1 \right\} \xi$$

where $\nabla_\perp^2 = \partial^2/\partial x^2 + \partial^2/\partial y^2$ and assuming $k = \omega^2/c$ Neglecting the leftmost term in equation 2.20 yields the well-known parabolic or paraxial approximation of the Helmholtz equation

$$2jk_0 \frac{\partial \xi'}{\partial z} = \nabla_\perp^2 \xi + k_0^2 \left\{ \left[\frac{n(x,y)}{n_0}\right]^2 - 1 \right\} \xi'.$$

The eigen solutions of equations can be expressed as

$$\xi(x, y, z) = u'(x, y) e^{-j\beta z}$$

$$\xi'(x, y, z) = u'(x, y) e^{-j\beta' z}$$

Substituting equations and back into equations provides the relationship between the eigenvalues as :

$$\beta = k_0 \left[1 - \left(1 + 2\frac{\beta'}{k_0}\right)^{1/2}\right]$$

The conclusion from equation is that eigenvalues can be obtained by solving the parabolic or Fresnel equation.

Equation is solved numerically by using an implicit finite difference scheme. However, a fundamental physical limitation exists resulting from the parabolic approximation to the Helmholtz equation. The effect of these limitations can be relaxed by using more accurate approximations, such as the Padé approximations found in the BeamPROP software used in this thesis. Another limitation results from the assumption of scalar waves, and prevents polarization effects from being considered. Several vector beam propagation techniques can be employed to overcome this problem, as described in. The third limitation is that BPM cannot account for backward reflections because the one-way equation on which it is based does not allow both positive and negative travelling waves. It is possible to introduce a bi-directional BPM algorithm, as is done in BeamPROP.

Numerical Methods

Both *spatial domain* methods, and *frequency (spectral) domain* methods are available for the numerical solution of the discretized master equation. Upon discretization into a grid, (using various centralized difference, Crank–Nicolson method, FFT-BPM etc.) and field values rearranged in a causal fashion, the field evolution is computed through iteration, along the propagation direction. The spatial domain method computes the field at the next step (in the propagation direction) by solving a linear equation, whereas the spectral domain methods use the powerful forward/inverse DFT algorithms. Spectral domain methods have the advantage of stability even in the presence of nonlinearity (from refractive index or medium properties), while spatial domain methods can possibly become numerically unstable.

Applications

BPM is a quick and easy method of solving for fields in integrated optical devices. It is typically used only in solving for intensity and modes within shaped (bent, tapered, terminated) waveguide structures, as opposed to scattering problems. These structures typically consist of isotropic optical materials, but the BPM has also been extended to be applicable to simulate the propagation of light in general anisotropic materials such as liquid crystals. This allows one to analyze e.g. the polarization rotation of light in anisotropic materials, the tunability of a directional coupler based on liquid crystals or the light diffraction in LCD pixels.

Limitations of BPM

The Beam Propagation Method relies on the slowly varying envelope approximation, and is inaccurate for the modelling of discretely or fastly varying structures. Basic implementations are also inaccurate for the modelling of structures in which light propagates in large range of angles and for devices with high refractive-index contrast, commonly found for instance in silicon photonics. Advanced implementations, however, mitigate some of these limitations allowing BPM to be used to accurately model many of these cases, including many silicon photonics structures.

The BPM method can be used to model bi-directional propagation, but the reflections need to be implemented iteratively which can lead to convergence issues.

Implementations

There are several simulation tools that implement BPM algorithms.

Bloch wave – MoM Method

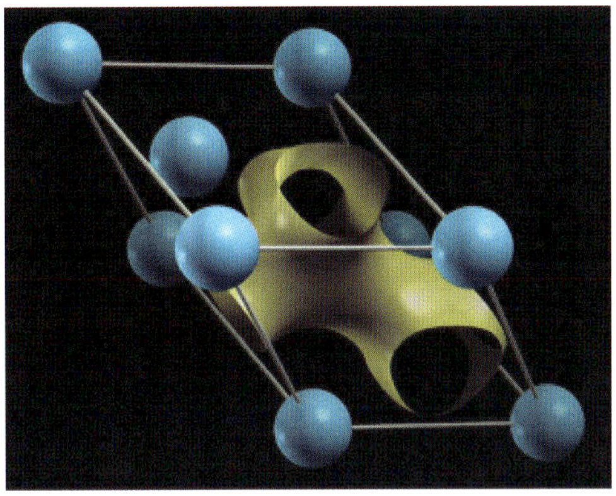

Bloch wave – MoM is a technique for determining the band structure of triply-periodic electromagnetic media such as photonic crystals. This technique uses the method of moments (MoM) in combination with a Bloch wave expansion of the electromagnetic field in the structure. This approach is very efficient in terms of the number of plane waves needed for good convergence and is analogous to the spectral domain MoM method commonly used for analyzing 2D periodic structures such as frequency selective surfaces (FSS). In the latter case, the electromagnetic field is expanded in terms of a plane wave spectrum. In both cases, the field is expanded in terms or a set of eigenfunction modes (either a Bloch wave in 3D or a plane wave spectrum in 2D), and an integral equation is enforced on the surface of the scatterers in each unit cell. In the FSS case, the unit cell is 2-dimensional and in the photonic crystal case, the unit cell is 3-dimensional.

Field Equations for 3D PEC Photonic Crystal Structures

The Bloch wave - MoM approach will be illustrated here for the case of perfectly electrically conducting (PEC) structures admitting only electric current sources, J. However, It can also be readily expanded to dielectric structures, using the well-known interior and exterior equivalent problems commonly employed in ordinary spatial domain method of moments formulations. In dielectric problems, there are twice as many unknowns - J & M - and also twice as many equations to enforce - continuity of tangential E & H - at the dielectric interfaces.

For PEC structures, the electric field E is related to the vector magnetic potential A via the well-known relation:

$$\mathbf{E} = -jk\eta\left[\mathbf{A} + \frac{1}{k^2}\nabla(\nabla \cdot \mathbf{A})\right]$$

and the vector magnetic potential is in turn related to the source currents via:

$$\nabla^2\mathbf{A} + k^2\mathbf{A} = -\mathbf{J}$$

where

$$k^2 = \omega^2(\mu\epsilon)$$

Bloch Wave Expansion of the Fields

To solve equations within the infinite periodic volume, we may assume a Bloch wave expansion for all currents, fields and potentials:

$$\mathbf{J}(x,y,z)=\sum_{mnp}\mathbf{J}(\alpha_m,\beta_n,\gamma_p)e^{j(\alpha_m x+\beta_n y+\gamma_p z)}$$

$$\mathbf{E}(x,y,z)=\sum_{mnp}\mathbf{E}(\alpha_m,\beta_n,\gamma_p)e^{j(\alpha_m x+\beta_n y+\gamma_p z)}$$

$$\mathbf{A}(x,y,z)=\sum_{mnp}\mathbf{A}(\alpha_m,\beta_n,\gamma_p)e^{j(\alpha_m x+\beta_n y+\gamma_p z)}$$

where for simplicity, we assume an orthogonal lattice in which α only depends on m, β only depends on n and γ only depends on p. With this assumption,

$$\alpha_m = k_0 \sin\theta_0 \cos\phi_0 + \frac{2m\pi}{l_x}$$

$$\beta_n = k_0 \sin\theta_0 \sin\phi_0 + \frac{2n\pi}{l_y}$$

$$\gamma_p = k_0 \cos\theta_0 + \frac{2p\pi}{l_z}$$

and,

$$k_0 = 2\pi / \lambda$$

where l_x, l_y, l_z are the unit cell dimensions in the x,y,z directions respectively, λ is the effective wavelength in the crystal and θ_0, φ_0 are the directions of propagation in spherical coordinates.

The quantity k in equations and comes originally from the time derivative in Maxwell's equations and is the *free space* propagation constant (actually, the propagation constant of whatever dielectric medium the metallic scatterers are embedded in), proportional to frequency as in equation. On the other hand, k_0 in the equations above comes from the *assumed Bloch wave solution* given by equations . As a result, it represents the propagation constant inside the periodic medium, inversely proportional to wavelength. These two k's, i.e. the free space propagation constant (proportional to frequency) and the propagation constant of the Bloch wave (inversely proportional to wavelength), are in general different thereby allowing for dispersion in the solution. The band diagram is essentially a plot of k as a function of k_0.

The Bloch wave expansions in equations are nothing more than exponential Fourier series multiplied by the cell-to-cell propagation factor: $e^{j(\alpha_0 x+\beta_0 y+\gamma_0 z)}$. The Bloch wave expansions are chosen since any field solution within an infinite periodic volume must have the same periodicity as the medium itself, or stated another way, the fields in neighboring cells must be identical up to a (real or complex) propagation factor. In passbands the propagation factor is an exponential function with purely imaginary argument and in the stop bands (or band gaps) it is a decaying exponential function whose argument has a real component.

The wave numbers α_0, β_0 and γ_0 satisfy the relations: $|\alpha_0|\leq\frac{\pi}{l_x}$, $|\beta_0|\leq\frac{\pi}{l_y}$, $|\gamma_0|\leq\frac{\pi}{l_z}$ and outside of these ranges, the bands are periodic.

The Bloch waves are periodic functions of space, with periods l_x, l_y, l_z and the bands are periodic functions of wavenumber, with periods: $\frac{2\pi}{l_x}$, $\frac{2\pi}{l_y}$ and $\frac{2\pi}{l_z}$

Integral Equation for PEC Media

Substituting equations yields the spectral domain Greens function relating the radiated electric field to its source currents:

$$\mathbf{E}(\alpha_m,\beta_n,\gamma_p)=\frac{jk\eta}{k^2-\alpha_m^2-\beta_n^2-\gamma_p^2}\mathbf{G}_{mnp}\,\mathbf{J}(\alpha_m,\beta_n,\gamma_p)$$

where,

$$\mathbf{G}_{mnp}=\begin{pmatrix} 1-\frac{\alpha_m^2}{k^2} & -\frac{\alpha_m\beta_n}{k^2} & -\frac{\alpha_m\gamma_p}{k^2} \\ -\frac{\alpha_m\beta_n}{k^2} & 1-\frac{\beta_n^2}{k^2} & -\frac{\beta_n\gamma_p}{k^2} \\ -\frac{\alpha_m\gamma_p}{k^2} & -\frac{\beta_n\gamma_p}{k^2} & 1-\frac{\gamma_p^2}{k^2} \end{pmatrix}$$

is the tensor Green's function in the spectral domain. Note that the spatial domain convolution has been transformed into a simple multiplication in the spectral domain, consistent with the convolution theorem for Fourier transforms.

With this equation for the electric field, the electric field boundary condition (requiring that the total tangential electric field be zero on the surface of PEC scatterer) becomes:

$$\sum_{mnp}\frac{1}{k^2-\alpha_m^2-\beta_n^2-\gamma_p^2}\mathbf{G}_{mnp}\,\mathbf{J}(\alpha_m,\beta_n,\gamma_p)e^{j(\alpha_m x+\beta_n y+\gamma_p z)}=\mathbf{0}$$

Since we are seeking characteristic modes (eigenmodes) of the structure, there is no impressed E-field on the RHS of this electric field integral equation (EFIE). Equation is not strictly correct however, since it is only the tangential components of electric field that are actually zero on the surface of the PEC scatterer. This inexactness will be resolved presently, when we test this equation with the electric current basis functions - defined as residing on the surface of the scatterer.

Method of Moments (MoM) Solution

As is usual in the method of moments, the source currents are now expanded in terms of a sum over some known set of basis functions with unknown weighting coefficients J_j:

$$\mathbf{J}(x,y,z)=\sum_j J_j\,\mathbf{J}_j(x,y,z)$$

Different structures will have different sets of basis functions for representing the currents on the elements and as in the ordinary spatial domain method of moments, the solution (in this case, the band diagram) *is a function of the set of basis functions used.*

Substituting and then testing the resulting equation with the *i*-th current basis function (i.e., dotting from the left and integrating over the domain of the *i*-th current basis function, thereby completing the quadratic form) produces the *i*-th row of the matrix eigenvalue equation for a 3-dimensional array of PEC scatterers as:

$$\sum_j J_j \left[\sum_{mnp} \frac{\mathbf{J}_i(-\alpha_m,-\beta_n,-\gamma_p)\mathbf{G}_{mnp}\,\mathbf{J}_j(\alpha_m,\beta_n,\gamma_p)}{k^2-\alpha_m^2-\beta_n^2-\gamma_p^2} \right] = \mathbf{0}$$

As in all MoM formulations, the reaction concept in electromagnetics was used in obtaining this equation. The electric field boundary/continuity conditions are "tested" (or enforced) by being integrated against electric current basis functions (for dielectric structures, the magnetic field continuity conditions are additionally tested by being integrated against magnetic current basis functions), and this is how the electric (and magnetic) field boundary conditions are converted into a matrix equation via the method of moments. This process is wholly analogous to that used to decompose a periodic function into its Fourier sine & cosine components, the only difference being that in this case the basis functions are not necessarily orthogonal, merely linearly independent.

This matrix equation is easy to implement and requires only that the 3D Fourier transform (FT) of the basis functions be computed, preferably in closed form, available on researchgate.net). In fact, computing bands of a 3D photonic crystal with this method is no more difficult than computing reflection and transmission from a 2D periodic surface using the spectral domain method . This is because equation is identical to the basic EFIE for a freestanding PEC FSS or Frequency selective surface), the only difference being the stronger singularity in 3D which significantly accelerates convergence of the triple sums, and of course the fact that the vectors are now 3-dimensional. As a result, an ordinary PC is sufficient to compute bands of many types of photonic crystals.

It's evident from that the EFIE could become singular whenever the free space wavenumber is exactly equal to one of the wave numbers in any of the 3 periodic coordinate directions. This can happen for example when the free space wavelength exactly equals the lattice spacing. This is a statistically rare occurrence in computational practice and corresponds to a propagation anomaly similar to a Wood's reflection anomaly for gratings.

Computing Bands

To compute bands of the crystal (i.e. k-k_0 diagrams), successive values of frequency (k) are tried - in conjunction with pre-selected values of propagation constant (k_0) and propagation direction (θ_0 & φ_0) - until a combination is found which drives the determinant of the matrix to zero. Equation has been used to compute bands in various types of doped and undoped photonic crystals, both available on researchgate.net). Not surprisingly, doping photonic crystals with defects provides a means for designing photonic passbands, in just the same way that doping semiconductors with chemical impurities provides a means for designing electronic passbands.

For many subsectional basis functions, such as those having a half-sine or triangular shape along a round wire, the FT of the basis function for negative wave numbers -α, -β, -γ is the complex conjugate of the basis function FT for positive wave numbers. As a result, the matrix in eqn. is Hermitian. And as a result of that, only half the matrix needs to be computed. And a second result is that the determinant is a purely real function of the real-valued wavenumber k. Zeroes generally occur at zero-crossings (inflection points, where curvature is zero), so a simple root-finding algorithm such as Newton's method is usually sufficient to find the roots to a very high degree of accuracy. If may still be useful however to plot the determinant as a function of k, to observe its behavior near the zeros.

As a matter of computational convenience, whenever the matrix is larger than 2x2 it's vastly more efficient to compute the determinant either by reducing the matrix to upper triangular form using QR decomposition or to compute the determinant by reducing to echelon form using Gaussian elimination, rather than trying to actually compute the determinant of the matrix directly.

Transfer-matrix Method

The transfer matrix method can be used for the analysis of the wave propagation of quantum particles, such as electrons [29,46,49,81,82,115–117,124,103,108,131,129,141] and of electromagnetic [39,123,124], acoustic, and elastic waves. Once this technique is developed for one type of wave, it can easily be applied to any other wave problem.

Transfer Matrices in Optics

In a real world analysis of a complex optical system, one of the most useful tricks is the introduction of a transfer matrix. In this problem, we'll see how to construct and use transfer matrices for normally incident electromagnetic waves in a large optical system.

Suppose we have an optical system which is translationally invariant in the y and z directions. We can write the electric field of a normally incident plane wave therefore

$$E = E + e^{in\omega x/c} + E - e^{-in\omega x/c}$$

The polarization of the wave does not concern us. Suppose that at some initial state, therefore, our electric field is described by the vector (E_{+1}, E_{-1}), and at some later time by the vector (E_{+2}, E_{-2}). (The vector refers to whether the wave is moving forwards/backwards, not the vector nature of E). If our system is linear, we then construct a 2 × 2 matrix T such that

$$\begin{pmatrix} E_{+2} \\ E_{-2} \end{pmatrix} = T \begin{pmatrix} E_{+1} \\ E_{-1} \end{pmatrix}$$

T is called a transfer matrix. To start off, let's calculate some sample transfer matrices.

a) Let $P(n; x)$ be the matrix representing propagation through a medium of index of refraction n for a distance x. Show that

$$P(n;\, x) = cos\frac{n\omega x}{c} + i\, sin\frac{n\omega x}{c}\sigma_3.$$

b) Let $B(n_1 \to n_2)$ be the matrix representing a wave propagating from a medium of index n_1 $(x < xo)$ to a medium with index n_2 $(x > xo)$. Show that

$$B(n_1 \to n_2) = \frac{1 + n_1 / n_2}{2} + \frac{1 - n_1 / n_2}{2}\sigma_1.$$

Now suppose we are considering a stack of N thin films of index of refraction n, each with thickness b. The films are separated by air ($n = 1$) of thickness a. The transfer matrix formalism is ideal to analyze this problem in a nearly-exact manner.

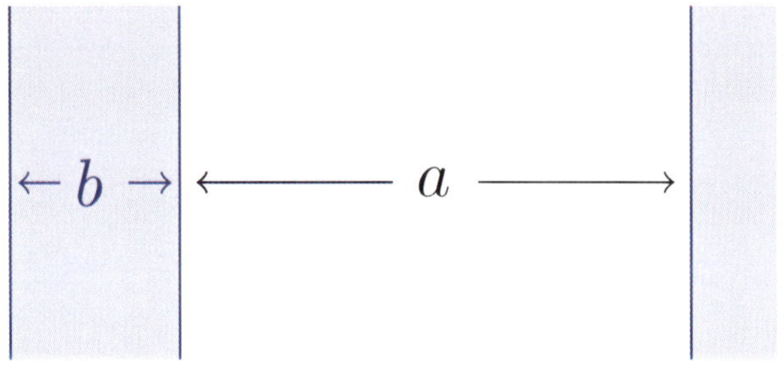

c) Let $\alpha = \omega a/c$ and $\beta = n\omega b/c$; let U be the transfer matrix for one unit of the thin film stack (as shown in the figure). Show that

$$U = \frac{1}{4n}\Big[\Big((n+1)^2 cos(\alpha+\beta) - (n-1)^2 cos(\alpha-\beta)\Big)$$
$$+ 2(n^2-1)\sin\beta\sin\alpha\sigma_1 - 2(n^2-1)\sin\beta\cos\alpha\sigma_2$$
$$+\Big((n+1)^2 \sin(\alpha+\beta) - (n-1)^2 \sin(\alpha-\beta) i\sigma_3\Big)\Big].$$

If we build a stack such that $\alpha = \beta = \pi/2$, compute T and show that

$$\frac{I_{trans}}{I_{inc}} = \frac{4}{\left(1 + \frac{1}{n^2}\right)^N},$$

Comment on this result, and on its possible usefulness in an application.

Transfer Matrix Approach: General Treatment of Multilayered Optical Materials

Consider a 3-layered material with an index of refraction profile:

$$n(x) = \begin{cases} n_1, & x < 0 \\ n_2, & 0 < x < d \\ n_3, & d < x \end{cases}$$

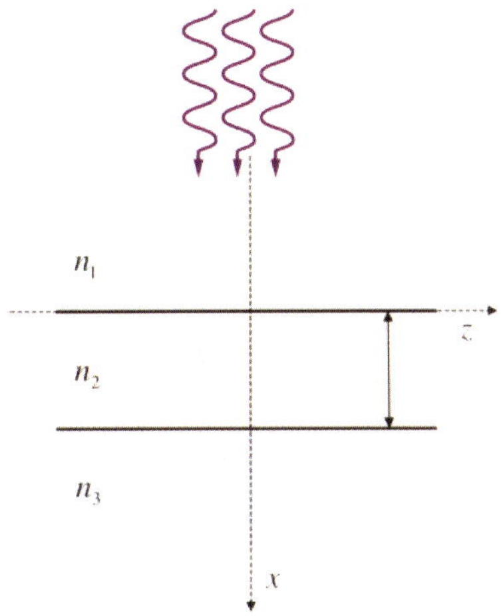

As discussed in the previous lecture, the z component of the wavevector $k_z = \beta$ does not change throughout the problem (consequence of phase continuity). This leads to a simple dependence for the evolution in *z* direction:

$$\vec{E} = \vec{E}(x)e^{i(\omega t - \beta z)}$$

However for *x* and *y* directions the scenario is more complicated as we have noticed in the previous lecture when we derived the reflection and transmission coefficients for the sing le interface. Now we would like to find a general way to relate any pair of electric and magnetic field amplitudes in any layer to those in any other layer.

As usual we would break our electric (and consequently magnetic) field components into *s*- and *p*-polarizations. Here we will only consider *s*-polarized electric field for simplicity (*y*-direction for *E* and *x*-*z* plane, perpendicular to wavevector for *H*).

For *s*-polarized field:

$$E(x) = \begin{cases} E_1 e^{ik_{1x}x} + E_1' e^{-ik_{1x}x}, & x < 0 \\ E_2 e^{ik_{2x}x} + E_2' e^{-ik_{2x}x}, & 0 < x < d \\ E_3 e^{ik_{3x}(x-d)} + E_1' e^{-ik_{3x}(x-d)}, & x > d \end{cases}$$

Where E_1, E_2, E_3 are the amplitudes of components of the electric field propagating forward and E'_1, E'_2, E'_3 are the amplitudes of the reflected components propagating backward.

If we were to relate the amplitudes in the layer 1 to the layer 3, we will have to consider propagation across two interfaces 1-2 and 2-3 and the simple propagation through the homogeneous layer 2.

The effect of propagation through a medium of index n_2 and thickness *d* is captured by the propagation matrix:

$$P_2 = \begin{pmatrix} e^{ik_{2x}d} & 0 \\ 0 & e^{-ik_{2x}d} \end{pmatrix}$$

Which is just simple phase accumulation for a plane wave propagating distance d in a homogeneous medium of index n_2.

The interface matrices can be derived from the boundary conditions akin to the way we treated a single interface:

$$\vec{E}_{1\|} = \vec{E}_{1\|} \Rightarrow E_{1y} + E'_{1y} = E_{2y} + E'_{2y}$$

$$H_{1\|} = H_{2\|} \Rightarrow H_{1z} + H'_{1z} = H_{2z} + H'_{2z} \Rightarrow n_1 \cos\theta_1 E_{1y} - n_1 \cos\theta_1 E'_{1y} = n_2 \cos\theta_2 E_{2y} - n_2 \cos\theta_2 E'_{2y}$$

Then can rewrite the equations above in a matrix form:

$$\left.\begin{array}{c} E_{1y} + E'_{1y} = E_{2y} + E'_{2y} \\ n_1 \cos\theta_1 E_{1y} - n_1 \cos\theta_1 E'_{1y} = n_2 \cos\theta_2 E_{2y} - n_2 \cos\theta_2 E'_{2y} \end{array}\right\} \Rightarrow$$

$$\begin{pmatrix} 1 & 1 \\ n_1 \cos\theta_1 & -n_1 \cos\theta_1 \end{pmatrix} \begin{pmatrix} E_{1y} \\ E'_{1y} \end{pmatrix} = \begin{pmatrix} 1 & 1 \\ n_2 \cos\theta_2 & -n_2 \cos\theta_2 \end{pmatrix} \begin{pmatrix} E_{2y} \\ E'_{2y} \end{pmatrix} \Rightarrow D_1 \begin{pmatrix} E_{1y} \\ E'_{1y} \end{pmatrix} = D_2 \begin{pmatrix} E_{2y} \\ E'_{2y} \end{pmatrix}$$

$$D_2^{-1} D_1 \begin{pmatrix} E_{1y} \\ E'_{1y} \end{pmatrix} = \begin{pmatrix} E_{2y} \\ E'_{2y} \end{pmatrix} \Rightarrow D_{12} \begin{pmatrix} E_{1y} \\ E'_{1y} \end{pmatrix} = \begin{pmatrix} E_{2y} \\ E'_{2y} \end{pmatrix}; D_{21} = D_2^{-1} D_1$$

Then the matrix D_{12} can be simply found as:

$$D_{12} = D_2^{-1} D_1 = \begin{pmatrix} 1 & 1 \\ n_2 \cos\theta_2 & -n_2 \cos\theta_2 \end{pmatrix}^{-1} \begin{pmatrix} 1 & 1 \\ n_1 \cos\theta_1 & -n_1 \cos\theta_1 \end{pmatrix} = \begin{pmatrix} \frac{1}{2} & \frac{1}{2n_2 \cos\theta_2} \\ \frac{1}{2} & -\frac{1}{2n_2 \cos\theta_2} \end{pmatrix} \begin{pmatrix} 1 & 1 \\ n_1 \cos\theta_1 & -n_1 \cos\theta_1 \end{pmatrix}$$

$$D_{12} = \begin{pmatrix} \frac{n_2 \cos\theta_2 + n_1 \cos\theta_1}{2n_2 \cos\theta_2} & \frac{n_2 \cos\theta_2 - n_1 \cos\theta_1}{2n_2 \cos\theta_2} \\ \frac{n_2 \cos\theta_2 - n_1 \cos\theta_1}{2n_2 \cos\theta_2} & \frac{n_2 \cos\theta_2 + n_1 \cos\theta_1}{2n_2 \cos\theta_2} \end{pmatrix}$$

Then in order to get find the connection between the electric field amplitudes in layer 1 and layer 3 we need to do the following:

$$D_{23} P_2 D_{12} \begin{pmatrix} E_1 \\ E'_1 \end{pmatrix} = \begin{pmatrix} E_3 \\ E'_3 \end{pmatrix}$$

Example

As an illustration, consider a single layer of glass with a refractive index n and thickness d suspended in air at a wave number k (in air). In glass, the wave number is $k' = nk$. The transfer matrix is

$$M = \begin{pmatrix} \cos k'd & \sin(k'd)/k' \\ -k' \sin k'd & \cos k'd \end{pmatrix}.$$

The amplitude reflection coefficient can be simplified to

$$r = \frac{(1/n - n)\sin(k'd)}{(n + 1/n)\sin(k'd) + 2i\cos(k'd)}.$$

This configuration effectively describes a Fabry–Pérot interferometer or etalon: for $k'd = 0, \pi, 2\pi, \cdots$, the reflection vanishes.

Acoustic Waves

It is possible to apply the transfer-matrix method to sound waves. Instead of the electric field *E* and its derivative *F*, the displacement *u* and the stress $\sigma = Cdu/dz$, where C is the p-wave modulus, should be used.

Abeles Matrix Formalism

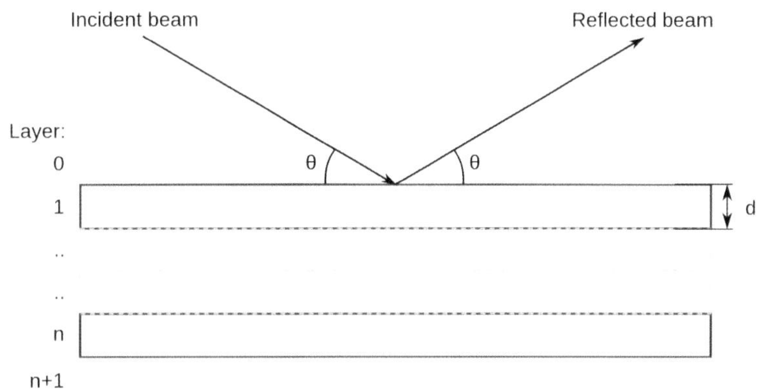

Reflection from a stratified interface

The Abeles matrix method is a computationally fast and easy way to calculate the specular reflectivity from a stratified interface, as a function of the perpendicular momentum transfer, Q_z:

$$Q_z = \frac{4\pi}{\lambda}\sin\theta = 2k_z$$

where θ is the angle of incidence/reflection of the incident radiation and λ is the wavelength of the radiation. The measured reflectivity depends on the variation in the scattering length density (SLD) profile, $\rho(z)$, perpendicular to the interface. Although the scattering length density profile is normally a continuously varying function, the interfacial structure can often be well approximated by a slab model in which layers of thickness (d_n), scattering length density (ρ_n) and roughness ($\sigma_{n,n+1}$) are sandwiched between the super- and sub-phases. One then uses a refinement procedure to minimise the differences between the theoretical and measured reflectivity curves, by changing the parameters that describe each layer.

In this description the interface is split into *n* layers. Since the incident neutron beam is refracted by each of the layers the wavevector, *k*, in layer *n*, is given by:

$$k_n = \sqrt{k_z^2 - 4\pi(\rho_n - \rho_0)}$$

The Fresnel reflection coefficient between layer n and $n+1$ is then given by:

$$r_{n,n+1} = \frac{k_n - k_{n+1}}{k_n + k_{n+1}}$$

Since the interface between each layer is unlikely to be perfectly smooth the roughness/diffuseness of each interface modifies the Fresnel coefficient and is accounted for by an error function.

$$r_{n,n+1} = \frac{k_n - k_{n+1}}{k_n + k_{n+1}} \exp(-2k_n k_{n+1} \sigma_{n,n+1}^{2})$$

A phase factor, β, is introduced, which accounts for the thickness of each layer.

$$\beta_0 = 0$$

$$\beta_n = ik_n d_n$$

where $i^2 = -1$. A characteristic matrix, c_n is then calculated for each layer.

$$c_n = \begin{bmatrix} \exp(\beta_n) & r_{n,n+1}\exp(\beta_n) \\ r_{n,n+1}\exp(-\beta_n) & \exp(-\beta_n) \end{bmatrix}$$

The resultant matrix is defined as the product of these characteristic matrices

$$M = \prod_n c_n$$

from which the reflectivity is calculated as:

$$R = \left| \frac{M_{10}}{M_{00}} \right|^2$$

Gerchberg–Saxton Algorithm

The standard Gerchberg–Saxton (GS) algorithm is normally used to find the phase (measured on two different parallel planes) of a propagating optical field (usually far-field propagation), given that the irradiance information on those planes is known. This is mostly used to calculate the modulation function of a phase mask so that when illuminated by a plane wave, it produces a known far-field irradiance distribution, or the equivalent, to calculate the phase mask to be used in a Fourier optical system so the desired pattern is obtained on the image plane. There are some extensions of the GS algorithm that can be used when the transformations that describe the optical setup are non-unitary, for example the Yang-Gu algorithm, but these are usually demonstrated using nonunitary translational-invariant optical systems.

The Gerchberg-Saxton (GS) algorithm was proposed in 1972 by R. W. Gerchberg and W. O.

Saxton to solve the problem of phase retrieval of a field at two different planes, when at those, only the field amplitudes are known and given that the fields are related by a Fourier transform. Besides its original purpose, it has also been extensively used to calculate phase-only diffractive optical elements (DOE) to be used as beam-shapers or in imaging applications where the optical field's propagation from the DOE to the image plane and vice versa are known and Fourier like, such as systems involving far-field propagation or in a well-corrected Fourier system.

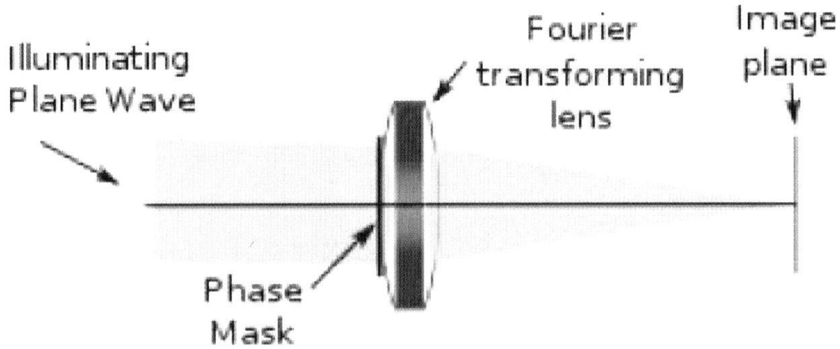

Figure: Diagram of a typical Fourier system.

There are also some applications related to beam-shaping and imaging, where it is of interest to apply some constrains to the obtained DOE, such as discretizing its phase to simplify the fabrication. Here, some modifications of the standard GS algorithm have been proposed and analyzed, but the optical systems involved are usually simulated by a Fourier transform too. Besides, there are some extensions of the GS algorithm that can be used when the transformations that describe the optical setup are non-unitary (and even translational-variant), for example the Yang-Gu algorithm but these are usually demonstrated using simple non-unitary and translational-invariant optical systems such a (paraxial) Fourier system that includes an aperture, and where a method to characterize more complex systems is not addressed. In addition, some other modifications of the GS algorithm where the object (phase mask) and image plane are not related by a Fourier transform but by a Fresnel or a Fractional Fourier transform have been proposed but these algorithms have been developed for and applied to simple systems involving ideal lenses and free space propagation.

A first approximation to handle non-ideal systems, is to regard them as paraxial (translational-invariant) optical systems with aberrations. Here the GS algorithm still works, provided that the adequate compensation phase is added to the calculated phase mask. However, in applications where the optical setup is a translational-variant and non paraxial Fourier-like system that cannot be simulated by ideal lenses and free space propagation, the GS algorithm or any of its derivatives cannot be used, because they do not addresses the problem of how to generate the optical transformations that characterize the system adequately.

What it is proposed in this work is to use a mixed approach, where raytracing and the Huygens-Fresnel principle are used to characterize an arbitrarily complex optical setup, making it possible to obtain the adequate direct and inverse propagation functions to be used with the GS algorithm to calculate the phase mask needed to produce an arbitrary irradiance distribution in the image plane of the setup. As an example, the method is tested using a highly translational-variant multi-lens Fourier-like system.

The Implemented GS Algorithm

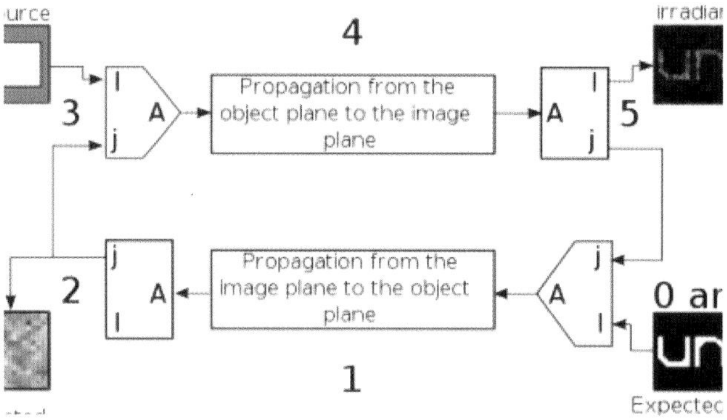

Figure: Block diagram of the algorithm used.

A block diagram of the algorithm used is presented in figure. The steps performed (that are basically the same ones used in the standard GS algorithm), are the following:

0. At the beginning, a field with an amplitude given by the square root of the expected irradiance and a constant phase is taken.
1. The field is propagated from the image-plane to the object-plane.
2. The amplitude information is discarded, leaving only the phase information (for the phase mask).
3. The amplitude and phase of the illumination field are added to the phase information to obtain the resulting object field.
4. The field is propagated from the object plane to the image plane.
5. The resulting reconstructed image (square of the field amplitude) is compared with the expected one. By using the correlation between both images as a criterion, a decision is taken to finish the process or continue iterating.
6. The phase from the reconstructed image is combined with the field amplitude obtained from the expected irradiance, and the process is repeated from step 1.

The algorithm is implemented using two propagation functions equivalent for Steps 1 and 4. The first propagation function is obtained by locating point sources on the object (phase mask) plane, and their corresponding contributions are obtained by raytracing through the optical system until the image plane. For the second one, the process is done backward and the point sources are located on the image plane, and their corresponding contributions on the object plane are obtained by raytracing in the opposite direction. The result of all the raytracing propagation are saved, so they can be used again in each iteration of the GS implementation.

Pseudo-code Algorithm

```
Let:
 FT - forward Fourier transform
 IFT - inverse Fourier transform
```

```
 i - the imaginary unit, √-1 (square root of -1)
 exp - exponential function (exp(x)=e^x)
 Target and Source be the Target and Source Amplitude planes respectively
 A, B, C & D be complex planes with the same dimension as Target and
Source
 Amplitude - Amplitude-extracting function:
   e.g. for complex z = x + iy, amplitude(z) = sqrt(x·x + y·y)
        for real x, amplitude(x) = |x|
 Phase - Phase extracting function:
   e.g. Phase(z) = arctan(y/x)
end Let


Gerchberg-Saxton Algorithm(Source, Target, Retrieved_Phase)
 A = IFT(Target)
 while error criterion is not satisfied
   B = Amplitude(Source) * exp(i*Phase(A))
   C = FT(B)
   D = Amplitude(Target) * exp(i*Phase(C))
   A = IFT(D)
 end while
 Retrieved_Phase = Phase(A)
end Gerchberg-Saxton Algorithm
```

This is just one of the many ways to implement the GS algorithm. Aside from optimizations, others may start by performing a forward Fourier Transform to the source distribution.

Adaptive-additive Algorithm

The adaptive-additive (AA) algorithm of Soifer *et al.* an iterative numerical technique which explores the space of degenerate phase profiles, $\Phi^f(\vec{\rho})$, to find a phase modulation of the incident laser beam encoding any desired intensity profile in the focal plane. To facilitate calculation and fabrication, both the input and output planes are discretized into $M \times M$ square arrays of pixels. Optimal spatial resolution requires pixels in the focal plane to be one half-wavelength on a side

$\delta \quad \lambda/2$. The number, M, of pixels on a side then depends on the desired dimensions of the trapping array. Lengths in the input and focal planes are related by Eqs., so that the corresponding pixel size in the input plane is $\delta^{in} = \lambda f/(M\delta^{f}) = 2f/M$. If δ^{in} is inconveniently small, then L_1 and L_2 can be chosen so that a more amenable pixel size in the input plane corresponds to δ^{in} in the focal plane.

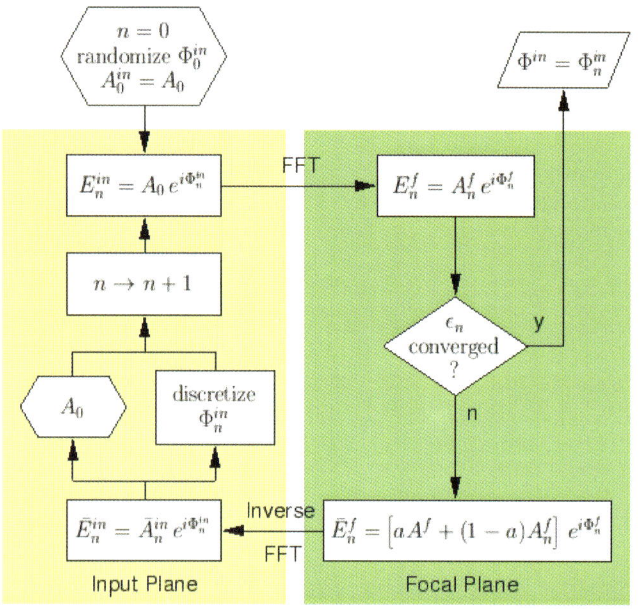

Figure: Flow chart for the adaptive-additive algorithm. The phase modulation, $\Phi_n^{in}(\vec{r})$, can be quantized into discrete steps with every iteration, as shown, or after the algorithm has converged.

The AA algorithm, depicted in Fig. starts with an arbitrary initial guess for $\Phi_1^{in}(\vec{r})$ and an initial input wavefront $E_1^{in}(\vec{r}) = E_0(\vec{r})\exp[i\Phi_1^{in}(\vec{r})]$. The Fourier transform of this wavefront is the starting estimate for the output electric field: $E_1^{f}(\vec{\rho}) = F\{E_1^{in}(\vec{r})\} = A_1^{f}(\vec{\rho})\exp[i\Phi_1^{f}(\vec{\rho})]$. The corresponding intensity in the output plane, $I_1^{f}(\vec{\rho}) = |A_1^{f}(\vec{\rho})|^2$ is unlikely to be a good rendition of the desired intensity pattern, $I^{f}(\vec{\rho}) = |A^{f}(\vec{\rho})|^2$. The error,

$$\epsilon_1 \equiv \frac{1}{M^2}\sum_{i=1}^{M^2}\left[I^{f}(\vec{\rho}_i) - I_1^{f}(\vec{\rho}_i)\right]^2,$$

is reduced by mixing a proportion, a, of the desired amplitude into the field in the focal plane:

$$\bar{E}_1^{f}(\vec{\rho}) = [aA^{f}(\vec{\rho}) + (1-a)A_1^{f}(\vec{\rho})]\exp[i\Phi_1^{f}(\vec{\rho})].$$

Inverse transforming $\bar{E}_1^{f}(\vec{\rho})$ yields the corresponding field in the input plane, $\bar{E}_1^{in}(\vec{r}) = \bar{A}_1^{in}(\vec{r})\exp[i\Phi_2^{in}(\vec{r})]$. At this point, the amplitude in the input plane no longer matches the actual laser profile, so we replace $\bar{A}_1^{in}(\vec{r})$ with $A_0(\vec{r})$. The result is an improved estimate for the input field: $E_2^{in}(\vec{r}) = E_0(\vec{r})\exp[i\Phi_2^{in}(\vec{r})]$. This completes one iteration of the AA algorithm. Subsequent iterations lead to monotonically improving estimates, $\Phi_n^{in}(\vec{r})$, for the desired phase modulation . The cycle is repeated until the error, ϵ_n, in the n-th iteration converges to within an acceptable tolerance: $(\epsilon_n - \epsilon_{n-1})/\epsilon_n < \chi$.

The phase and amplitude fields are computed as arrays of double-precision numbers, and their Fourier transforms calculated with fast Fourier transform (FFT) routines. Starting from random input phases, $\Phi_1^{in}(\vec{r}_i)$, uniformly distributed in the range 0 to 2π, the AA algorithm typically requires eight iterations to converge within $\chi = 10^{-6}$ of an acceptably accurate local minimum of ϵ_n using an intermediate value for the mixing parameter, $a = 0.5$.

Practical Considerations

The AA algorithm generates phase profiles, $\Phi^{in}(\vec{r})$, that vary continuously between 0 and 2π. Actually creating a phase element with continuously varying phase delay is difficult; usually only a small number of discrete levels are available. Discretizing the output of the adaptive-additive algorithm necessarily introduces errors. These can be minimized by integrating the discretization step into the AA algorithm itself, as shown in Fig. although this can lead to problems with convergence.

Pseudo-code Algorithm

- Define input amplitude and random phase
- Forward Fourier Transform
- Separate transformed amplitude and phase
- Compare transformed amplitude/intensity to desired output amplitude/intensity
- Check convergence conditions
- Mix transformed amplitude with desired output amplitude and combine with transformed phase
- Inverse Fourier Transform
- Separate new amplitude and new phase
- Combine new phase with original input amplitude
- Loop back to Forward Fourier Transform

References

- Physical-optics-514000: accessscience.com, Retrieved 26 March 2018
- Interference-physics: britannica.com, Retrieved 29 May 2018
- Diffraction, science, Retrieved 29 May 2018
- Polarization-physics: britannica.com, britannica.com, Retrieved 11 July 2018
- The algorithm, pseudo-code-algorithm, adaptive-additive-algorithm: liquisearch.com, Retrieved 18 July 2018

2

Understanding Geometrical Optics

Geometrical optics is a branch of optics, which describes the propagation of light as rays. Some of the optical phenomena addressed under geometrical optics include reflection, refraction and total internal reflection. This chapter has been carefully written to provide an easy understanding of such phenomena. It also includes topics such as specular reflection, diffused reflection, signal reflection, reflection phase change and refractive index, besides others.

Geometrical optics deals with the propagation of light in a straight line and phenomena such as reflection, refraction, polarization, etc. A ray of light gives the direction of propagation of light. In the absence of an obstacle, the rays advance in a straight line without changing direction. When light meets a surface separating two transparent media, reflection and refraction occur and the light rays bend.

Every real system experiences diffraction effects, so geometric optics is necessarily an approximation. However, the simplicity arising from treating only rays which move in straight lines affords many uses.

Reflection at Smooth Surfaces

A light ray is reflected by a smooth surface in accordance with the 2 laws of reflection:

1. The angle of incidence is always equal to the angle of reflection.
2. The incident ray, the reflected ray, and the normal to the reflecting surface are coplanar. The images formed by a plane mirror are equidistant from the distance between the object and the mirror.

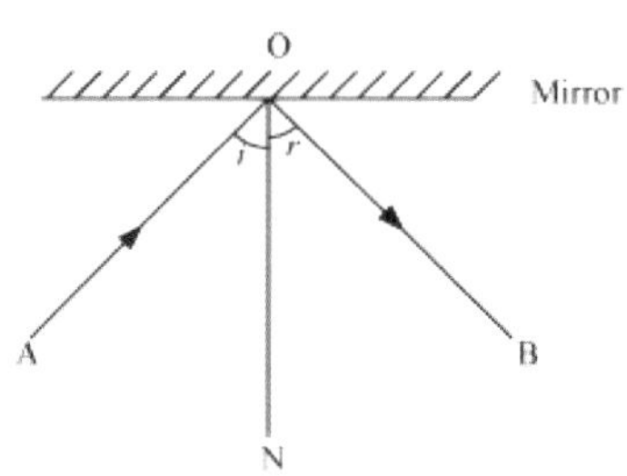

Example

An object is placed in front of the mirror. If the mirror is moved away from the object through a distance of x by how much distance will the image move?

The answer will be added shortly.

Our final answer is $2x$.

A ray of light falls on a plane mirror. Show that if the mirror is tilted through an angle of θ, then the reflected ray tilts through an angle of 2θ.

The laws of reflection are the same for plane and curved surfaces. A normal can be drawn from any point of the curved surface by first drawing the tangent plane from that point and then drawing the line perpendicular to that plane. Angles of incidence and reflection are defined from this normal. The angle of incidence is equal to angle of reflection.

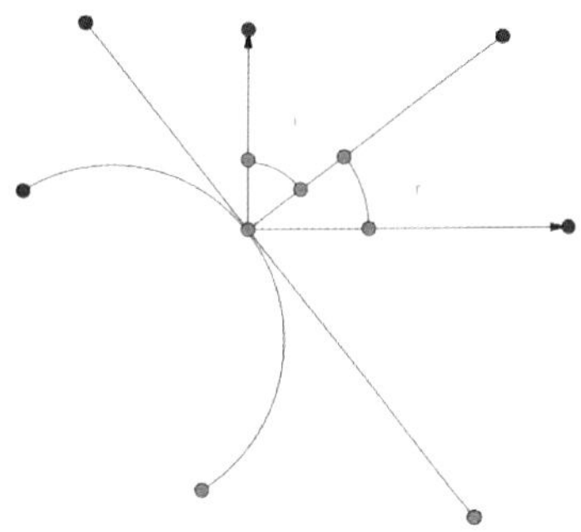

Spherical Mirrors

A spherical mirror is a part cut from a hollow sphere. The spherical mirrors are generally constructed from glass. One surface of the glass is silvered. If reflection takes place at the convex surface, it is called a convex mirror and if it takes place on the concave surface, it is called a concave mirror.

Terms used in spherical mirrors:

- Center of curvature (C): it is the center of the sphere from which the mirror is made.
- Focus (f): it is the point at which the rays meet (either virtually or physically). Although we have not verified, from further exploration we will understand that $R = 2f$.
- Pole (P): it is the surface of the mirror on which light falls.
- Principle axis (PA): it is the imaginary line which is perpendicular to the pole.
- Paraxial rays: These are rays parallel and close to the principal axis.
- u : This is the distance between the object and the pole and is sometimes referred to as the object distance.
- v : This is the distance between the pole and the image and is sometimes referred to as the image distance.

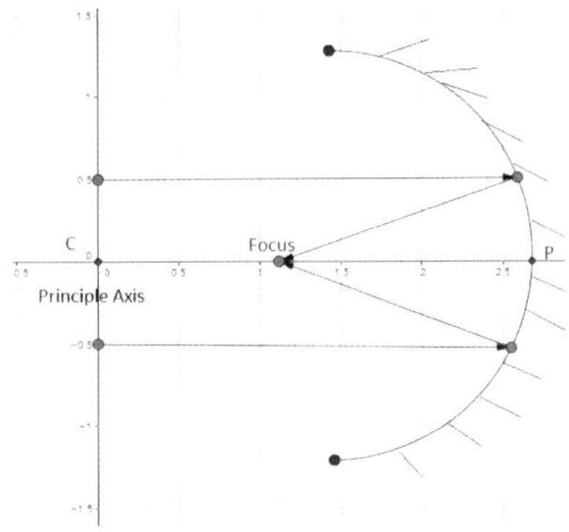

Here $(0, 0)$ is the C and the principal axis is the x-axis.

Sign Conventions

Cartesian Sign Conventions

- The pole is taken to be the origin.
- The direction in which the light moves is taken as the positive direction.
- The height above the principal axis is taken as positive while the height below it is taken as negative.

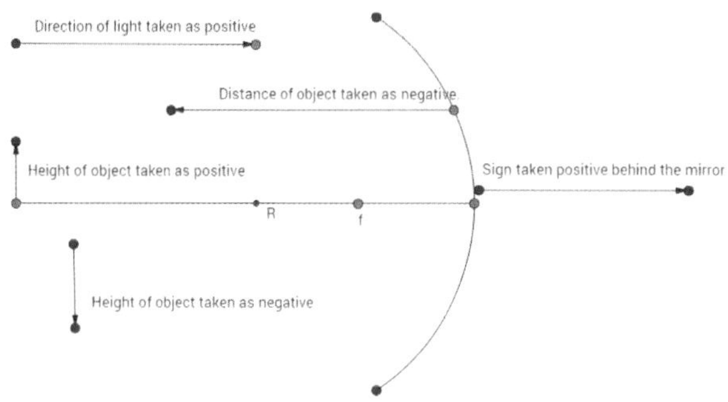

Mirror Formula

Theorem

The main formula used for solving problems on mirrors is

$$\frac{1}{v}+\frac{1}{u}=\frac{1}{f}.$$

While solving problems, do not forget to apply sign conventions and make sure you always take the object distance, u as *negative only*. the examples below which will make you comfortable with solving the problems based on the mirror formula.

Example

If the distance between the object and the mirror is 10cm and its focal length is 5cm given it is a convex mirror, find the distance between the mirror and the object.

We are given $u=-10\text{cm}$ and $f=5\text{cm}$

Applying the mirror formula, we get

$$\frac{1}{v}+\frac{1}{u}=\frac{1}{f}$$

$$\frac{1}{v}+\frac{1}{-10}=\frac{1}{5}$$

$$\frac{1}{v}=\frac{3}{10}$$

$$\Rightarrow v=3.33\text{ cm}.$$

THEOREM

The magnification of a mirror is given by $\frac{h_i}{h_0}$, where h_i is height of the image and h_0 is height of the object.

Magnification is also given by $-\frac{v}{u}$.

Again we should remember not to forget applying the sign conventions.

If magnification is $-ve$ it means the formed image is inverted.

Example

If the image formed by a spherical mirror is half the size of the object and it is virtual, then find the distance between the mirror and the image given that the object's distance is 16cm.

We are given $m = \frac{h_i}{h_0} = \frac{1}{2} = \frac{-v}{u}$.

So, $u = -2v = -16 \Rightarrow v = 8\text{cm}$.

Refraction

When light passes from one medium to another and undergoes a change in speed and direction, it is called refraction. The light does reflect but continues on its path with a change in its direction.

When light moves from a rarer to denser medium, i.e. moves from a light to heavy medium, the light tends to bend towards the normal while the opposite happens when light travels from a denser to rarer medium.

For the concept of refraction, one important thing to know is to learn about Snell's law.

This law tells us that

$$\mu_1 \sin i = \mu_2 \sin r,$$

where μ_1 is the medium's refractive index from which light travels and μ_2 is the medium's refractive index to which light goes. The angle of incidence is i and the angle of refraction is r.

Generally, we deal with a situation where light goes from air to another medium and since the refractive index of air is 1, we get the equation

$$\sin i = \mu \sin r$$

THEOREM

The lens formula is given by

$$\frac{1}{v} - \frac{1}{u} = \frac{1}{f}.$$

Again it is important to use the sign conventions, or the answers are going to be wrong.

Most importantly, the sign conventions are going to remain the same as that of the mirrors, but here we have two focal lengths rather than one in the case of mirrors. So the focal length of concave is $-Ve$ by convention and that of convex is $+Ve$.

Example

If the image formed by a convex lens is $20\,\text{cm}$ from the lens and the radius of curvature is $30\,\text{cm}$ then find the distance of the object from the lens.

We have $R = 30\,\text{cm} \Rightarrow f = 15$. Using the lens formula, we have $\frac{1}{+20} - \frac{1}{u} = \frac{1}{15}$. Remember not to put the negative sign of u before solving it:

$$\frac{1}{20} - \frac{1}{15} = \frac{1}{u}$$
$$\frac{3-4}{60} = \frac{1}{u}$$
$$\frac{-1}{60} = \frac{1}{u}$$
$$u = -60\ \text{cm}.$$

Theorem

The lens formula we used above was a result of using a biconvex or biconcave lens. These lenses have equal radii of curvature and equal foci. However, the same cannot be used for a lens having different radii of curvature. So a new method of finding the focus, u and v was developed, called the lens maker formula.

$$= \left(\frac{\mu_2}{\mu_1} - 1\right)\left(\frac{1}{R_1} - \frac{1}{R_2}\right).$$

Here F is the resultant focal length, μ_2 is the refractive index of the object in which light enters, and μ_1 is the refractive index from which light travels to reach the object. Normally μ_1 is ignored when light travels from air because it has a refractive index of 1, so μ_1 becomes 1. R_1 is the radius of curvature of the 1st lens, and R_2 is the radius of curvature of the 2nd lens.

Atmospheric Refraction

We know that the earth is surrounded by a layer of air called the atmosphere. The density of air is not the same everywhere. The refractive index of air depends on its density, i.e. the higher the density of air, the greater its refractive index. The changes in refractive index can give rise to many phenomena.

For example, a rising current of hot air makes the objects viewed through it flicker and sway. You might have seen that on a summer day; hot air seems to be rising above the surface of the roads.

Early Sunrise and Late Sunset

Consider an oblique ray from a heavenly body such as the sun or a star. While travelling through the atmosphere, it continuously moves into regions of higher refractive index. So, it continuously bends towards the normal, resulting in a path similar to the figure given below.

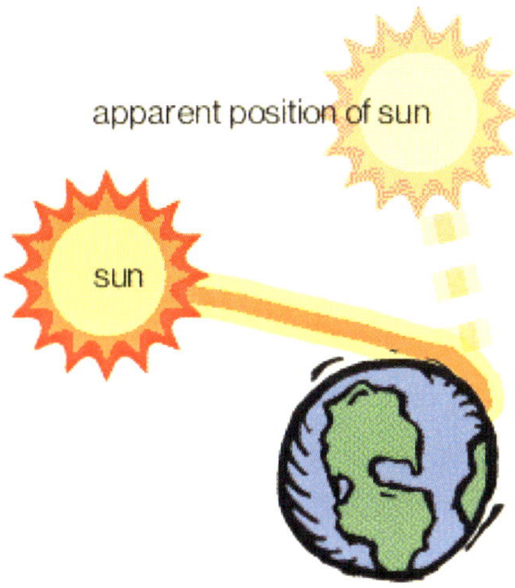

Since we see the object, that is the sun now, in the direction of the ray incident on the eye, the heavenly body appears higher than its actual position.

Now, consider a situation when the sun is just below the horizon. Rays of light coming from it get bent such that they seem to be coming from above the horizon (to the observer). Thus, when the sun's position is just below the horizon, the sun is visible to us. So, at sunrise, we see the sun before it actually comes to the horizon. At sunset, we see it even after it has dipped below the horizon. The fact is that it increases daylight by 4 minutes everyday (2 minutes at sunrise and 2 minutes at sunset).

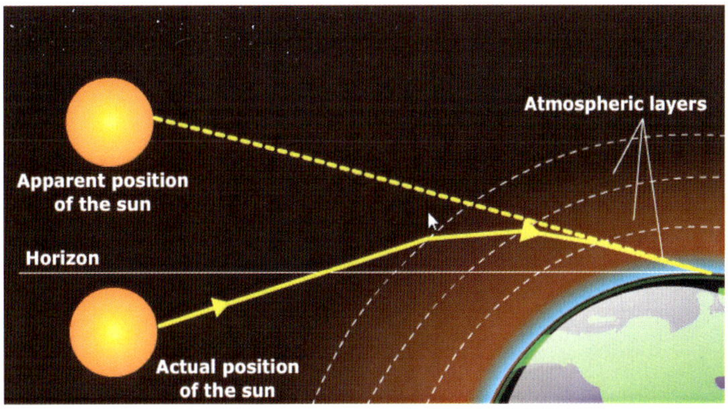

You might also have noticed that the sun looks oval/flattened at sunrise and sunset. The rays from the lower regions of the sun travel a bit greater distance than those rays which are from the middle and upper regions. So, they bend more. As a result, the image of the lower region gets shifted upwards more than that of the upper region. This makes the sun appear like a flattened circle or an oval.

Similarly, *twinkling of stars* is also a result of atmospheric refraction.

Now, we can define atmospheric refraction as follows:

Atmospheric refraction is a phenomenon, where light/any electromagnetic wave deviates from its rectilinear path because of passing through *atmosphere* where there are variations in the air density and refractive index.

Total Internal Reflection

When light travels from a denser to rarer medium with an angle greater than the critical angle, the ray of light does not deviate in its path or does not refract, but it undergoes a reflection known as total internal reflection.

The critical angle differs from medium to medium.

It is given by the formula

$$\mu = \frac{1}{\sin \theta_c}.$$

This is very useful as it is used in fiber glasses where total internal reflection helps in fast movement of wavelengths.

Optical fibers are devices used for guiding light in many applications, most notably for fast communication. A fiber consists of a glass cylinder surrounded by a wall covered in a special coating.

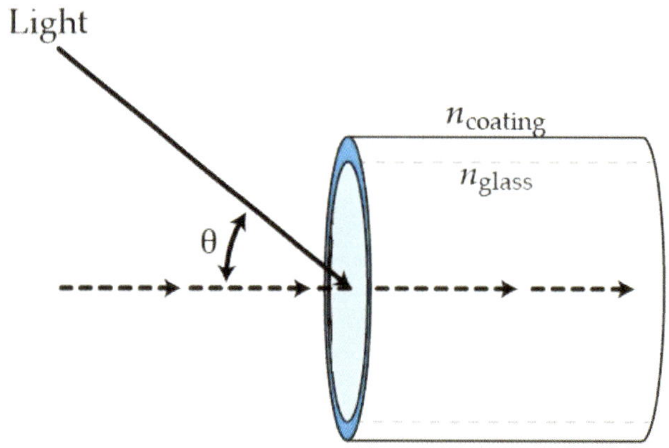

The fibers work on a principle called *total internal reflection*: light enters the fiber at an angle such that it does not get transmitted through the wall of the fiber when it hits the inside of the wall. Therefore, the refraction index of the glass part of the fiber has to be higher than that of its coating.

What is the maximum entering angle in degrees a light ray can pass from the air to the glass fiber for the total internal reflection to occur?

Details and Assumptions:

- Measure the entering angle from the axis of the fiber.
- Use the following refraction indexes: $n_{air} = 100, n_{glass} = 1.50$, and $n_{coating} = 1.46$.

Shifting of Image and Object

When viewing an object from different media the refraction of light often creates a shifting of the image. Because of this, for example, when viewing a pencil through a thick glass slab there is a shift of the image.

This shift of image is given by

$$\mu = \frac{(\text{real height})}{(\text{apparent height})}$$

This is useful when viewing an object from water where the light bends only once.

However, when viewing an object from a glass slab, the light rays bend twice: once when going to the glass and again when coming out from the glass. So here we use

$$\Delta x = t\left(1 - \frac{1}{\mu}\right)$$

where Δx is the shift in the image and t is the thickness of the glass slab.

As a mathematical study, geometrical optics emerges as a short-wavelength limit for solutions to hyperbolic partial differential equations. In this short-wavelength limit, it is possible to approximate the solution locally by

$$u(t,x) \approx a(t,x)e^{i(k\cdot x - \omega t)}$$

where k,ω satisfy a dispersion relation, and the amplitude $a(t,x)$ varies slowly. More precisely, the leading order solution takes the form

$$a_0(t,x)e^{i\varphi(t,x)/\varepsilon}$$

The phase $\varphi(t,x)/\varepsilon$ can be linearized to recover large wavenumber $\varepsilon k := \nabla_x \varphi$, and frequency $\omega := -\partial_t \varphi$. The amplitude a_0 satisfies a transport equation. The small parameter ε enters the scene due to highly oscillatory initial conditions. Thus, when initial conditions oscillate much faster than the coefficients of the differential equation, solutions will be highly oscillatory, and transported along rays. Assuming coefficients in the differential equation are smooth, the rays will be too. In other words, refraction does not take place. The motivation for this technique comes from studying the typical scenario of light propagation where short wavelength light travels along rays that minimize (more or less) its travel time. Its full application requires tools from microlocal analysis.

A Simple Example

Starting with the wave equation for $(t,x) \in \mathbb{R} \times \mathbb{R}^n$

$$L(\partial_t, \nabla_x)u := \left(\frac{\partial^2}{\partial t^2} - c(x)^2\Delta\right)u(t,x) = 0,\ u(0,x) = u_0(x),\ u_t(0,x) = 0$$

assume an asymptotic series solution of the form

$$u(t,x) \sim a_\varepsilon(t,x)e^{i\varphi(t,x)/\varepsilon} = \sum_{j=0}^{\infty} i^j \varepsilon^j a_j(t,x)e^{i\varphi(t,x)/\varepsilon}.$$

Check that

$$L(\partial_t,\nabla_x)(e^{i\varphi(t,x)/\varepsilon})a_\varepsilon(t,x) = e^{i\varphi(t,x)/\varepsilon}\left(\left(\frac{i}{\varepsilon}\right)^2 L(\varphi_t,\nabla_x\varphi)a_\varepsilon + \frac{2i}{\varepsilon}V(\partial_t,\nabla_x)a_\varepsilon + \frac{i}{\varepsilon}(a_\varepsilon L(\partial_t,\nabla_x)\varphi) + L(\partial_t,\nabla_x)a_\varepsilon\right)$$

with

$$V(\partial_t,\nabla_x) := \frac{\partial\varphi}{\partial t}\frac{\partial}{\partial t} - c^2(x)\sum_j \frac{\partial\varphi}{\partial x_j}\frac{\partial}{\partial x_j}$$

Plugging the series into this equation, and equating powers of ε, the most singular term $O(\varepsilon^{-2})$ satisfies the eikonal equation (in this case called a dispersion relation),

$$0 = L(\varphi_t,\nabla_x\varphi) = (\varphi_t)^2 - c(x)^2(\nabla_x\varphi)^2.$$

To order ε^{-1}, the leading-order amplitude must satisfy a transport equation

$$2Va_0 + (L\varphi)a_0 = 0$$

With the definition $k := \nabla_x\varphi,\ \omega := -\varphi_t$, the eikonal equation is precisely the dispersion relation that results by plugging the plane wave solution $e^{i(k\cdot x-\omega t)}$ into the wave equation. The value of this more complicated expansion is that plane waves cannot be solutions when the wave-speed c is non-constant. However, it can be shown that the amplitude a_0 and phase φ v are smooth, so that on a local scale there are plane waves.

To justify this technique, the remaining terms must be shown to be small in some sense. This can be done using energy estimates, and an assumption of rapidly oscillating initial conditions. It also must be shown that the series converges in some sense.

Reflection

"Reflection is a change in direction of a wave upon striking the interface between two materials."

When a ray of light strikes a plane mirror, the light ray reflects off the mirror. Reflection involves a change in direction of the light ray. The convention used to express the direction of a light ray is to indicate the angle which the light ray makes with a normal line drawn to the surface of the mirror. The angle of incidence is the angle between this normal line and the incident ray; the angle of reflection is the angle between this normal line and the reflected ray.

In optics, it relates to an image of an object.

Reflectional Symmetry

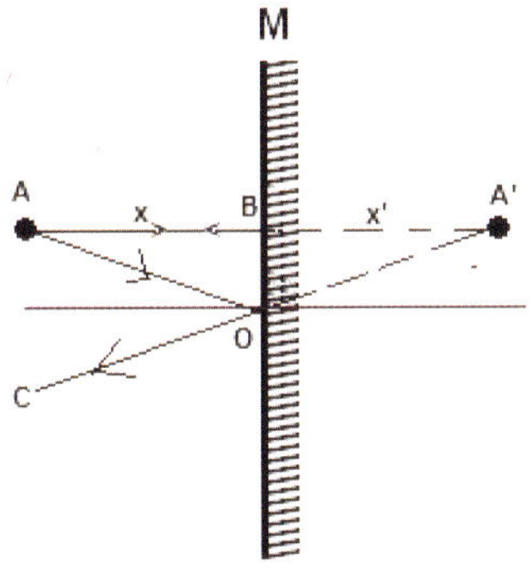

Look at the above diagram. An object A is placed in front of a mirror M. Consider rays AB and AO from A. As per law of reflection, the ray AB is reflected back as BA and ray AO reflected as OC.

Retracing these rays behind the mirror they appear to intersect at A'. This is where you find A', an image of A and it appears to be at the same distance behind as much as the object is before the mirror. That is the measures x and x' are same, though you cannot actually measure x'. The creation of image A' is actually called as the reflection of A.

Reflection of Light

Light rays are basically wave motions and they are capable of being reflected by an interface in between two mediums. Such interfaces are called Light Reflectors.

The most common Light reflector is the one we use many times a day, which is nothing but a plane mirror. But it is not only a plane mirror reflects light. Even a curved mirror does that though the nature of reflection of light by curved mirrors is different. Let us discuss this part separately at a later point.

Now for the present, let us concentrate on reflection of light by a plane mirror described by the following diagram.

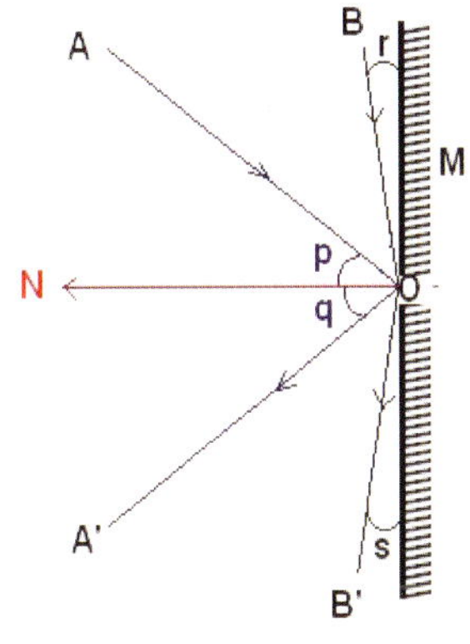

M is a plane mirror. It is a plane glass but on one side (the hatching show the surface) a suitable abrasive is coated to give the reflection effect to the glass, that is to make the glass to be a plane mirror. It also acts as an interface to separate the mediums on the left and right to that.

Let O be a any point on the plane mirror. The line NO is the perpendicular or better called as 'principal' to the plane mirror at O. Suppose a light ray AO hits the plane mirror at O, it is 'reflected to the same medium as ray OA'. This is called Reflection of light.

Types of Reflection

Diffuse Reflection

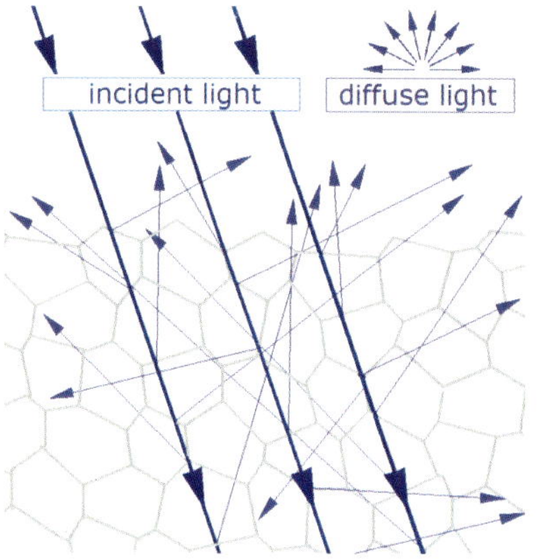

General scattering mechanism which gives diffuse reflection by a solid surface

When light strikes the surface of a (non-metallic) material it bounces off in all directions due to multiple reflections by the microscopic irregularities *inside* the material (e.g. the grain boundaries of a polycrystalline material, or the cellor fiber boundaries of an organic material) and by its surface, if it is rough. Thus, an ‹image' is not formed. This is called *diffuse reflection*. The exact form of the reflection depends on the structure of the material. One common model for diffuse reflection is Lambertian reflectance, in which the light is reflected with equal luminance (in photometry) or radiance (in radiometry) in all directions, as defined by Lambert's cosine law.

The light sent to our eyes by most of the objects we see is due to diffuse reflection from their surface, so that this is our primary mechanism of physical observation.

Retroreflection

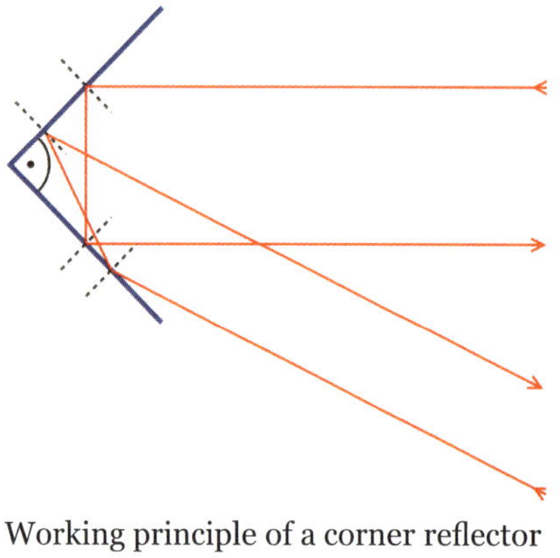

Working principle of a corner reflector

Some surfaces exhibit *retroreflection*. The structure of these surfaces is such that light is returned in the direction from which it came.

When flying over clouds illuminated by sunlight the region seen around the aircraft's shadow will appear brighter, and a similar effect may be seen from dew on grass. This partial retro-reflection is created by the refractive properties of the curved droplet's surface and reflective properties at the backside of the droplet.

Some animals' retinas act as retroreflectors as this effectively improves the animals' night vision. Since the lenses of their eyes modify reciprocally the paths of the incoming and outgoing light the effect is that the eyes act as a strong retroreflector, sometimes seen at night when walking in wildlands with a flashlight.

A simple retroreflector can be made by placing three ordinary mirrors mutually perpendicular to one another (a corner reflector). The image produced is the inverse of one produced by a single mirror. A surface can be made partially retroreflective by depositing a layer of tiny refractive spheres on it or by creating small pyramid like structures. In both cases internal reflection causes the light to be reflected back to where it originated. This is used to make traffic signs and automobile license plates reflect light mostly back in the direction from which it came. In this application perfect retroreflection is not desired, since the light would then be directed back into the headlights of an oncoming car rather than to the driver's eyes.

Multiple Reflections

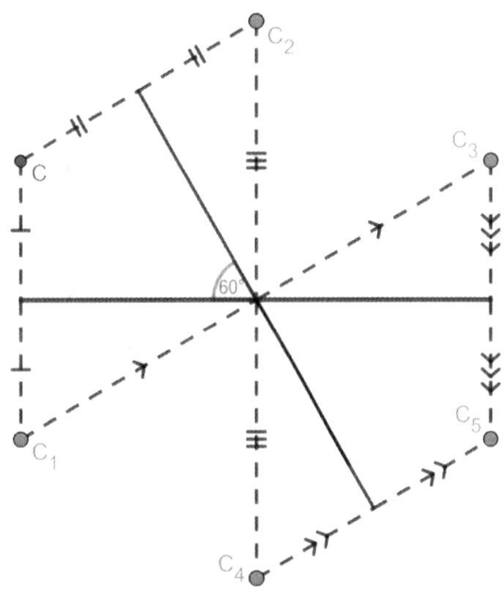

Multiple reflections in two plane mirrors at a 60° angle.

When light reflects off a mirror, one image appears. Two mirrors placed exactly face to face give the appearance of an infinite number of images along a straight line. The multiple images seen between two mirrors that sit at an angle to each other lie over a circle. The center of that circle is located at the imaginary intersection of the mirrors. A square of four mirrors placed face to face give the appearance of an infinite number of images arranged in a plane. The multiple images seen between four mirrors assembling a pyramid, in which each pair of mirrors sits an angle to each other, lie over a sphere. If the base of the pyramid is rectangle shaped, the images spread over a section of a torus.

Note that these are theoretical ideals, requiring perfect alignment of perfectly smooth, perfectly flat perfect reflectors that absorb none of the light. In practice, these situations can only be approached but not achieved because the effects of any surface imperfections in the reflectors propagate and magnify, absorption gradually extinguishes the image, and any observing equipment (biological or technological) will interfere.

Complex Conjugate Reflection

In this process (which is also known as phase conjugation), light bounces exactly back in the direction from which it came due to a nonlinear optical process. Not only the direction of the light is reversed, but the actual wavefronts are reversed as well. A conjugate reflector can be used to remove aberrations from a beam by reflecting it and then passing the reflection through the aberrating optics a second time.

Other Types of Reflection

Neutron Reflection

Materials that reflect neutrons, for example beryllium, are used in nuclear reactors and nuclear weapons. In the physical and biological sciences, the reflection of neutrons off of atoms within a material is commonly used to determine the material's internal structure.

Sound Reflection

Sound diffusion panel for high frequencies

When a longitudinal sound wave strikes a flat surface, sound is reflected in a coherent manner provided that the dimension of the reflective surface is large compared to the wavelength of the sound. Note that audible sound has a very wide frequency range (from 20 to about 17000 Hz), and thus a very wide range of wavelengths (from about 20 mm to 17 m). As a result, the overall nature of the reflection varies according to the texture and structure of the surface. For example, porous materials will absorb some energy, and rough materials (where rough is relative to the wavelength) tend to reflect in many directions—to scatter the energy, rather than to reflect it coherently. This leads into the field of architectural acoustics, because the nature of these reflections is critical to the auditory feel of a space. In the theory of exterior noise mitigation, reflective surface size mildly detracts from the concept of a noise barrier by reflecting some of the sound into the opposite direction. Sound reflection can affect the acoustic space.

Seismic Reflection

Seismic waves produced by earthquakes or other sources (such as explosions) may be reflected by layers within the Earth. Study of the deep reflections of waves generated by earthquakes has allowed seismologists to determine the layered structure of the Earth. Shallower reflections are used in reflection seismology to study the Earth's crust generally, and in particular to prospect for petroleum and natural gas deposits.

Law of Reflection

Let us first study what is law of reflection with respect to a single light ray. Look at the following diagram.

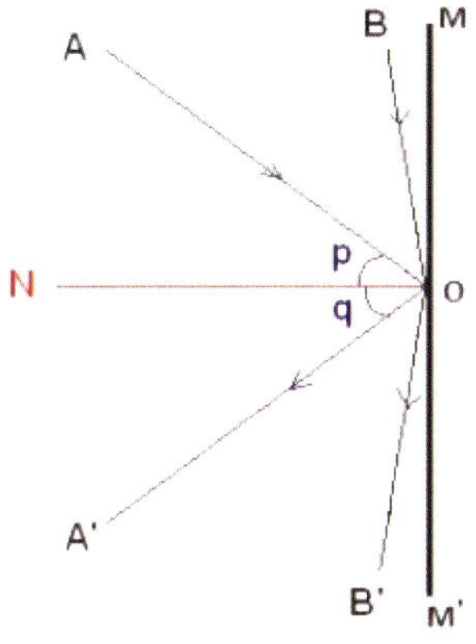

Let MM' be a plane mirror which is absolutely smooth. A light ray AO strikes the mirror at O (the incident ray), and it is reflected as ray OA', called as reflected ray. ON is the perpendicular to MM' at O. It is also called as the normal to the surface at the point of reflection.

The angle made by the incident ray with the normal ON is p and q be the angle made by the reflected ray with the same normal. These are called as 'Angle of Incidence' and the 'Angle of Reflection' respectively.

According to the Law of Reflection, Angle of incidence is always equal to angle of reflection.

Law of Reflection Definition

Now, let us consider a set of rays assuming a plane mirror with a smooth and perfect surface. The situation is explained in the following diagram:

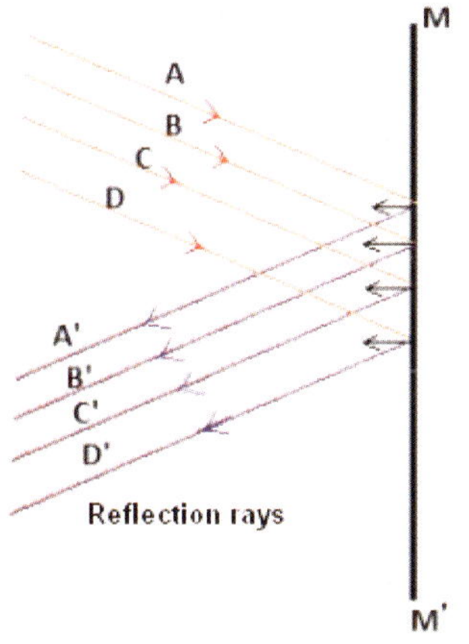

Let ABCD is a set of rays striking a mirror MM'. Obviously, the rays of parallel to each other. When each individual ray strikes the mirror it is reflected as per law of reflection, that is the angle of incidence and the angle of reflection are same for each ray. Thus the set of incident rays are reflected as a set of rays. In other words, a beam of light is reflected as a beam of light if the mirror is plane and its reflecting surface is smooth. This type of reflection of a set of rays is called Specular Reflection.

Thus, observing at this diagram we can Define Law of Reflection as:

When light falls on a surface which is plane, the angle with which the light falls on the surface is equal to the angle with which the light reflects back and the incident ray, reflected ray and the normal ray lies in the same plane.

Does the Law of Reflection Hold for Diffuse Reflection?

What happens if the reflecting surface is rough or curved? Does the law of reflection hold for diffuse reflection? Does the law of reflection hold for curved mirrors? Diffused Reflection takes place when the incident light falls on the rough surface.

Let us study the situation again considering each ray in a set of rays through the following diagram.

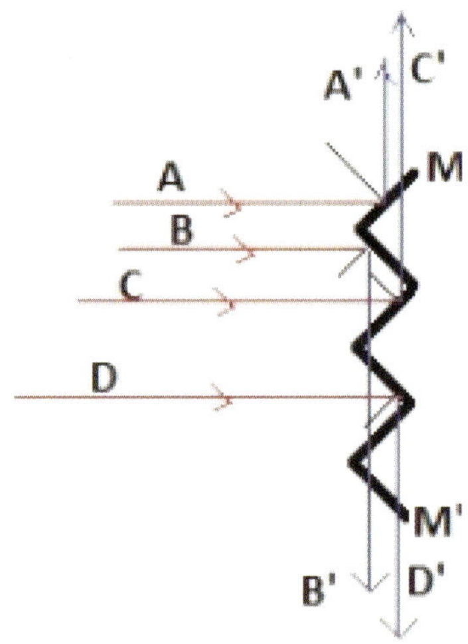

Consider a reflecting surface MM' which is highly exaggerated rough for the purpose of study. A set of rays ABCD strike the reflecting surface. But when you consider the rays individually, the infinitesimally small reflecting surface on which each ray falls is plane. Hence as far as that ray is concerned the law of reflection holds good and the reflection obeys the law that angle of incidence is equal to the angle of reflection. For example, the ray A is reflected as A'. Overall, the set of rays ABCD is reflected as A'B'C'D' as shown in the diagram.

It may be noticed that while the incident rays are parallel, the reflected rays emerge in different directions. This is called scatter reflection or diffuse reflection. Thus the law of reflection still holds in case of diffuse reflection but the net is result different from specular reflection.

Does the Law of Reflection Hold for Curved Mirrors?

Now let us consider the case of a curved mirror. As an example a convex mirror is shown in the following diagram.

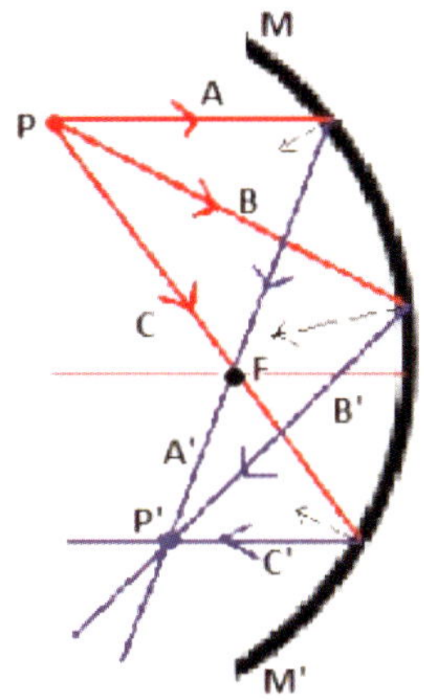

MM' is a concave mirror and rays A, B and C from an object P emerging out in different directions. Each of these is reflected following the law of reflection at the respective striking points. Thus the law of reflection holds good for curved mirrors also.

The corresponding reflected rays are A', B' and C' respectively and we find these rays meet at a point P' to create an image of the object P.

An incident ray, that is parallel to the principal axis of the mirror is reflected through a point F on the principal axis and an incident ray through F is reflected parallel to the principal axis. This is called two rules of a curved mirror and simplifies the job of drawing a ray diagram. The point F is called the Focal point of the mirror.

Hence, we can Conclude that Law of reflection does not hold for Curved mirrors as the angle of incidence is different from the angle of reflection.

Angle of Incidence

The Angle of incidence can be defined as the angle with which a light ray strikes a reflecting surface. But the angle of strike is measured as the angle between the ray and the normal to the surface at the point of incidence of the ray. It has a relation with the angle of reflection and also with the type of reflecting surface.

Angle of Incidence Definition

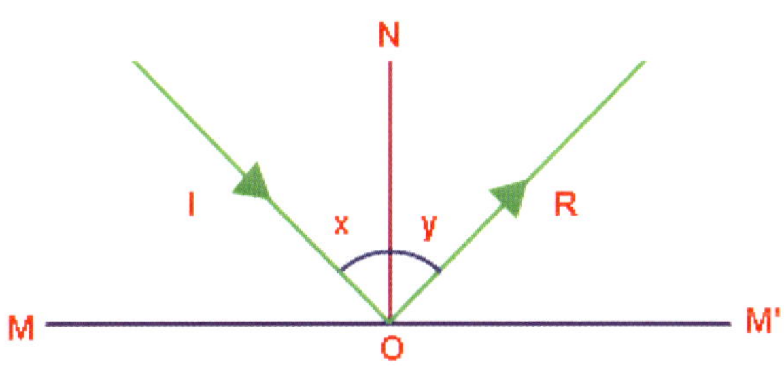

The ray I from a light source strikes the mirror MM' at point O This ray is called as incident ray. ON is the perpendicular to the mirror MM' at O which is also called the normal to the mirror at O. The angle 'x' made by the ray I, with the normal ON is called the 'angle of incidence'.

The incident ray, is reflected by the mirror when it strikes the mirror at O and the ray after reflection 'R' is called as 'reflected ray'. The angle 'y' made by the reflected ray with the same ray is called 'angle of reflection'.

Angle of Incidence and Reflection

As per law of reflection the angle of incidence is always equal to angle of reflection. This can be proved mathematically and also by a simple experiment. There are many mathematical proofs but the simplest proof is based on Fermat's Principle.

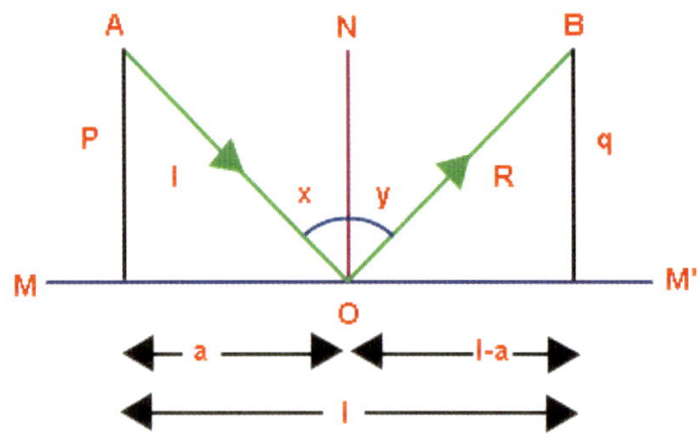

Look at the above diagram. Let a right ray AO from a source A strike a plane mirror at O and reflected as ray OB. Let B be a point on ray OB and let ON be the normal to the mirror at O. The angle of incidence is 'x' and angle of reflection is 'y', as shown. Let 't' be the time taken for the ray to travel the distance AO + OB and 'c' be the velocity of light, which is a constant.

The time 't' can be worked out as,

$$t = \frac{AO}{c} + \frac{OB}{c} \; or,$$

$$t = \frac{\sqrt{a^2 + p^2}}{c} + \frac{\sqrt{(l-a)^2 + q^2}}{c}$$

As per Fermat's principle light takes the path of least time. Therefore, the derivative $\frac{dt}{da}$ must be 0.

$$Thus\ is\ 0 = \frac{1}{c}\left[\frac{a}{\sqrt{a^2 + p^2}} + \frac{-(l-a)}{\sqrt{(l-a)^2 + q^2}}\right]$$

$$\text{or, } \frac{a}{\sqrt{a^2 + p^2}} = \frac{-(l-a)}{\sqrt{(l-a)^2 + q^2}}$$

But as per trigonometry, this is same as $sin(x) = sin(y)\ or\ x = y$.

Hence,

The Angle of Incidence = Angle of Reflection

The law of reflection can also be proved experimentally. The arrangement for the experiment is described below.

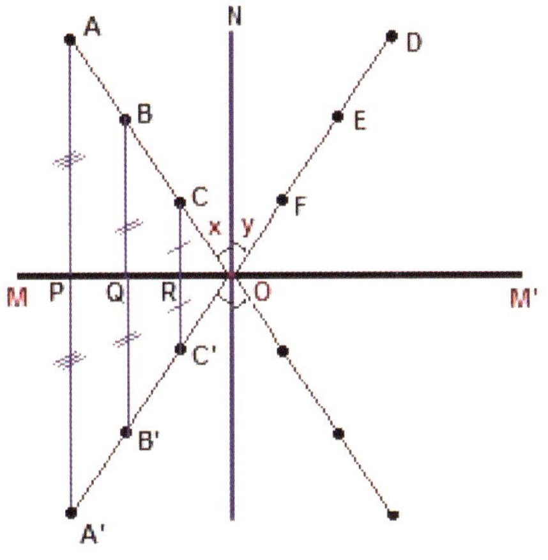

Take a white card board sheet and fix the same firmly on a board. Draw a thick line MM' and mark the midpoint O. Draw a perpendicular ON at O. Place a plain mirror on MM' with its center at O. Take three pins and fix them vertically on the sheet at positions C, B and A on one side of ON such that the pins lie on a straight line when viewed from A. Now move to the other side of the perpendicular ON and fix three more pins at F, E and D in such way that the images A' B' C' of the pins at A, B, C and the pins at D, E, F are all in one straight line. This is the most important point of the entire experiment.

Now remove the mirror and the pins carefully marking the points where they have been fixed. At this stage we will not be able to locate the positions of A', B' and C'. Draw a perpendicular AP from A and extend that line. Join DO and extend that too. Let A' be the point of intersection of these extended line. A' is nothing but the position of the image that appeared when the mirror was there. Same way locate the positions of B' and C'.

If you join AO and measure the angles AON and DON, you will find them to be equal. Thus, it proves the angle of incidence equals angle of reflection. Also you will notice that the following measures to be equal. AP = PA', BQ = QB' and CR = RC', which proves the image distance equals the object distance.

Angle of Reflection

Referring to the same diagram, the angle A'ON between the reflected ray OA' and the principal axis NO is called as 'Angle of Reflection'.

The Angle of reflection is the angular reflection formed by a reflected ray and a perpendicular to the surface at the point of reflection. The Angle of Reflection depends on the Angle of Incidence.

Angle of Reflection Formula

The Angle of incidence always equals the angle of reflection and the distance of image always equals the distance of object are called Angle of reflection formula.

if θ_i and θ_r are the angle of incidence and angle of reflection respectively, then

$$\sin\theta_i = \sin\theta_r$$

or

$$\theta_i = \theta_r$$

This is the Angle of Reflection Formula.

Grazing Angle

Now, in the same diagram, we see another ray BO which falls on the mirror and reflected as as a ray OB'. In this case, it is easier to perceive the angles at which the incident ray and the reflected ray with the surface of the mirror rather than to its normal. They are angles 'r' and 's' and these are called 'grazing angles'. The angles 'r' and's' are congruent because of the fact they are the difference of 90° – the angle of incidence or reflection as the case may be.

Specular Reflection

Specular reflection is a type of surface reflectance often described as a mirror-like reflection of light from the surface. In specular reflection, the incident light is reflected into a single outgoing direction. The name specular is derived from the Latin word speculum, meaning mirror.

Specular reflection obeys the Laws of Reflection and the light is reflected from a smooth surface at a definite angle. Specular reflection occurs for all polished and smooth surfaces, like mirrors. The angle of reflection in specular reflection is equal to the angle of incidence. Rough or unpolished surfaces are not good specular reflectors as they do not reflect light at the same angle as the angle of incidence.

Mirrors that are used for specular reflection are usually made of metals such as aluminum or silver. Specular reflection is measured using incidence reflectometers by scanning a variable wavelength source.

If a specular reflection light ray goes straight into a camera, it is as if the camera is seeing the light source directly, even though it has bounced off of an object. The camera is seeing the light of the light source, not the color of the object. If you have a white light source, the specular reflection will be white. If you have a red light source, the specular reflection will be red. Therefore, to model specular reflection you need to specify the color of the light source in your lighting model.

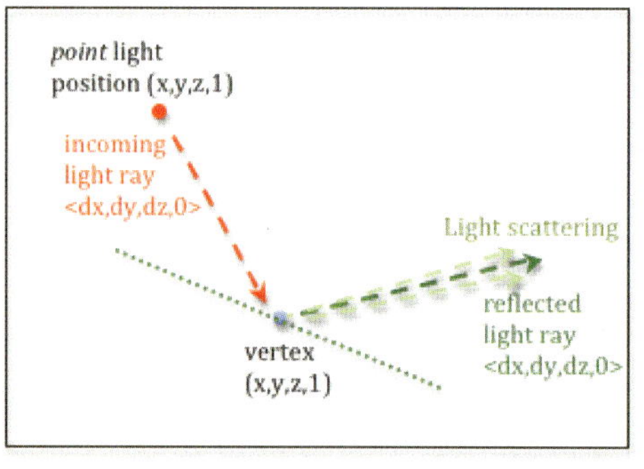

Specular light scattering.

We assume there is a small amount of light scattering around the reflection vector, but not much. The scattering is centered around the reflection vector as shown in the diagram to the left.

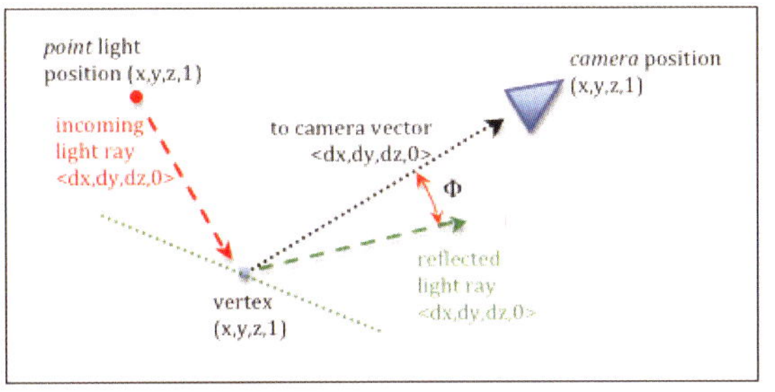

Specular light intensity.

If we calculate a vector from a fragment to the camera, and then calculate the angle between it and the reflection vector, we can use the angle to estimate the amount of specular reflection the camera receives. If the angle is zero the camera is receiving the entire reflection light ray. If the angle is very small, then some percentage of the reflection ray is going into the camera. If the angle is large, then no light from the reflection is entering the camera. If the angle is between small and large, we need a way to calculate some percentage of reflected light that is entering the camera. We used a cosine function to calculate diffuse light percentages, but the cosine function is too broad for the a focused specular reflection. However, if we raise the cosine function to a power, the curve collapses around the Y axis. Experiment with various exponents in the plot of the $\cos(angle)^{exp}$ function below. By using various exponents for this equation, we can simulate various amounts of light scattering around the specular reflection vector. If the exponent is large, such as 100, then the $\cos(angle)^{exp}$ shrinks around the Y axis and only very small angles will return a significant percentage value. If the exponent is small, such as 1.0, a broad amount of light around the reflection ray will be simulated.

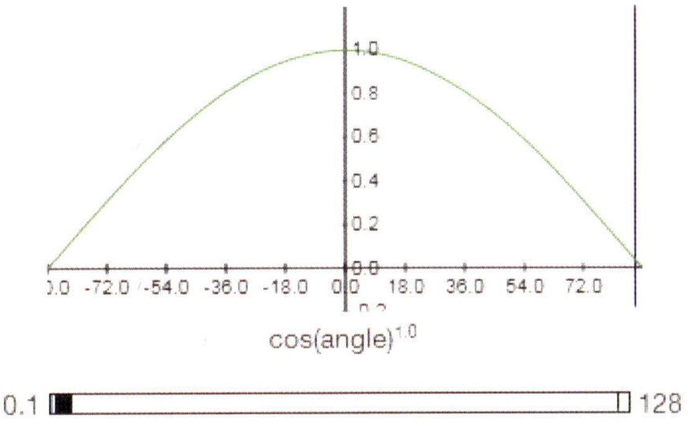

The Math for Specular Reflection

To calculate specular reflection, we need two vectors:

- A vector from the fragment location to the camera.
- A reflected light vector

The angle between these two vectors determines the amount of specular reflection.

To calculate a vector from the fragment position to the camera, subtract the head of the vector (the camera) from the tail (the fragment location). As we previously discussed, we are performing these calculations in "camera space" and the camera is located at the global origin, (0,0,0). Therefore, a vector to the camera is:

$$to\_camera[0] = 0 - fragment\_position[0]$$
$$to\_camera[1] = 0 - fragment\_position[1]$$
$$to\_camera[2] = 0 - fragment\_position[2]$$

We can calculate this vector by simply multiplying the fragment location by -1.

Calculating the Reflection Vector

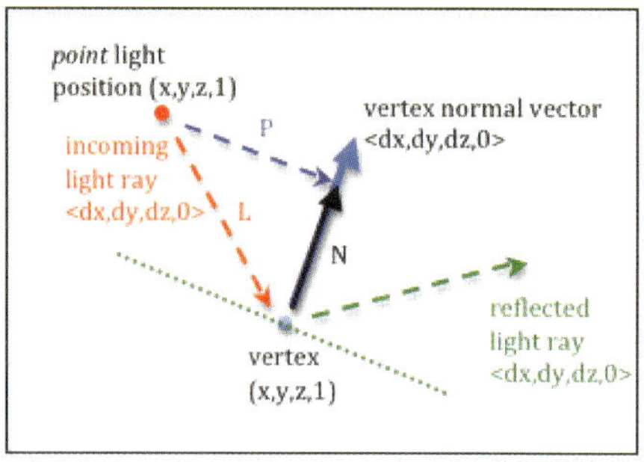

Calculating the reflection vector

To calculate the reflection vector, we need to set up several intermediate vectors. Please study the diagram to the right before proceeding with this discussion. The position of the light source can be projected onto the vertex's normal vector. This is the "p" vector in the diagram. It can be shown that a vector in the direction of the normal vector that has a length to the projected point can be calculated by taking the dot product of -L and a normalized vertex normal vector. The dot product calculates the angle between the two vectors, but only when the vectors have unit length. Notice that the normal vector is normalized, but the -L vector is not. This provides the correct length for the N vector to the projected point. To summarize, to calculate N we:

- Normalize the vertex's normal vector to make it one unit in length. Let's call it n.
- Calculate the dot product of -L and n. Let's call this value *s*.
- Scale n by *s* to create the vector N, which is in the direction of the vertex's normal vector and has a length to the projected point of the light source.
- From simple vector addition, L + N === P.

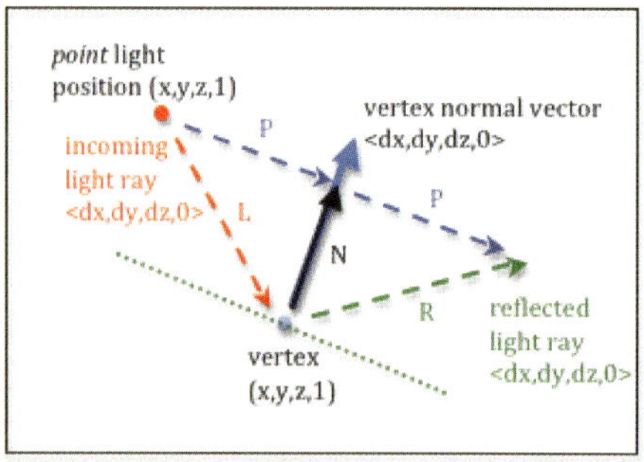

Calculating the reflection vector

Now we know the vector P. Observe that from simple vector addition, the reflection vector R is equal to N + P. We can substitute the calculations for N and P into this equation like this:

$$R = N + PR$$
$$= n * dot\_product(n, -L) + (L + N)$$
$$R = n * dot\_product(n, -L) + (L + n * dot\_product(n, -L))$$
$$R = 2 * n * dot\_product(n, -L) + L$$

If you flip the direction of L and follow the same logic, you get:

$$R = 2 * n * dot\_product(n, L) - L \; // \; \textit{When L goes from the vertex to the light sourc}$$

The Percentage of a Reflection Ray a Camera Receives

We have a vector from the vertex to the camera and we have a reflection vector. The angle between these two vectors indicates how much of the reflection ray goes into the camera. The cosine of this angle, which is a percentage between 0.0 and 1.0 is raised to some power to simulate a focused bean of light around the reflection ray. The exponent is typically called the "shininess" exponent. The larger the exponent, the more shiny an object appears. Shiny objects have small, focused specular reflection. Dull objects have large, spread out specular reflection.

Specular Reflection in Shader Programs

Please study the following *shader programs*. Then compare the programs to the comments below.

Vertex Shader

```
// Vertex Shader
precision mediump int;
precision mediump float;


// Scene transformations
uniform mat4 u_PVM_transform; // Projection, view, model transform
uniform mat4 u_VM_transform;  // View, model transform


// Light model
uniform vec3 u_Light_position;
uniform vec3 u_Light_color;
uniform float u_Shininess;
```

```
// Original model data
attribute vec3 a_Vertex;
attribute vec3 a_Color;
attribute vec3 a_Vertex_normal;

// Data (to be interpolated) that is passed on to the fragment shader
varying vec3 v_Vertex;
varying vec4 v_Color;
varying vec3 v_Normal;

void main() {

  // Perform the model and view transformations on the vertex and pass
this
  // location to the fragment shader.
  v_Vertex = vec3( u_VM_transform * vec4(a_Vertex, 1.0) );

  // Perform the model and view transformations on the vertex's normal
vector
  // and pass this normal vector to the fragment shader.
  v_Normal = vec3( u_VM_transform * vec4(a_Vertex_normal, 0.0) );

  // Pass the vertex's color to the fragment shader.
  v_Color = vec4(a_Color, 1.0);

  // Transform the location of the vertex for the rest of the graphics
pipeline
  gl_Position = u_PVM_transform * vec4(a_Vertex, 1.0);
}
```

- This is the same vertex shader program used for diffuse lighting in the previous lesson, with the addition of two new uniform variables: u_Light_color and u_Shininess. These values enhance our lighting model.

- All calculations in the *fragment shader* will be done in "camera space," so the vertex data is transformed by the *model* and *camera* transformations, but not the *projection* transformation.

Fragment Shader

The fragment shader calculates a reflection vector and then determines if any reflected light should be used to color the pixel.

```
// Fragment shader program
precision mediump int;
precision mediump float;

// Light model
uniform vec3 u_Light_position;
uniform vec3 u_Light_color;
uniform float u_Shininess;

// Data coming from the vertex shader
varying vec3 v_Vertex;
varying vec4 v_Color;
varying vec3 v_Normal;

void main() {
  vec3 to_light;
  vec3 vertex_normal;
  vec3 reflection;
  vec3 to_camera;
  float cos_angle;
  vec3 specular_color;
  vec3 object_color;
  vec3 color;

  // Calculate a vector from the fragment location to the light source
  to_light = u_Light_position - v_Vertex;
```

```
  to_light = normalize( to_light );

  // The vertex's normal vector is being interpolated across the primi-
tive
  // which can make it un-normalized. So normalize the vertex's normal
vector.
  vertex_normal = normalize( v_Normal );

  // Calculate the reflection vector
  reflection = 2.0 * dot(vertex_normal,to_light) * vertex_normal - to_
light;

  // Calculate a vector from the fragment location to the camera.
  // The camera is at the origin, so negating the vertex location gives
the vector
  to_camera = -1.0 * v_Vertex;

  // Calculate the cosine of the angle between the reflection vector
  // and the vector going to the camera.
  reflection = normalize( reflection );
  to_camera = normalize( to_camera );
  cos_angle = dot(reflection, to_camera);
  cos_angle = clamp(cos_angle, 0.0, 1.0);
  cos_angle = pow(cos_angle, u_Shininess);

  // If this fragment gets a specular reflection, use the light's color,
  // otherwise use the objects's color
  specular_color = u_Light_color * cos_angle;
  object_color = vec3(v_Color) * (1.0 - cos_angle);
  color = specular_color + object_color;

  gl_FragColor = vec4(color, v_Color.a);
}
```

- The angle between a vector to the camera and a reflection vector is calculated. The calculations are exactly as described in the math section above.
- It is important to note that the cosine of the angle is clamped to a value between 0.0 and 1.0 before is it raised to the shininess exponent. The angle can be negative if the reflection vector and the vector to the camera have an angle greater than 90 degrees. Raising a number to a power can make a value positive. For example, $(-2)^2$ is +4. (If the cosine of the angle is negative, it will be clamped to zero, which means 0% of reflected light gets to the camera, which is the physical result we want.)
- If the value of the cos(angle) is one, we want the color of the light source for this fragment. If the value of the cos(angle) is zero, we want the color of the object. If the cos(angle) is some percentage between zero and one, we take this percentage of the light's color, and the left-over percentage, (1.0 - cos_angle), of the color of the object. "Color is additive" so we can simply add the two color values together.

Law of Reflection

The law of reflection describes the angle of reflected light: the angle of incident light is the same as the angle of the reflected light.

The law of reflection arises from diffraction of a plane wave with small wavelength on a flat boundary: when the boundary size is much larger than the wavelength, then electrons of the boundary are seen oscillating exactly in phase only from one direction – the specular direction. If a mirror becomes very small compared to the wavelength, the law of reflection no longer holds, and the behavior of light is more complicated.

Vector Formulation

The law of reflection can also be equivalently expressed using linear algebra. The direction of a reflected ray is determined by the vector of incidence and the surface normal vector. Given an incident direction $\hat{\mathbf{d}}_{\mathrm{i}}$ from the surface to the light source and the surface normal direction $\hat{\mathbf{d}}_{\mathrm{n}}$ the specularly reflected direction $\hat{\mathbf{d}}_{\mathrm{s}}$ (all unit vectors) is:

$$\hat{\mathbf{d}}_{\mathrm{s}} = 2\left(\hat{\mathbf{d}}_{\mathrm{n}} \cdot \hat{\mathbf{d}}_{\mathrm{i}}\right)\hat{\mathbf{d}}_{\mathrm{n}} - \hat{\mathbf{d}}_{\mathrm{i}},$$

where $\hat{\mathbf{d}}_{\mathrm{n}} \cdot \hat{\mathbf{d}}_{\mathrm{i}}$ is a scalar obtained with the dot product. Different authors may define the incident and reflection directions with different signs. Assuming these Euclidean vectors are represented in column form, the equation can be equivalently expressed as a matrix-vector multiplication:

$$\hat{\mathbf{d}}_{\mathrm{s}} = \mathbf{R}\,\hat{\mathbf{d}}_{\mathrm{i}},$$

where $\mathbf{R}$ is the so-called Householder transformation matrix, defined as:

$$\mathbf{R} = \mathbf{I} - 2\hat{\mathbf{d}}_{\mathrm{n}}\hat{\mathbf{d}}_{\mathrm{n}}^{\mathrm{T}};$$

T denotes transposition and $\mathbf{I}$ is the identity matrix.

Reflectivity

Reflectivity is the ratio of the power of the reflected wave to that of the incident wave. It is a function of the wavelength of radiation, and is related to the refractive index of the material as expressed by Fresnel's equations. In regions of the electromagnetic spectrum in which absorption by the material is significant, it is related to the electronic absorption spectrum through the imaginary component of the complex refractive index. The electronic absorption spectrum of an opaque material, which is difficult or impossible to measure directly, may therefore be indirectly determined from the reflection spectrum by a Kramers-Kronig transform. The polarization of the reflected light depends on the symmetry of the arrangement of the incident probing light with respect to the absorbing transitions dipole moments in the material.

Measurement of specular reflection is performed with normal or varying incidence reflection spectrophotometers (*reflectometer*) using a scanning variable-wavelength light source. Lower quality measurements using a glossmeter quantify the glossy appearance of a surface in gloss units.

Consequences

Internal Reflection

When light is propagating in a material and strikes an interface with a material of lower index of refraction, some of the light is reflected. If the angle of incidence is greater than the critical angle, total internal reflection occurs: all of the light is reflected. The critical angle can be shown to be given by

$$\theta_{crit} = \arcsin\left(\frac{n_2}{n_1}\right).$$

Polarization

When light strikes an interface between two materials, the reflected light is generally partially polarized. However, if the light strikes the interface at Brewster's angle, the reflected light is *completely* linearly polarized parallel to the interface. Brewster's angle is given by

$$\theta_B = \arctan\left(\frac{n_2}{n_1}\right).$$

Reflected Images

The image in a flat mirror has these features:

- It is the same distance behind the mirror as the object is in front.
- It is the same size as the object.
- It is the right way up (erect).
- It is reversed.
- It is *virtual*, meaning that the image appears to be behind the mirror, and cannot be projected onto a screen.

The reversal of images by a plane mirror is perceived differently depending on the circumstances. In many cases, the image in a mirror appears to be reversed from left to right. If a flat mirror is mounted on the ceiling it can appear to reverse *up* and *down* if a person stands under it and looks up at it. Similarly a car turning *left* will still appear to be turning *left* in the rear view mirror for the driver of a car in front of it. The reversal of directions, or lack thereof, depends on how the directions are defined. More specifically a mirror changes the handedness of the coordinate system, one axis of the coordinate system appears to be reversed, and the chirality of the image may change. For example, the image of a right shoe will look like a left shoe.

Examples

Esplanade of the Trocadero in Paris after rain. The layer of water exhibits specular reflection, reflecting an image of the Eiffel Tower and other objects.

A classic example of specular reflection is a mirror, which is specifically designed for specular reflection.

In addition to visible light, specular reflection can be observed in the ionospheric reflection of radiowaves and the reflection of radio- or microwave radar signals by flying objects. The measurement technique of x-ray reflectivityexploits specular reflectivity to study thin films and interfaces with sub-nanometer resolution, using either modern laboratory sources or synchrotron x-rays.

Non-electromagnetic waves can also exhibit specular reflection, as in acoustic mirrors which reflect sound, and atomic mirrors, which reflect neutral atoms. For the efficient reflection of atoms from a solid-state mirror, very cold atoms and/or grazing incidence are used in order to provide significant quantum reflection; ridged mirrors are used to enhance the specular reflection of atoms. Neutron reflectometry utilizes specular reflection to study material surfaces and thin film interfaces in an analogous fashion to x-ray reflectivity.

Diffuse Reflection

Diffuse reflection is the property of light to scatter when reflecting off a surface. Light travels in a straight line, and when it strikes an object, its expected primary angle of reflection is called a specular reflection. In addition, the light also reflects in all possible angles and directions. The degree or magnitude of this diffused reflection is dependent upon the characteristics of the reflecting substance and surface.

Some surfaces, such as a highly polished metal mirror, reflect light at nearly 100% specular efficiency. Other surfaces, such as those of some crystals or liquids, may most or all of it to pass through the surface and medium. Most objects of the world, including the above examples, reflect light both specularly and diffusely to varying degrees. Diffuse reflection is the primary property of light that allows human eyes to see an object.

The main determinant of reflection is the surface's absorption of light. Polished surfaces such as white marble stone, or the randomly irregular fibers of paper, reflect diffuse light with nearly equal efficiency. Black objects tend to absorb more light. Some naturally occurring as well as human made things emit light and overwhelm any diffuse reflection from their surface to distinguish them as objects.

Diffuse Reflection of Ambient Light

If we assume a uniform intensity of ambient light, Ia, then the intensity of the diffuse reflection at any point on the surface: I = ka * Ia, where ka = the ambient diffuse coefficient of reflection with: 0.0 <= ka <= 1.0. For colored surfaces there are three ka values, one for red, green, and blue. We are approximating the ambient diffuse part of the reflection intensity function R(j,q,l) by the constant ka. For a dull, non-reflecting surface ka is close to 0.0 and for a highly reflective surface, such as a mirror, ka is close to 1.0.

Diffuse Reflection of Point Light Sources

Consider a point source and assume the source is far from the surface so that incident light rays are parallel. Note: if the light source is near the surface then the light rays are not parallel as below:

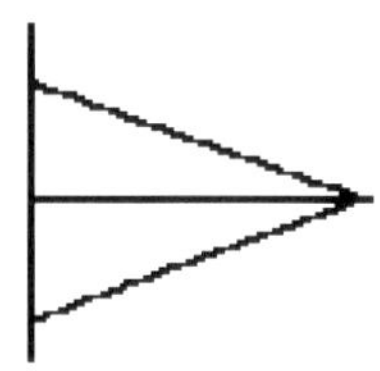

For a point light far from the surface, we use Lambert's cosine law which is as follows: the intensity of reflected light is f(angle of illumination). This can be qualitatively seen if we look at the two cases below:

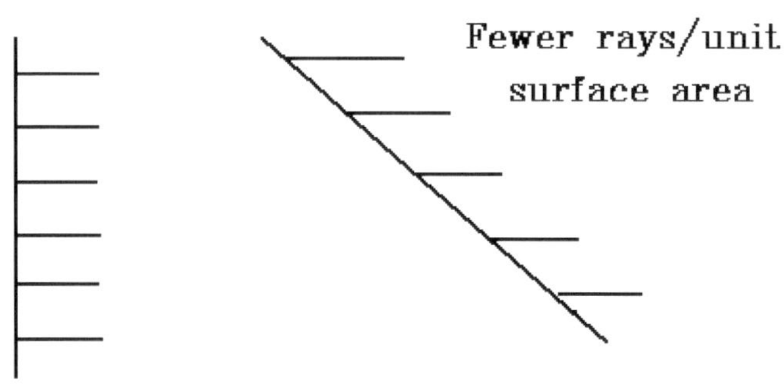

Define N = the unit normal to the surface and L = the unit vector which is the direction to the light source.

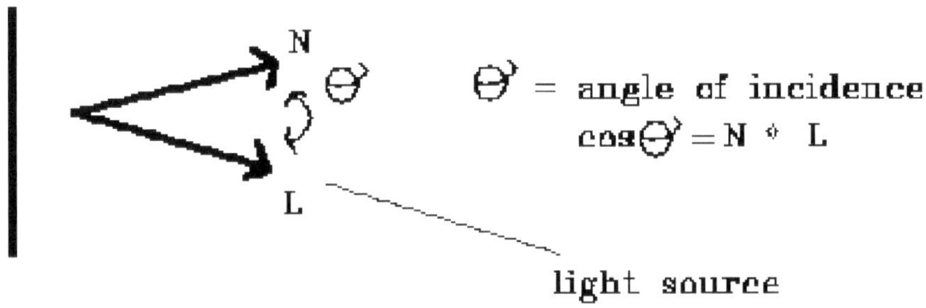

Then Lambert's law states that the reflected light is proportional to cos q (with q <= 90 else the light is behind the surface). Let Ip="intensity" of point light source, then the brightness is $f(1/d^2)$ where d="distance" of light source from the surface. Some models use a d distance dependence rather than d^2, and since most light sources are not really point sources, this works. Some models define a "distant" light source, such as the sun, which have no distance dependence.

Then I = [(kd * Ip) / (D)] * (N�L) with D = the distance function (< = 1 for a distant light source, d or d^2 for other models and kd = the diffuse coefficient of reflection with: 0.0 <= kd <= 1.0. For colored surfaces there are three kd values, one for red, green, and blue. Some models also add a constant term d_0 to prevent d the denominator from going to zero. Frequently ka is set to be the same as kd. Note that the vectors N and L are always normalized.

Then including ambient light we get a simple two term illumination model:

I = ka * Ia + [(kd * Ip) / (d)] * (N�L)

If we consider color then we should have an explicit wavelength dependence. But we usually make the approximation of sampling the color space at only three colors, i.e., "red", "green", and "blue", without really defining what wavelengths these are. Then we will have 3 equations like the above: one for Red, Green, and Blue. For example, a blue surface might have $kd^R = kd^G = 0$, $Kd^B = 0.8$. Note that the intensity, I, should always have a value between 0.0 and 1.0. It may be necessary to clamp the computed values so that they remain in the valid range.

Colored Objects

Up to this point white objects have been discussed, which do not absorb light. But the above scheme continues to be valid in the case that the material is absorbent. In this case, diffused rays will lose some wavelengths during their walk in the material, and will emerge colored.

Diffusion affects the color of objects in a substantial manner because it determines the average path of light in the material, and hence to which extent the various wavelengths are absorbed. Red ink looks black when it stays in its bottle. Its vivid color is only perceived when it is placed on a scattering material (e.g. paper). This is so because light's path through the paper fibers (and through the ink) is only a fraction of millimeter long. However, light from the bottle has crossed several centimeters of ink and has been heavily absorbed, even in its red wavelengths.

And, when a colored object has both diffuse and specular reflection, usually only the diffuse component is colored. A cherry reflects diffusely red light, absorbs all other colors and has a specular reflection which is essentially white (if the incident light is white light). This is quite general, because, except for metals, the reflectivity of most materials depends on their refractive index, which varies little with the wavelength (though it is this variation that causes the chromatic dispersion in a prism), so that all colors are reflected nearly with the same intensity. Reflections from different origin, instead, may be colored: metallic reflections, such as in gold or copper, or interferential reflections: iridescences, peacock feathers, butterfly wings, beetle elytra, or the antireflection coating of a lens.

Importance for Vision

Looking at one's surrounding environment, the vast majority of visible objects are seen primarily by diffuse reflection from their surface. This holds with few exceptions, such as glass, reflective liquids, polished or smooth metals, glossy objects, and objects that themselves emit light: the Sun, lamps, and computer screens (which, however, emit *diffuse* light). Outdoors it is the same, with perhaps the exception of a transparent water stream or of the iridescent colors of a beetle. Additionally, Rayleigh scattering is responsible for the blue color of the sky, and Mie scattering for the white color of the water droplets of clouds.

Light scattered from the surfaces of objects is by far the primary light which humans visually observe.

Interreflection

Diffuse interreflection is a process whereby light reflected from an object strikes other objects in the surrounding area, illuminating them. Diffuse interreflection specifically describes light reflected from objects which are not shiny or specular. In real life terms what this means is that light is reflected off non-shiny surfaces such as the ground, walls, or fabric, to reach areas not directly in view of a light source. If the diffuse surface is colored, the reflected light is also colored, resulting in similar coloration of surrounding objects.

In 3D computer graphics, diffuse interreflection is an important component of global illumination. There are a number of ways to model diffuse interreflection when rendering a scene. Radiosity and photon mapping are two commonly used methods.

Signal Reflection

Signal reflection is the process of bouncing a signal from a medium that does not totally absorb it. It may occur in copper cables for electrical signals, and in optical fiber for laser or optical signals. Signal reflection may also occur in outdoor metallic surfaces for electromagnetic (EM) waves. The EM waves travel through most open spaces and are not visible.

The study of signal reflection is used in specialized applications. Signal as sound may be reflected by a rigid surface and return to a receiver as a sound navigation and ranging (SONAR) signal. The ground-penetrating radar makes use of the principle that different radio frequencies and different ground materials will produce different amounts of signal absorption and reflection. In impedance matching, the goal is to make sure that most of the signal reaches the destination or load. The source impedance usually has to match the destination or load impedance for a given sub-band of radio frequencies.

In analog transmission cables, signal reflection is experienced as an echo when there is mismatch in the audio. Most of the issues in audio transmission have been resolved by using digitized audio in the form of Internet protocol data packets. Any signal reflection will be seen as data errors and eliminated with error correction schemes. Crosstalk, which used to be the unwanted induction of one analog signal from one cable to another, is also eliminated by using digital audio like the voice over Internet protocol (VoIP) packets in a digital subscriber line.

The Bergeron diagram shows the resulting voltages and currents when reflected electrical energy-combines with the incident energy. For best signal integrity, there should be minimal reflection, which is achieved by impedance matching. In some cases, the addition of resistive components that absorb electrical energy may totally eliminate reflection, while in other cases complex impedances formed by the series-parallel combinations of inductors and capacitors may provide the solution. The presence of distributed inductance and capacitance that is dependent on frequency makes the design of good impedance matching circuits very challenging.

Other specialized signal reflection methods include optical ranging where timed beam of light is allowed to reflect to a ranging target. Given the speed of light and the time it takes to receive the reflection, the distance to target can be computed. In radio detection and ranging (RADAR), the target reflects radio signals when the radar equipment sends a burst of radio frequency. The equipment waits for any reflected signal and computes the distance based on the delay between the transmission of radio frequency burst and reception of reflected signal, and the speed of radio waves in air.

Total Internal Reflection

Total internal reflection, in physics, complete reflection of a ray of light within a medium such as water or glass from the surrounding surfaces back into the medium. The phenomenon occurs if the angle of incidence is greater than a certain limiting angle, called the critical angle. In general, total internal reflection takes place at the boundary between two transparent media when a ray of light in a medium of higher index of refraction approaches the other medium at an angle of incidence greater than the critical angle. For a water-air surface the critical angle is 48.5°. Because indices of refraction depend on wavelength, the critical angle (and hence the angle of total internal reflection) will vary slightly with wavelength and, therefore, with colour. At all angles less than the critical angle, both refraction and reflection occur in varying proportions.

Glass prisms can be shaped to produce total internal reflection and as such are employed in binoculars, periscopes, telescopes, and other optical instruments. Light rays may be conducted over long, twisting paths by multiple total internal reflection in glass or plastic rods or fibres.

Two Requirements for Total Internal Reflection

Total internal reflection (TIR) is the phenomenon that involves the reflection of all the incident light off the boundary. TIR only takes place when both of the following two conditions are met:

- The light is in the more dense medium and approaching the less dense medium.
- The angle of incidence is greater than the so-called critical angle.

Total internal reflection will not take place unless the incident light is traveling within the more optically dense medium towards the less optically dense medium. TIR will happen for light traveling from water towards air, but it will not happen for light traveling from air towards water. TIR would happen for light traveling from water towards air, but it will not happen for light traveling from water (n=1.333) towards crown glass (n=1.52). TIR occurs because the angle of refraction reaches a 90-degree angle before the angle of incidence reaches a 90-degree angle. The only way for the angle of refraction to be greater than the angle of incidence is for light to bend away from the normal. Since light only bends away from the normal when passing from a more dense medium into a less dense medium, then this would be a necessary condition for total internal reflection.

Total internal reflection only occurs with large angles of incidence. Question: How large is large? Answer: larger than the critical angle. As mentioned above, the critical angle for the water-air boundary is 48.6 degrees. So for angles of incidence greater than 48.6-degrees, TIR occurs. But 48.6 degrees is the critical angle only for the water-air boundary. The actual value of the critical angle is dependent upon the two materials on either side of the boundary. For the crown glass-air boundary, the critical angle is 41.1 degrees. For the diamond-air boundary, the critical angle is 24.4 degrees. For the diamond-water boundary, the critical angle is 33.4 degrees. The critical angle is different for different media.

Reflection and Refraction

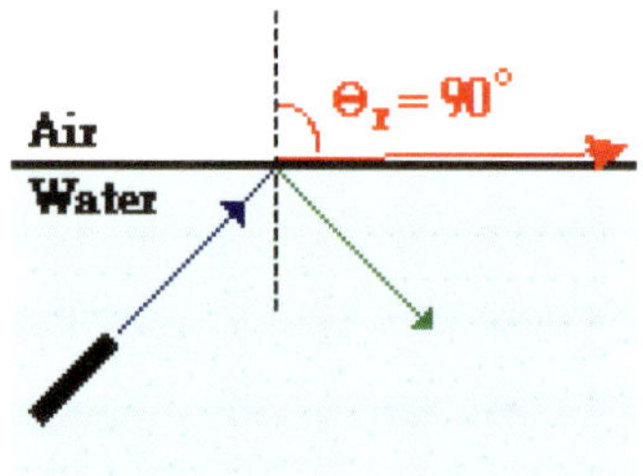

When the angle of incidence equal the critical angle, the angle of refraction is 90-degrees.

Total Internal Reflection

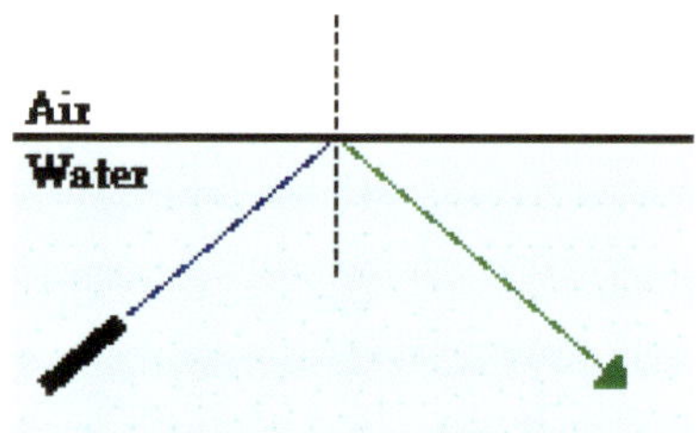

When the angle of incidence is greater than the critical angle, all the light undergoes reflection.

Optical Description

Total internal reflection of light can be demonstrated using a semi-circular block of glass or plastic. A "ray box" shines a narrow beam of light (a "ray") onto the glass medium. The semi-circular shape ensures that a ray pointing towards the centre of the flat face will hit the curved surface at a right angle; this will prevent refraction at the air/glass boundary of the curved surface. At the glass/air boundary of the flat surface, what happens will depend on the angle. If θ_c is the critical angle, then the following scenarios depict what will happen according to the size of the incident angle.

- If $\theta \le \theta_c$, the ray will split; some of the ray will reflect off the boundary, and some will refract as it passes through. This is not total internal reflection.
- If $\theta > \theta_c$, the entire ray reflects from the boundary. None passes through. This is called total internal reflection. TIR is the abbreviation.

This physical property makes optical fibers useful and prismatic binoculars possible. It is also what gives diamonds their distinctive sparkle, as diamond has an unusually high refractive index.

Critical Angle

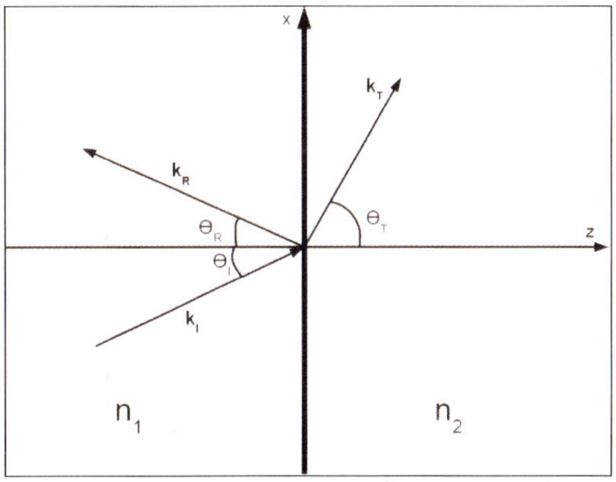

Illustration of Snell's law, $n_1 \sin\theta_i = n_2 \sin\theta_t$.

View up from directly under a diver. The above-water hemisphere is visible, compressed (as by a circular fisheye lens) into a circle (Snell's window) bounded by the critical angle. Everything outside the critical-angle circle is reflected from below the water (and is therefore dark).

The critical angle is the angle of incidence for which the angle of refraction is 90°. The angle of incidence is measured with respect to the normal at the refractive boundary. Consider a light ray passing from glass into air. The light emanating from the interface is bent towards the glass. When the incident angle is increased sufficiently, the transmitted angle (in air) reaches 90 degrees. It is at this point no light is transmitted into air. The critical angle θ_c is given by Snell's law,

$$n_1 \sin\theta_i = n_2 \sin\theta_t.$$

Rearranging Snell's Law, we get incidence

$$\sin\theta_i = \frac{n_2}{n_1}\sin\theta_t.$$

To find the critical angle, we find the value for θ_i when $\theta_t = 90°$ and thus $\sin\theta_t = 1$. The resulting value of θ_i is equal to the critical angle θ_c.

Now, we can solve for θ_i, and we get the equation for the critical angle:

$$\theta_c = \theta_i = \arcsin\left(\frac{n_2}{n_1}\right),$$

If the incident ray is precisely at the critical angle, the refracted ray is tangent to the boundary at the point of incidence. If for example, visible light were traveling through acrylic glass(with an index of refraction of approximately 1.50) into air (with an index of refraction of 1.00), the calculation would give the critical angle for light from acrylic into air, which is

$$\theta_c = \arcsin\left(\frac{1.00}{1.50}\right) = 41.8^{\circ}.$$

Light incident on the border with an angle less than 41.8° would be partially transmitted, while light incident on the border at larger angles with respect to normal would be totally internally reflected.

If the fraction n_2 / n_1 is greater than 1, then arcsine is not defined—meaning that total internal reflection does not occur even at very shallow or grazing incident angles.

So the critical angle is only defined when n_2 / n_1 is less than or equal to 1.

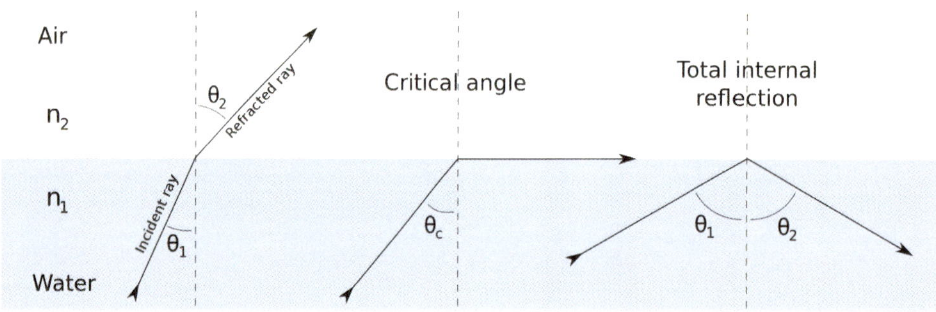

Refraction of light at the interface between two media, including total internal reflection.

Derivation of Evanescent Wave

An important side effect of total internal reflection is the appearance of an evanescent wave beyond the boundary surface. Essentially, even though the entire incident wave is reflected back into the originating medium, there is some penetration into the second medium at the boundary. The evanescent wave appears to travel along the boundary between the two materials, leading to the Goos-Hänchen shift.

If a plane wave, confined to the xz plane, is incident on a dielectric with an angle θ_I and wavevector $\mathbf{k_I}$ then a transmitted ray will be created with a corresponding angle of transmittance as shown in the figure above. The transmitted wavevector is given by:

$$\mathbf{k_T} = k_T \sin(\theta_T)\hat{x} + k_T \cos(\theta_T)\hat{z}$$

If $n_1 > n_2$, then $\sin(\theta_T) > 1$ since in the relation $\sin(\theta_T) = \frac{n_1}{n_2}\sin(\theta_I)$ obtained from Snell's law, $\frac{n_1}{n_2}\sin(\theta_I)$ is greater than one for $\theta_I > \theta_C$

As a result of this $\cos(\theta_T)$ becomes complex:

$$\cos(\theta_T) = \sqrt{1-\sin^2(\theta_T)} = \sqrt{\sin^2(\theta_T)-1}$$

The electric field of the transmitted plane wave is given by $\mathbf{E_T} = \mathbf{E_o}e^{i(\mathbf{k_T}\cdot\mathbf{r}-\omega t)}$ and so evaluating this further one obtains:

$$\mathbf{E_T} = \mathbf{E_o}e^{i(\mathbf{k_T}\cdot\mathbf{r}-\omega t)} = \mathbf{E_o}e^{i(xk_T\sin(\theta_T)+zk_T\cos(\theta_T)-\omega t)}$$

and

$$\mathbf{E_T} = \mathbf{E_o}e^{i(xk_T\sin(\theta_T)+zk_T i\sqrt{\sin^2(\theta_T)-1}-\omega t)}.$$

Using the fact that $k_T = \frac{\omega n_2}{c}$ and Snell's law, one finally obtains

$$\mathbf{E_T} = \mathbf{E_o}e^{-\kappa z}e^{i(kx-\omega t)}$$

where $\kappa = \frac{\omega}{c}\sqrt{(n_1\sin(\theta_I))^2 - n_2^2}$ a and $k = \frac{\omega n_1}{c}\sin(\theta_I)$.

This wave in the optically less dense medium is known as the evanescent wave. It is characterized by its propagation in the x direction and its exponential attenuation in the z direction. Although there is a field in the second medium, it can be shown that no energy flows across the boundary. The component of Poynting vector in the direction normal to the boundary is finite, but its time average vanishes. Whereas the other two components of Poynting vector (here x-component only) and their time averaged values are in general found to be finite (i.e. nonzero).

When a glass of water is held firmly, ridges making up the fingerprints are made visible by frustrated total internal reflection. Light tunnels from the glass into the ridges through the very short air gap.

It is important to note that the Goos-Hänchen effect only occurs for linearly polarized light. If the light is circularly or elliptically polarized, it will undergo the analogous Imbert–Fedorov effect.

Frustrated Total Internal Reflection

Under ordinary conditions, the evanescent wave transmits zero net energy across the interface. However, if a third medium with a higher refractive index than the low-index second medium is placed within less than several wavelengths distance from the interface between the first medium and the second medium, the evanescent wave will be different from the usual one, and it will pass energy across the second into the third medium. This process is called "frustrated" total internal reflection (FTIR) and is very similar to quantum tunneling. The quantum tunneling model is mathematically analogous if one thinks of the electromagnetic field as being the wave function of the photon. The low index medium can be thought of as a potential barrier through which photons can tunnel.

The transmission coefficient for FTIR is highly sensitive to the spacing between the third medium and the second medium (the function is approximately exponential until the gap is almost closed), so this effect has often been used to modulate optical transmission and reflection with a large dynamic range.

Phase Shift Upon Total Internal Reflection

An important aspect of total internal reflection is the appearance of a significant (rather than just 0° or 180°) phase shift upon reflection. Mathematically this is expressed by the Fresnel reflection coefficients acquiring an imaginary part. This resulting phase advance grows as the incidence angle increases beyond the critical angle, but unequally for the s and p polarization components of the incident wave.

This polarization dependent phase shift was discovered and quantitatively explained by Augustin-Jean Fresnel who employed it in his invention, the Fresnel rhomb, in which the angles of incidence are adjusted such that the two internal reflections cause a *relative* phase shift of 90° between the two polarizations of an incident wave. This performs the same function as a quarter-wave plate, but is more achromatic (phase shift has less dependence on wavelength). Either type of quarter-wave retarder may be used, for instance, to transform linear polarization to circular polarization (which Fresnel also discovered) and vice versa.

The polarization dependent phase shift is also the reason why TE and TM guided modes have different dispersion relations.

Applications

- Total internal reflection is the operating principle of optical fibers, which are used in endoscopes and telecommunications.
- Total internal reflection is the operating principle of automotive rain sensors, which control automatic windscreen/windshield wipers.
- Another application of total internal reflection is the spatial filtering of light.
- Prisms in binoculars use total internal reflection, rather than reflective coatings, to fold optical paths and show erect images.

- Some multi-touch screens use frustrated total internal reflection in combination with a camera and appropriate software to pick up multiple targets.
- Gonioscopy employs total internal reflection to view the anatomical angle formed between the eye's cornea and iris.
- A gait analysis instrument, CatWalk XT, uses frustrated total internal reflection in combination with a high speed camera to capture and analyze footprints of laboratory rodents.
- Optical fingerprinting devices use frustrated total internal reflection in order to record an image of a person's fingerprint without the use of ink.
- A total internal reflection fluorescence microscope uses the evanescent wave produced by TIR to excite fluorophores close to a surface. This is useful for the study of surface properties of biological samples.
- Total internal reflection is the operating principle of LED Light Panels. This technology utilizes LGPs (Light Guide Plates) as the vehicle for transmitting light over large areas. By etching a grid pattern on the second surface of the LGP, frustrated total internal reflection occurs allowing the light to escape the LGP as visible illumination.
- Diamonds shine brightly due to total internal reflection.
- Total internal reflection (TIR) are used on lighting industry to manage narrow distribution (NEMA classification) for floodlight.

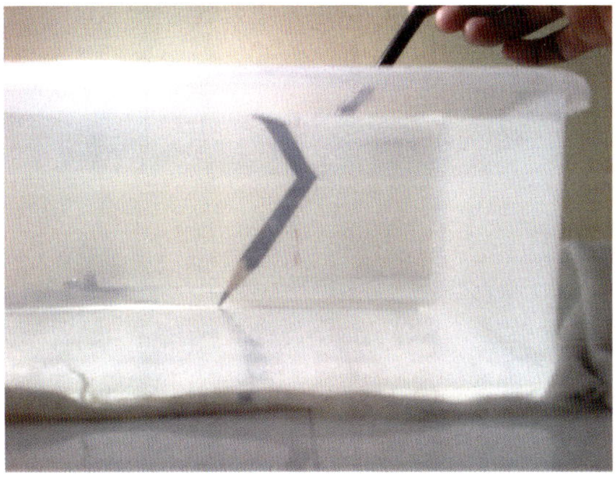

Mirror like effect

Examples in Everyday Life

Total internal reflection can be observed while swimming, when one opens one's eyes just under the water's surface. If the water is calm, the surface appears mirror-like (except directly overhead, in Snell's window).

Diamonds used for jewelry are normally shaped in such a way as to maximize the amount of light that is totally internally reflected from the back side of the diamond. The very high refractive index of diamond gives a small critical angle allowing almost all the light entering the diamond to be reflected back for optimized diamond cuts such as the brilliant.

Total internal reflection can be seen at the air-water boundary.

Light Piping and Optical Fibers

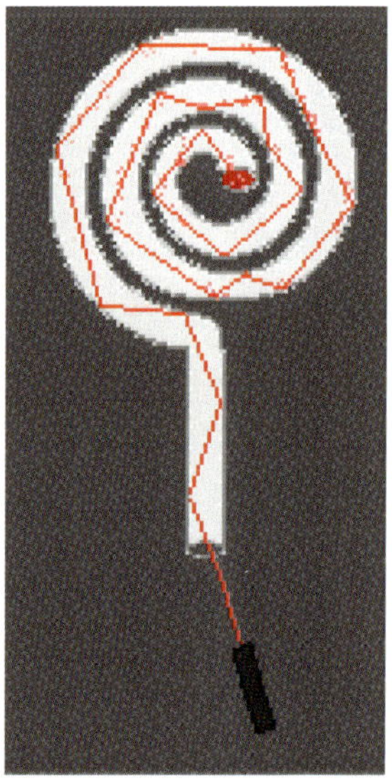

Total internal reflection is often demonstrated in a Physics class through a variety of demonstrations. In one such demonstration, a beam of laser light is directed into a coiled plastic *thing-a-ma jig*. The plastic served as a *light pipe*, directing the light through the coils until it finally exits out the opposite end. Once the light entered the plastic, it was in the more dense medium. Every time the light approached the plastic-air boundary, it is approaching at angles greater than the critical angle. The two conditions necessary for TIR are met, and all of the incident light at the plastic-air boundary stays internal to the plastic and undergoes reflection. And with the room lights off, every student becomes quickly aware of the ancient truth that Physics is better than drugs.

This demonstration helps to illustrate the principle by which optical fibers work. The use of a long strand of plastic (or other material such as glass) to pipe light from one end of the medium to the other is the basis for modern day use of optical fibers. Optical fibers are used in communication systems and micro-surgeries. Since total internal reflection takes place within the fibers, no incident energy is ever lost due to the transmission of light across the boundary. The intensity of the signal remains constant.

The laser beam stays internal to the water, continuously reflecting at each boundary.

Another common Physics demonstration involves the use of a large jug filled with water and a laser beam. The jug has a pea-sized hole drilled in its side such that when the cork is removed from the top of the jug, water begins to stream out the jug's side. The beam of laser light is then directed into the jug from the opposite side of the hole, through the water and into the falling stream. The laser light exits the jug through the hole but is still in the water. As the stream of water begins to fall as a projectile along a parabolic path to the ground, the laser light becomes trapped within the water due to total internal reflection. Being in the more dense medium (water) and heading towards a boundary with a less dense medium (air), and being at angles of incidence greater than the critical angle, the light never leaves the stream of water. In fact, the stream of water acts as a light pipe to pipe the laser beam along its trajectory. Once more, students viewing the demonstration are convinced of the fact that Physics is better than drugs.

Reflection Phase Change

Phase Change on Reflection

Idea: When light is reflected off a more dense medium with higher index of refraction, crests get reflected as troughs and troughs get reflected as crests. The wave is said undergo a 180 o change of phase on reflection. The net effect of the phase change is that the reflected ray "jumps ahead" by half a wave length. Thus, if one of the rays undergoes an odd number of phase changes, the conditions for constructive and destructive interference must be modified:

$$r_1 - r_2 = (m + \frac{1}{2})\lambda \ \ m = 1,2,.... constructive$$

$$r_1 - r_2 = (m)\lambda \ \ m = 1,2,... destructive$$

- The above formulas are valid whenever an odd number of phase changes occur. If an even number of phase changes occur then the original, unmodified formulas must be used.
- The phase change does not happen when light is reflected off a less dense medium, i.e. for total internal reflection.

The phase changes that take place upon reflection play an important part in thin film interference.

Interference Condition for Thin Films

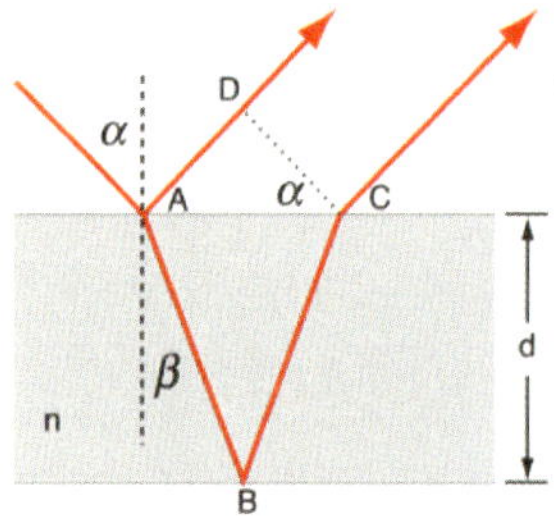

Optical pathlength difference Γ :

$$\Gamma = n(AB + BC) - AD$$

$$AB = \frac{d}{\cos\beta} \qquad AD = (2d\tan\beta)\sin\alpha$$

$$AD = 2d\tan\beta(n\sin\beta)$$

Collecting terms:

$$\Gamma = 2dn\left[\frac{1}{\cos\beta} - \tan\beta\sin\beta\right]$$

$$\Gamma = 2dn\left[\frac{1-\sin^2\beta}{\cos\beta}\right] = 2dn\cos\beta$$

Phase change = π: Destructive → $2dn\cos\beta = m\lambda$ ← Constructive : Phase change = 0

Phase change = π: Constructive → $2dn\cos\beta = (m - \frac{1}{2})\lambda$ ← Destructive : Phase change = 0

Refraction

Refraction is the bending of light (it also happens with sound, water and other waves) as it passes from one transparent substance into another.

This bending by refraction makes it possible for us to have lenses, magnifying glasses, prisms and rainbows. Even our eyes depend upon this bending of light. Without refraction, we wouldn't be able to focus light onto our retina.

Change of Speed Causes Change of Direction

Light refracts whenever it travels at an angle into a substance with a different refractive index (optical density).

This change of direction is caused by a change in speed. For example, when light travels from air into water, it slows down, causing it to continue to travel at a different angle or direction.

How much does light bend?

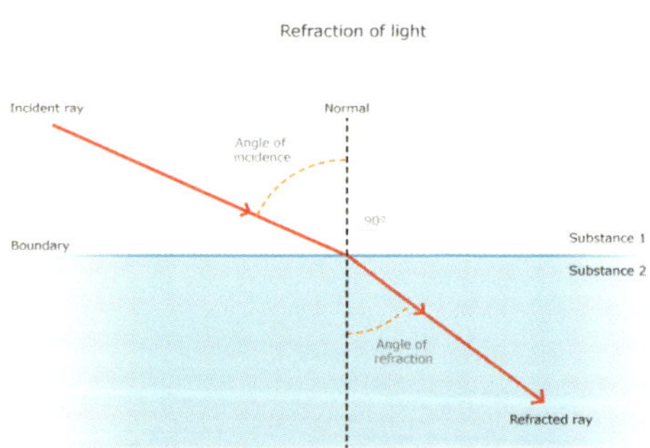

Refraction of light in water

When light travels from air into water, it slows down, causing it to change direction slightly. This change of direction is called refraction. When light enters a more dense substance (higher refractive index), it 'bends' more towards the normal line.

The amount of bending depends on two things:

- Change in speed – if a substance causes the light to speed up or slow down more, it will refract (bend) more.
- Angle of the incident ray – if the light is entering the substance at a greater angle, the amount of refraction will also be more noticeable. On the other hand, if the light is entering the new substance from straight on (at 90° to the surface), the light will still slow down, but it won't change direction at all.

Refractive index of some transparent substances

Substance	Refractive index	Speed of light in substance (x 1,000,000 m/s)	Angle of refraction if incident ray enters substance at 20°
Air	1.00	300	20
Water	1.33	226	14.9
Glass	1.5	200	13.2
Diamond	2.4	125	8.2

All angles are measured from an imaginary line drawn at 90° to the surface of the two substances. This line is drawn as a dotted line and is called the normal.

If light enters any substance with a *higher* refractive index (such as from air into glass) it slows down. The light bends *towards* the normal line.

If light travels enters into a substance with a *lower* refractive index (such as from water into air) it speeds up. The light bends *away* from the normal line.

A higher refractive index shows that light will slow down and change direction more as it enters the substance.

Lenses

A lens is simply a curved block of glass or plastic. There are two kinds of lens.

A biconvex lens is thicker at the middle than it is at the edges. This is the kind of lens used for a magnifying glass. Parallel rays of light can be focused in to a focal point. A biconvex lens is called a converging lens.

Refraction of light through a converging lens

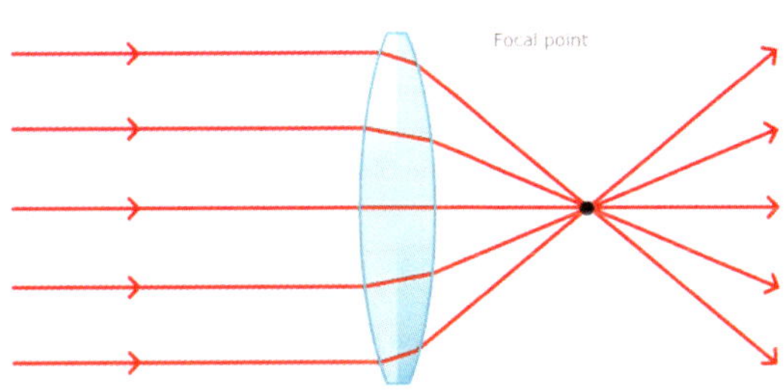

Each light ray entering a converging (convex) lens refracts inwards as it enters the lens and inwards again as it leaves. These refractions cause parallel light rays to spread out, travelling directly away from an imaginary focal point.

A biconcave lens curves is thinner at the middle than it is at the edges. Light rays refract outwards (spread apart) as they enter the lens and again as they leave.

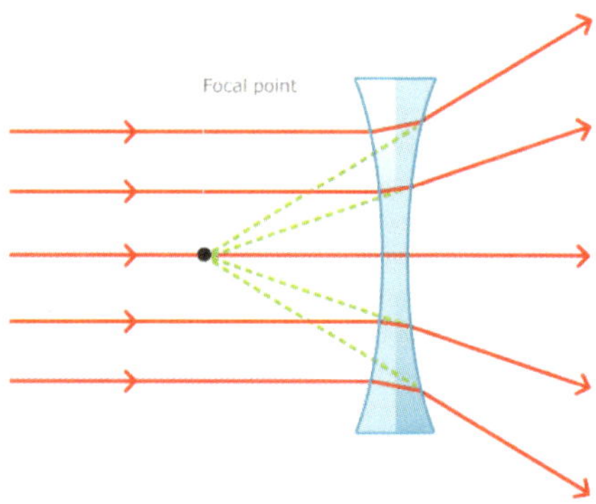

Refraction of light through a diverging lens

Each light ray entering a diverging (concave) lens refracts outwards as it enters the lens and outwards again as it leaves. These refractions cause parallel light rays to spread out, travelling directly away from an imaginary focal point.

Refraction Can Create a Spectrum

Isaac Newton performed a famous experiment using a triangular block of glass called a prism. He used sunlight shining in through his window to create a spectrum of colours on the opposite side of his room.

When white light shines through a prism, each colour refracts at a slightly different angle. Violet light refracts slightly more than red light. A prism can be used to show the seven colours of the spectrum that make up white light.

This experiment showed that white light is actually made of all the colours of the rainbow. These seven colours are remembered by the acronym ROY G BIV – red, orange, yellow, green, blue, indigo and violet.

Newton showed that each of these colours cannot be turned into other colours. He also showed that they can be recombined to make white light again.

The explanation for the colours separating out is that the light is made of waves. Red light has a longer wavelength than violet light. The refractive index for red light in glass is slightly different than for violet light. Violet light slows down even more than red light, so it is refracted at a slightly greater angle.

The refractive index of red light in glass is 1.513. The refractive index of violet light is 1.532. This slight difference is enough for the shorter wavelengths of light to be refracted more.

Rainbows

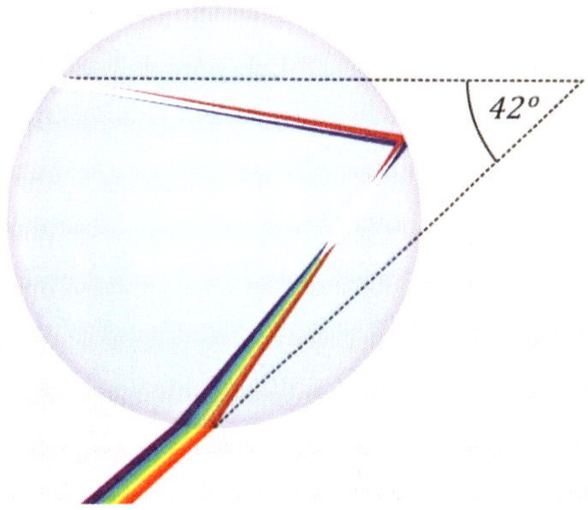

A rainbow is formed when light enters each water droplet, and the different colours bend (refract) at slightly different angles. They reflect off the inside of the drop before refracting again as they leave. The shorter wavelengths refract more.

A rainbow is caused because each colour refracts at slightly different angles as it enters, reflects off the inside and then leaves each tiny drop of rain.

A rainbow is easy to create using a spray bottle and the sunshine. The centre of the circle of the rainbow will always be the shadow of your head on the ground.

The secondary rainbow that can sometimes be seen is caused by each ray of light reflecting twice on the inside of each droplet before it leaves. This second reflection causes the colours on the secondary rainbow to be reversed. Red is at the top for the primary rainbow, but in the secondary rainbow, red is at the bottom.

Refractive Index

The refractive index or index of refraction (n_D) is one of the physicochemical properties of substances (optical medium). It is equal to velocity (*c*) of light of a given wavelength in empty space (vacuum) divided by its velocity (*v*) in a substance according to following equation:

$$n_D = \frac{c}{v}$$

When light travels at an angle between two materials, it bends according to their refractive indices. In order to reflect, light must be on the wider side of the critical angle.

The refractive index can provide information for us about the behavior of light. When light passes through the different substances its velocity decreases by increasing of the refractive index of these substances. It can be due to interaction between molecules of components in substrate and effect of these interactions on light. Also in the most substrates, the refractive index decreases by increasing of the temperature. The interaction between molecules decreases as the temperature increases.

The refractive index of different substrates measures with refractometers. There are four main types of refractometers: traditional handheld refractometers, digital handheld refractometers, laboratory or Abbe refractometers, and inline process refractometers. There is also the Rayleigh Refractometer used (typically) for measuring the refractive indices of gases.A sodium lamp may be used to provide the light source at a known wave-length (589.6 nm) although many instruments are corrected for daylight use.

In most of substrates which we know the refractive index have a positive value (more than zero) but in 1968 Veselago shows that there are substrates with negative permittivity and negative permeability. In these substrates refractive index have a negative value.

Maxwell's equations relate the permittivity and the permeability to the refractive index as follows:

$$n = \pm \sqrt{s\mu}$$

The sign of the index is often taken as positive. However, Veselago showed that if a substrate has both negative permeability and negative permittivity, this convention must be reversed and we must pick out the *negative* sign of the square root.

The Speed of Light

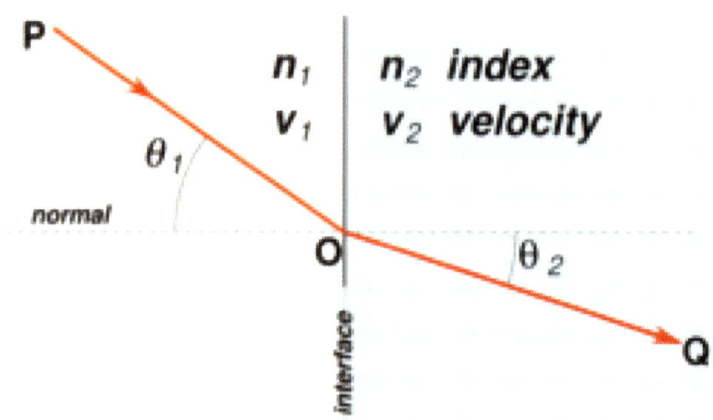

Refraction of light at the interface between two media of different refractive indices, with $n_2 > n_1$. The velocity is lower in the second medium ($v_2 < v_1$), therefore the angle of refraction θ_2 is less than the angle of incidence θ_1; that is, the ray in the higher-index medium is closer to the normal.

The speed of all electromagnetic radiation in vacuum is the same, approximately 3×10^8 meters per second, and is denoted by *c*. Therefore, if *v* is the phase velocity of radiation of a specific frequency in a specific material, the refractive index is given by

$$n = \frac{c}{v} \ .$$

This number is typically greater than one: the higher the index of the material, the more the light is slowed down. However, at certain frequencies (e.g., X-rays), n will actually be smaller than one. This does not contradict the theory of relativity, which holds that no information-carrying signal can ever propagate faster than c, because the phase velocity is not the same as the group velocity or the signal velocity, same as group velocity except when the wave is passing through an absorptive medium.

Sometimes, a "group velocity refractive index," usually called the *group index* is defined:

$$n_g = \frac{c}{v_g}$$

where v_g is the group velocity. This value should not be confused with n, which is always defined with respect to the phase velocity.

At the microscale, an electromagnetic wave's phase velocity is slowed in a material because the electric field creates a disturbance in the charges of each atom (primarily the electrons) proportional (a $y = kx$ relationship) to the permittivity. The charges will, in general, oscillate slightly out of phase with respect to the driving electric field. The charges thus radiate their own electromagnetic wave that is at the same frequency but with a phase delay. The macroscopic sum of all such contributions in the material is a wave with the same frequency but shorter wavelength than the original, leading to a slowing of the wave's phase velocity. Most of the radiation from oscillating material charges will modify the incoming wave, changing its velocity. However, some net energy will be radiated in other directions. If the refractive indices of two materials are known for a given frequency, then one can compute the angle by which radiation of that frequency will be refracted as it moves from the first into the second material from Snell's law.

Negative Refractive Index

Recent research has also demonstrated the existence of negative refractive index, which can occur if ε and μ are *simultaneously* negative. Not thought to occur naturally, it can be achieved with so called metamaterials. It offers the possibility of perfect lenses and other exotic phenomena such as a reversal of Snell's law.

References

- Ehrlich, Robert (1997). Why toast lands jelly-side down: zen and the art of physics demonstrations. Princeton, New Jersey, USA: Princeton University Press. p. 182. ISBN 0-691-02891-5. Retrieved 9 February 2012
- Geometrical-optics: brilliant.org, Retrieved 28 March 2018
- Axelrod, D. (1 April 1981). "Cell-substrate contacts illuminated by total internal reflection fluorescence". The Journal of Cell Biology. 89 (1): 141–145. doi:10.1083/jcb.89.1.141. PMC 2111781 . PMID 7014571
- Geometrical-optics, reflection-at-smooth-surfaces: brilliant.org, Retrieved 30 June 2018

- What-is-diffuse-reflection: wisegeek.com, Retrieved 30 June 2018
- OD, By Emily Bruce, OD, Rodney Bendure, OD, Sarah Krein, OD, and Nathan Lighthizer,. "Zoom in on Gonioscopy". Retrieved 2017-09-04
- Law-of-reflection, light: physics.tutorvista.com, Retrieved 10 June 2018
- Scott M. Juds (1988). Photoelectric sensors and controls: selection and application. CRC Press. p. 29. ISBN 978-0-8247-7886-6. Archived from the original on 2018-01-14
- Angle-of-incidence: light: physics.tutorvista.com, Retrieved 12 July 2018

3

Wave Interference

Wave interference is the phenomenon in which two waves superimpose to generate a wave with a lower, greater or the same amplitude. The aim of this chapter is to explore the fundamental concepts of constructive wave interference, destructive wave interference, Young's interference experiment and thin-film interference.

Wave interference is the phenomenon that occurs when two waves meet while traveling along the same medium to form a resultant wave of greater, lower or the same amplitude.

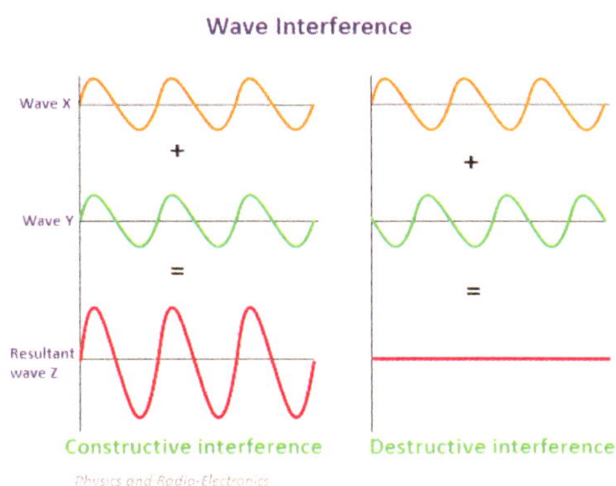

Interference usually refers to the interaction of waves which have constant phase difference and same or nearly same frequency.

Interference effects can be observed with all types of waves, for example, radio waves, light waves, sound waves, surface water waves or matter waves.

The Principle of Superposition

The task of determining the shape of the resultant demands that the principle of superposition is applied. The principle of superposition is sometimes stated as follows:

When two waves interfere, the resulting displacement of the medium at any location is the algebraic sum of the displacements of the individual waves at that same location.

In the cases above, the summing the individual displacements for locations of complete overlap was made out to be an easy task - as easy as simple arithmetic:

Displacement of Pulse 1	Displacement of Pulse 2	=	Resulting Displacement
+1	+1	=	+2
-1	-1	=	-2
+1	-1	=	0
+1	-2	=	-1

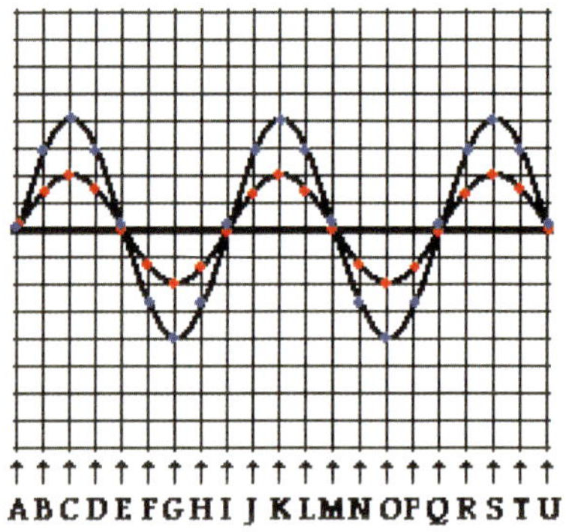

In actuality, the task of determining the complete shape of the entire medium during interference demands that the principle of superposition be applied for every point (or nearly every point) along the medium. As an example of the complexity of this task, consider the two interfering waves at the right. A snapshot of the shape of each individual wave at a particular instant in time is shown. To determine the precise shape of the medium at this given instant in time, the principle of superposition must be applied to several locations along the medium. A short cut involves measuring the displacement from equilibrium at a few strategic locations. Thus, approximately 20 locations have been picked and labeled as A, B, C, D, etc. The actual displacement of each individual wave can be counted by measuring from the equilibrium position up to the particular wave. At position A, there is no displacement for either individual wave; thus, the resulting displacement of the medium at position will be 0 units. At position B, the smaller wave has a displacement of approximately 1.4 units (indicated by the red dot); the larger wave has a displacement of approximately 2 units (indicated by the blue dot). Thus, the resulting displacement of the medium will be approximately 3.4 units. At position C, the smaller wave has a displacement of approximately 2 units; the larger wave has a displacement of approximately 4 units; thus, the resulting displacement of the medium will be approximately 6 units. At position D, the smaller wave has a displacement of approximately 1.4 units; the larger wave has a displacement of approximately 2 units; thus, the resulting displacement of the medium will be approximately 3.4 units. This process can be repeated for every position. When finished, a dot (done in green below) can be marked on the *graph* to note the displacement of the medium at each given location. The actual shape of the medium can then be sketched by estimating the position between the various marked points and sketching the wave. This is shown as the green line in the diagram below.

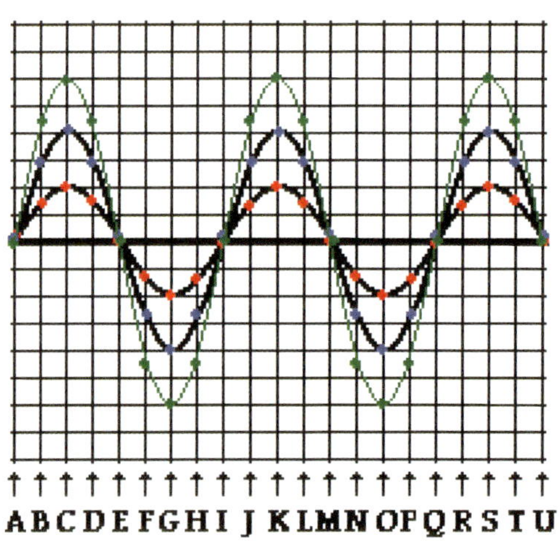

Interference of Waves

Let us consider two simple harmonic progressive waves of the same frequency travelling in the same direction. Let a_1 and a_2 be the amplitudes of the waves and ϕ the phase difference between them at any point in the medium. If the displacements due to these waves at that point be y_1 and y_2 then

$$Y_1 = a_1 \sin wt$$

and

$$Y_2 = a_2 \sin(wt + \phi)$$

And $\frac{w}{2\pi}$ is the frequency of each wave. By the principle of superposition, the net displacement at the given point is given by:

$$Y = Y_1 + Y_2$$

$$= a_1 \sin wt + a_2 \sin(wt + \phi)$$
$$= a_1 \sin wt + a_2 \sin wt \cos\phi + a_2 \cos wt \sin\phi$$
$$= \sin wt (a_1 + a_2 \cos\phi) + \cos wt (a_2 \sin\phi)$$

Now let

$$a_1 + a_2 \cos\phi = R\cos\theta)$$

$$a_1 + a_2 \cos\phi = R\sin\theta$$

Here R and θ are the newly generated constants. Then,

$$Y = R\sin wt \cos\theta + R\cos wt \sin\theta$$

Or

$$Y = R\sin(wt + \theta)$$

This equation is similar to the previous equations. Hence the resultant displacement at that point is due to a similar wave of amplitude R. To determine R we square (c) and (d) and add.

The intensity is directly proportional to the square of the amplitude. So the resultant intensity I is represented by:

$$I = a_1^2 + a_2^2 + 2a_1a_2 \cos\phi$$

And phase angle,

$$\phi = \tan^{-1}\left(\frac{a_2 \sin\phi}{a_1 + a_2 \cos\phi}\right)$$

Thus the resultant intensity at any point depends on the phase difference ϕ between the two waves at that point. If ϕ is equivalent to x path difference, then remember

$$\frac{\Phi}{2\pi} = \frac{x}{\lambda} = \frac{\Delta t}{T}$$

The above equations are called Wave Interference Equation.

Between Two Plane Waves

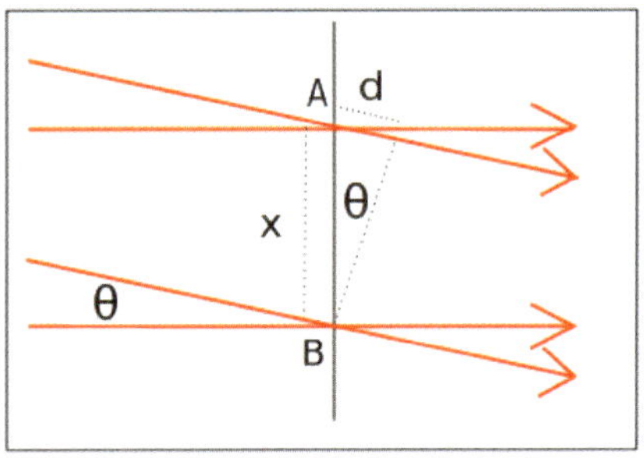

Geometrical arrangement for two plane wave interference

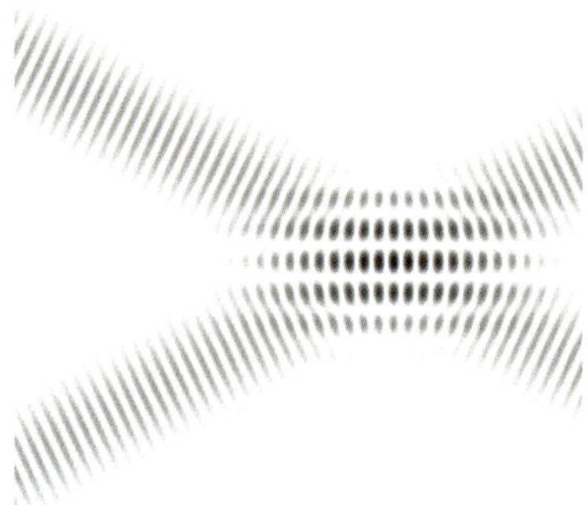

Interference fringes in overlapping plane waves

A simple form of interference pattern is obtained if two plane waves of the same frequency intersect at an angle. Interference is essentially an energy redistribution process. The energy which is lost at the destructive interference is regained at the constructive interference. One wave is travelling horizontally, and the other is travelling downwards at an angle θ to the first wave. Assuming that the two waves are in phase at the point B, then the relative phase changes along the x-axis. The phase difference at the point A is given by

$$\Delta\varphi = \frac{2\pi d}{\lambda} = \frac{2\pi x \sin\theta}{\lambda}$$

It can be seen that the two waves are in phase when

$$\frac{x \sin\theta}{\lambda} = 0, \pm 1, \pm 2, \ldots,$$

and are half a cycle out of phase when

$$\frac{x \sin\theta}{\lambda} = \pm\frac{1}{2}, \pm\frac{3}{2}, \ldots$$

Constructive interference occurs when the waves are in phase, and destructive interference when they are half a cycle out of phase. Thus, an interference fringe pattern is produced, where the separation of the maxima is

$$d_f = \frac{\lambda}{\sin\theta}$$

and d_f is known as the fringe spacing. The fringe spacing increases with increase in wavelength, and with decreasing angle θ.

The fringes are observed wherever the two waves overlap and the fringe spacing is uniform throughout.

Between Two Spherical Waves

Optical interference between two point sources that have different wavelengths and separations of sources.

A point source produces a spherical wave. If the light from two point sources overlaps, the interference pattern maps out the way in which the phase difference between the two waves varies in space. This depends on the wavelength and on the separation of the point sources. The figure to the right shows interference between two spherical waves. The wavelength increases from top to bottom, and the distance between the sources increases from left to right.

When the plane of observation is far enough away, the fringe pattern will be a series of almost straight lines, since the waves will then be almost planar.

Multiple Beams

Interference occurs when several waves are added together provided that the phase differences between them remain constant over the observation time.

It is sometimes desirable for several waves of the same frequency and amplitude to sum to zero (that is, interfere destructively, cancel). This is the principle behind, for example, 3-phase power and the diffraction grating. In both of these cases, the result is achieved by uniform spacing of the phases.

It is easy to see that a set of waves will cancel if they have the same amplitude and their phases are spaced equally in angle. Using phasors, each wave can be represented as $Ae^{i\varphi_n}$ for N waves from $n=0$ to $n=N-1$, where

$$\varphi_n - \varphi_{n-1} = \frac{2\pi}{N}.$$

To show that

$$\sum_{n=0}^{N-1} Ae^{i\varphi_n} = 0$$

one merely assumes the converse, then multiplies both sides by $0\, e^{i\frac{2\pi}{N}}$

The Fabry–Pérot interferometer uses interference between multiple reflections.

A diffraction grating can be considered to be a multiple-beam interferometer; since the peaks which it produces are generated by interference between the light transmitted by each of the elements in the grating.

Optical Interference

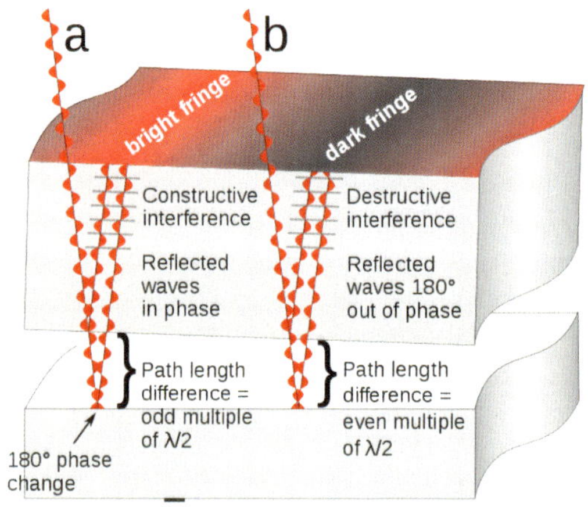

Creation of interference fringes by an optical flat on a reflective surface. Light rays from a monochromatic source pass through the glass and reflect off both the bottom surface of the flat and the supporting surface. The tiny gap between the surfaces means the two reflected rays have different path lengths. In addition the ray reflected from the bottom plate undergoes a 180° phase reversal. As a result, at locations *(a)* where the path difference is an odd multiple of λ/2, the waves reinforce. At locations *(b)* where the path difference is an even multiple of λ/2 the waves cancel. Since the gap between the surfaces varies slightly in width at different points, a series of alternating bright and dark bands, *interference fringes*, are seen.

Because the frequency of light waves (~10^{14} Hz) is too high to be detected by currently available detectors, it is possible to observe only the intensity of an optical interference pattern. The intensity of the light at a given point is proportional to the square of the average amplitude of the wave. This can be expressed mathematically as follows. The displacement of the two waves at a point r is:

$$U_1(\mathbf{r},t) = A_1(\mathbf{r})e^{i[\varphi_1(\mathbf{r})-\omega t]}$$

$$U_2(\mathbf{r},t) = A_2(\mathbf{r})e^{i[\varphi_2(\mathbf{r})-\omega t]}$$

where A represents the magnitude of the displacement, ϕ represents the phase and ω represents the angular frequency.

The displacement of the summed waves is

$$U(\mathbf{r},t) = A_1(\mathbf{r})e^{i[\varphi_1(\mathbf{r})-\omega t]} + A_2(\mathbf{r})e^{i[\varphi_2(\mathbf{r})-\omega t]}$$

The intensity of the light at r is given by

$$I(\mathbf{r}) = \int U(\mathbf{r},t)U^*(\mathbf{r},t)dt \propto A_1^2(\mathbf{r}) + A_2^2(\mathbf{r}) + 2A_1(\mathbf{r})A_2(\mathbf{r})\cos[\varphi_1(\mathbf{r}) - \varphi_2(\mathbf{r})]$$

This can be expressed in terms of the intensities of the individual waves as

$$I(\mathbf{r}) = I_1(\mathbf{r}) + I_2(\mathbf{r}) + 2\sqrt{I_1(\mathbf{r})I_2(\mathbf{r})}\cos[\varphi_1(\mathbf{r}) - \varphi_2(\mathbf{r})]$$

Thus, the interference pattern maps out the difference in phase between the two waves, with maxima occurring when the phase difference is a multiple of 2π. If the two beams are of equal intensity, the maxima are four times as bright as the individual beams, and the minima have zero intensity.

The two waves must have the same polarization to give rise to interference fringes since it is not possible for waves of different polarizations to cancel one another out or add together. Instead, when waves of different polarization are added together, they give rise to a wave of a different polarization state.

Light Source Requirements

The discussion above assumes that the waves which interfere with one another are monochromatic, i.e. have a single frequency—this requires that they are infinite in time. This is not, however, either practical or necessary. Two identical waves of finite duration whose frequency is fixed over that period will give rise to an interference pattern while they overlap. Two identical waves which consist of a narrow spectrum of frequency waves of finite duration, will give a series of fringe patterns of slightly differing spacings, and provided the spread of spacings is significantly less than the average fringe spacing, a fringe pattern will again be observed during the time when the two waves overlap.

Conventional light sources emit waves of differing frequencies and at different times from different points in the source. If the light is split into two waves and then re-combined, each individual light wave may generate an interference pattern with its other half, but the individual fringe patterns generated will have different phases and spacings, and normally no overall fringe pattern will be observable. However, single-element light sources, such as sodium- or mercury-vapor lamps have emission lines with quite narrow frequency spectra. When these are spatially and colour filtered, and then split into two waves, they can be superimposed to generate interference fringes. All interferometry prior to the invention of the laser was done using such sources and had a wide range of successful applications.

A laser beam generally approximates much more closely to a monochromatic source, and it is much more straightforward to generate interference fringes using a laser. The ease with which interference fringes can be observed with a laser beam can sometimes cause problems in that stray reflections may give spurious interference fringes which can result in errors.

Normally, a single laser beam is used in interferometry, though interference has been observed using two independent lasers whose frequencies were sufficiently matched to satisfy the phase requirements.

White light interference in a soap bubble. The iridescence is due to thin-film interference.

It is also possible to observe interference fringes using white light. A white light fringe pattern can be considered to be made up of a 'spectrum' of fringe patterns each of slightly different spacing. If all the fringe patterns are in phase in the centre, then the fringes will increase in size as the wavelength decreases and the summed intensity will show three to four fringes of varying colour. Young describes this very elegantly in his discussion of two slit interference. Since white light fringes are obtained only when the two waves have travelled equal distances from the light source, they can be very useful in interferometry, as they allow the zero path difference fringe to be identified.

Optical Arrangements

To generate interference fringes, light from the source has to be divided into two waves which have then to be re-combined. Traditionally, interferometers have been classified as either amplitude-division or wavefront-division systems.

In an amplitude-division system, a beam splitter is used to divide the light into two beams travelling in different directions, which are then superimposed to produce the interference pattern. The Michelson interferometer and the Mach-Zehnder interferometer are examples of amplitude-division systems.

In wavefront-division systems, the wave is divided in space—examples are Young's double slit interferometer and Lloyd's mirror.

Interference can also be seen in everyday phenomena such as iridescence and structural coloration. For example, the colours seen in a soap bubble arise from interference of light reflecting off the front and back surfaces of the thin soap film. Depending on the thickness of the film, different colours interfere constructively and destructively.

Applications

Optical Interferometry

Interferometry has played an important role in the advancement of physics, and also has a wide range of applications in physical and engineering measurement.

Thomas Young's double slit interferometer in 1803 demonstrated interference fringes when two small holes were illuminated by light from another small hole which was illuminated by sunlight. Young was able to estimate the wavelength of different colours in the spectrum from the spacing of the fringes. The experiment played a major role in the general acceptance of the wave theory of light. In quantum mechanics, this experiment is considered to demonstrate the inseparability of the wave and particle natures of light and other quantum particles (wave–particle duality). Richard Feynman was fond of saying that all of quantum mechanics can be gleaned from carefully thinking through the implications of this single experiment.

The results of the Michelson–Morley experiment are generally considered to be the first strong evidence against the theory of a luminiferous aether and in favor of special relativity.

Interferometry has been used in defining and calibrating length standards. When the metre was defined as the distance between two marks on a platinum-iridium bar, Michelson and Benoît used interferometry to measure the wavelength of the red cadmium line in the new standard, and also showed that it could be used as a length standard. Sixty years later, in 1960, the metre in the new SI system was defined to be equal to 1,650,763.73 wavelengths of the orange-red emission line in the electromagnetic spectrum of the krypton-86 atom in a vacuum. This definition was replaced in 1983 by defining the metre as the distance travelled by light in vacuum during a specific time interval. Interferometry is still fundamental in establishing the calibration chain in length measurement.

Interferometry is used in the calibration of slip gauges (called gauge blocks in the US) and in coordinate-measuring machines. It is also used in the testing of optical components.

Radio Interferometry

The Very Large Array, an interferometric array formed from many smaller telescopes, like many larger radio telescopes.

In 1946, a technique called astronomical interferometry was developed. Astronomical radio interferometers usually consist either of arrays of parabolic dishes or two-dimensional arrays of omni-directional antennas. All of the telescopes in the array are widely separated and are usually connected together using coaxial cable, waveguide, optical fiber, or other type of transmission line. Interferometry increases the total signal collected, but its primary purpose is to vastly increase the resolution through a process called Aperture synthesis. This technique works by superposing (interfering) the signal waves from the different telescopes on the principle that waves that coincide with the same phase will add to each other while two waves that have opposite phases will cancel each other out. This creates a combined telescope that is equivalent in resolution (though not in sensitivity) to a single antenna whose diameter is equal to the spacing of the antennas furthest apart in the array.

Acoustic Interferometry

An acoustic interferometer is an instrument for measuring the physical characteristics of sound wave in a gas or liquid. It may be used to measure velocity, wavelength, absorption, or impedance. A vibrating crystal creates the ultrasonic waves that are radiated into the medium. The waves strike a reflector placed parallel to the crystal. The waves are then reflected back to the source and measured.

Quantum Interference

If a system is in state ψ, its wavefunction is described in Dirac or bra–ket notation as:

$$|\psi\rangle = \sum_i |i\rangle\langle i|\psi\rangle = \sum_i |i\rangle \psi_i$$

where the $|i\rangle$ s specify the different quantum "alternatives" available (technically, they form an eigenvector basis) and the ψ_1 are the probability amplitude coefficients, which are complex numbers.

The probability of observing the system making a transition or quantum leap from state ψ to a new state φ is the square of the modulus of the scalar or inner product of the two states:

$$\text{prob}(\psi \Rightarrow \varphi) = |\langle\psi|\varphi\rangle|^2 = |\sum_i \psi_i^* \varphi_i|^2$$

$$= \sum_{ij} \psi_i^* \psi_j \varphi_j^* \varphi_i = \sum_i |\psi_i|^2 |\varphi_i|^2 + \sum_{ij;i\neq j} \psi_i^* \psi_j \varphi_j^* \varphi_i$$

where $\psi_i = \langle i|\psi\rangle$ (as defined above) and similarly $\varphi_i = \langle i|\varphi\rangle$ are the coefficients of the final state of the system. is the complex conjugate so that $\psi_i^* = \langle\psi|i\rangle$, etc.

Now let's consider the situation classically and imagine that the system transited from $|\psi\rangle\, to\, |\varphi\rangle$ via an intermediate state $|i\rangle$. Then we would *classically* expect the probability of the two-step transition to be the sum of all the possible intermediate steps. So we would have

$$\text{prob}(\psi \Rightarrow \varphi) = \sum_i \text{prob}(\psi \Rightarrow i \Rightarrow \varphi)$$

$$= \sum_i |\langle\psi|i\rangle|^2 |\langle i|\varphi\rangle|^2 = \sum_i |\psi_i|^2 |\varphi_i|^2,$$

The classical and quantum derivations for the transition probability differ by the presence, in the quantum case, of the extra terms $\sum_{ij;i\neq j} \psi_i^* \psi_j \varphi_j^* \varphi_i$; these extra quantum terms represent *interference* between the different $i \neq j$ intermediate "alternatives". These are consequently known as the quantum interference terms, or cross terms. This is a purely quantum effect and is a consequence of the non-additivity of the probabilities of quantum alternatives.

The interference terms vanish, via the mechanism of quantum decoherence, if the intermediate state $|i\rangle$ is measured or coupled with the environment.

Constructive Wave Interference

The wave interference is said to be a constructive wave interference if the crest of a wave meets the crest of another wave of the same frequency at the same point.

or

Constructive wave interference is a wave interference which occurs when in phase waves meets with each other and forms a new wave with greater amplitude.

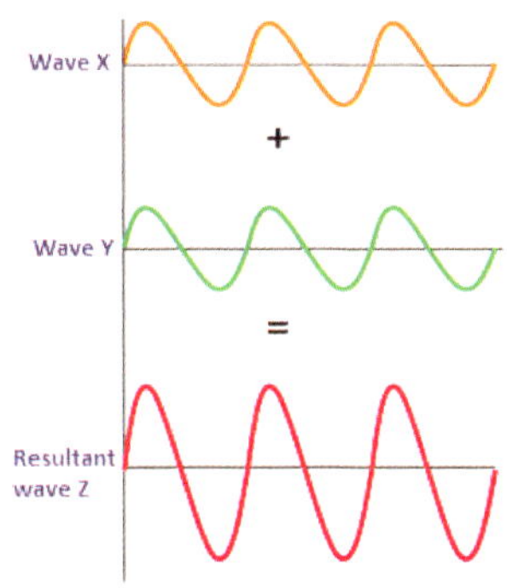

Constructive interference

When two in phase waves having the same frequency meets at the same place at the same time, the amplitudes of the two waves are simply added together. The resultant wave amplitude is greater than that of the original wave.

At these points, the maximum intensity of light is produced because intensity is directly proportional to the square of the amplitude of the wave.

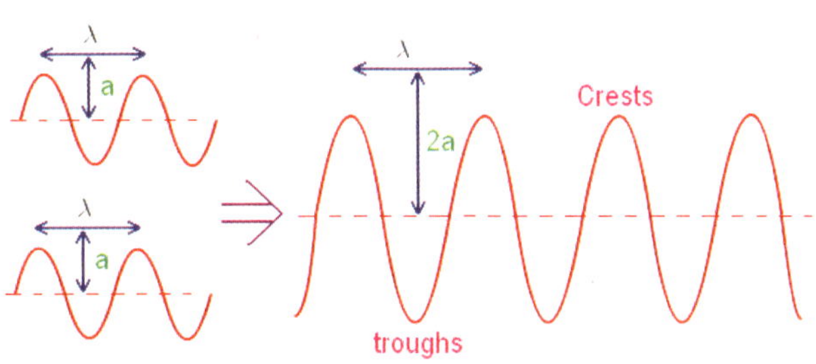

Crests and trough reinforce

The mathematical representation of the above statement is,

$$I \propto A^2$$

The images below show the effects of constructive interference between two waves with the same amplitude and frequency (ω) described by the equations:

$$y_1 = \cos(kx - wt) \text{ and } y_2 = \cos(kx - wt + \delta)$$

where δ is the phase difference between the waves, k is the wave number, x is the wave position and t is time.

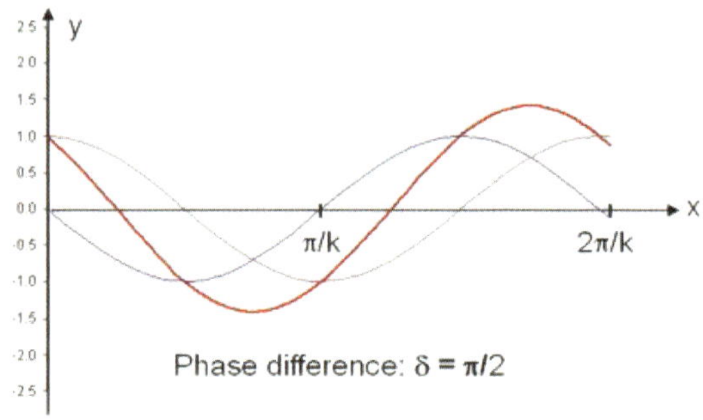

In the image on the left, the phase difference is $\delta = \pi/2$ or 90 degrees. The two waves (thin lines, shown in blue and purple) interfere, with the resulting wave (shown in red) equal to the sum of these two waves. Note that the nodes of the two original waves (where the waves cross the y=0 axis) do not occur at the same values of x, and the nodes of the resulting wave occur when $y_1 + y_2 = 0$.

In the image on the right, the phase difference is δ = 0, so that the two waves (shown in blue) interfere constructively and the amplitude of the resulting wave (shown in red) is the sum of the individual amplitudes. The nodes of the final wave occur at the same locations as the nodes of the individual waves.

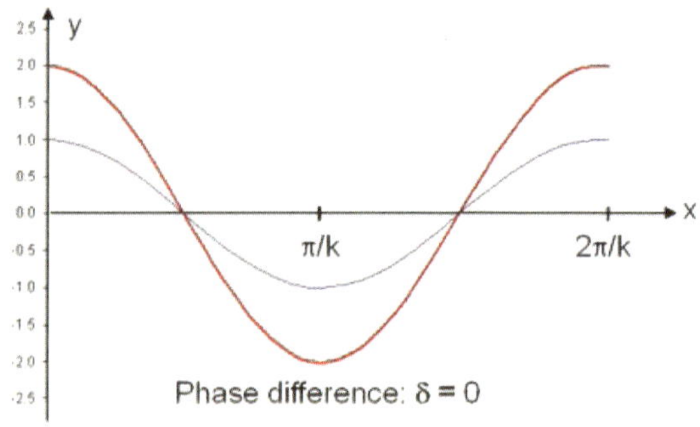

For interference of light waves, such as in Young's two-slit experiment, bands of bright and dark lines will appear. The bright regions occur whenever an integer number of waves constructively interfere.

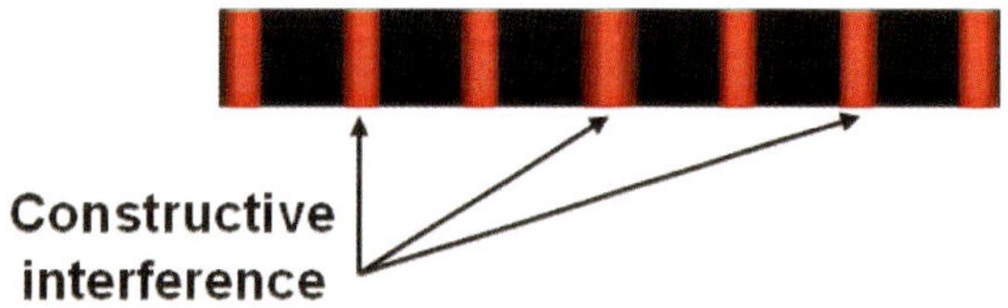

The alternating bright and dark bands resulting from two-slit intererference.

Constructive Interference Equation

The constructive interference is seen when the phase difference between the two waves is,

$$\phi = 0, 2\pi, 4\pi, 6\pi.......2\,n\,\pi.$$

Then $\cos\phi = +1$

$$A_{max} = a_1 + a_2$$

The phase difference between two waves an even multiple of

$\phi = 2\,n\,\pi$. Here n = 0,1,2.....

The path difference between two waves is an even multiple of $\frac{\lambda}{2}$.

$$\delta = 2n\frac{\lambda}{2}.$$

The time interval between two waves is even multiple of $\frac{T}{2}$.

$$\theta = 2\,n\frac{T}{2},\ where\ n = 0, 1, 2, 3...........$$

The resultant amplitude of wave is equal to the sum of amplitudes of individual waves

$$A = a_1 + a_2$$

The resultant intensity is more than the sum of intensities of individual waves

$$I_{max} = I_1 + I_2 + 2\sqrt{(I_1 I_2)}$$

$$= \sqrt{I_1} + \sqrt{I_2}^{\,2}$$

Intensity is more than the given intensity $I > I_1 + I_2$.

Constructive Interference Example

These are the few examples of constructive interference:

- Constructive interference takes place in light.
- Constructive interference takes place in sound.
- The rogue waves in the ocean are the example of constructive interference.
- When two speakers are put one meter apart and run we get constructive interference between the sound waves.
- Beats are the example of constructive interference.

Consider two sound waves traveling in the same medium as shown in the below figure. The sound wave X amplitude is 1 unit and the sound wave Y amplitude is also 1 unit. When these two sound waves having the same frequency meets at the same time at the same place, the amplitudes of the two sound waves are added together. Therefore, the resultant sound wave (Z) amplitude is greater than that of the original wave.

$$I.e. Z = X + Y = 1 + 1 = 2 \text{ units}$$

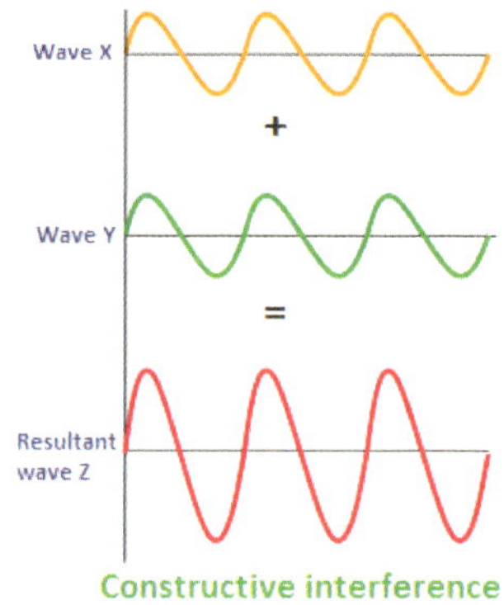

The constructive interference is pretty normal and natural.

Example:

The constructive wave interference occurs naturally. However, we can also create the constructive interference by using two speakers.

If we place two speakers side-by-side, point them in the same direction, and play the same frequency, we get a constructive interference.

As we know, sound means the vibrations that travel through the air. Both speaker 1 and speaker 2 pushes the air forwards. Hence, the vibrations travel in the forward direction (same direction) through the air.

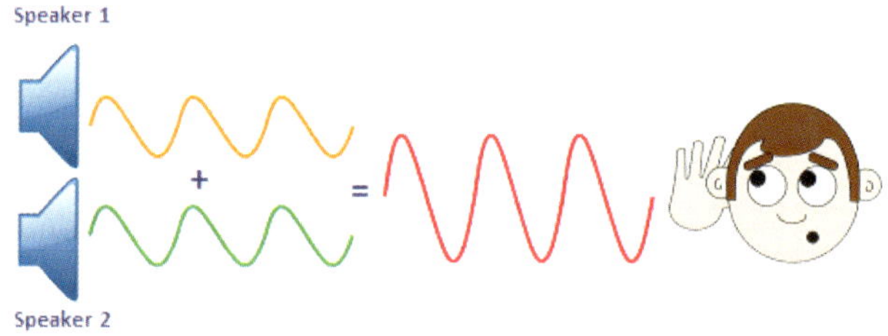

(Amplitude means the maximum extent of a vibration from the point of equilibrium.)

If we stand in front of the two speakers, the vibrations produced by two speakers will be added up in the air. As a result, we hear a sound which is much louder than the individual speakers would produce.

In other words, if we stand in front of the two speakers, the amplitudes of the two sound waves are added up. As a result, we hear a sound which is much louder than the individual speakers would produce.

This is nothing but a constructive interference.

Destructive Wave Interference

The wave interference is said to be a destructive wave interference if the crest of a wave meets the trough of another wave of the same frequency.

or

Destructive wave interference is a wave interference which occurs when out of phase waves meets with each other and forms a new wave with lower amplitude or zero amplitude.

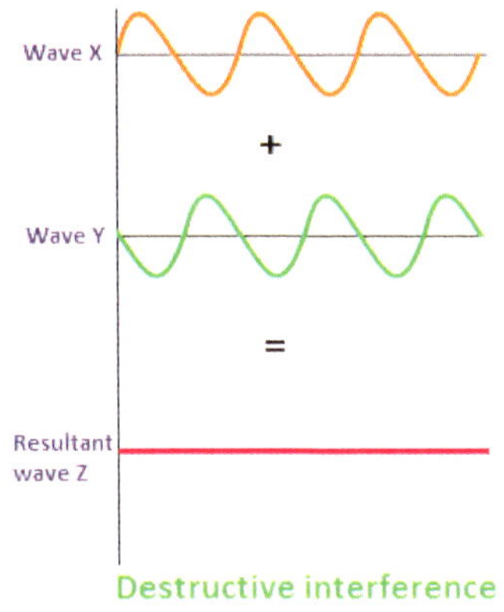

The images below show the effects of destructive interference between two waves with the same amplitude and frequency (ω) described by the equations:

$$y_1 = \cos(kx - wt) \text{ and } y_2 = \cos(kx - wt + \delta)$$

where δ is the phase difference between the waves, k is the wave number, x is the wave position and t is time.

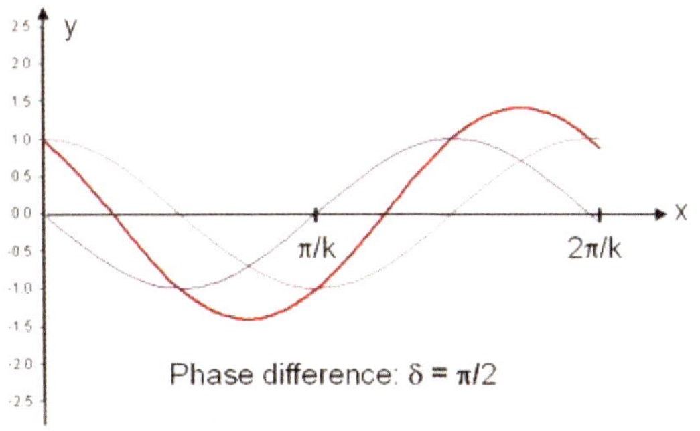

In the image on the left, the phase difference is $\delta = \pi/2$ or 90 degrees. The two waves (thin lines, shown in blue and purple) interfere, with the resulting wave (shown in red) equal to the sum of these two waves. Note that the nodes of the two original waves (where the waves cross the y=0 axis) do not occur at the same values of x, and the nodes of the resulting wave occur when $y_1 + y_2 = 0$.

In the image on the right, the phase difference is $\delta = \pi$, so that the two waves (shown in blue and purple) interfere destructively and the amplitude of the resulting wave (shown in red) is equal to zero.

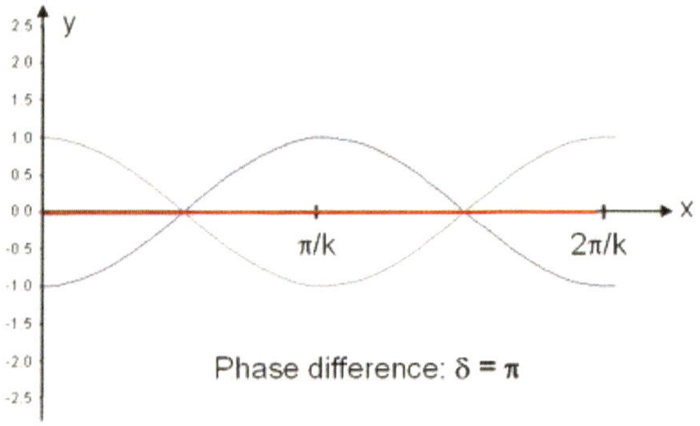

For interference of light waves, such as in Young's two-slit experiment, bands of bright and dark lines will appear. The dark regions occur whenever the waves destructively interfere.

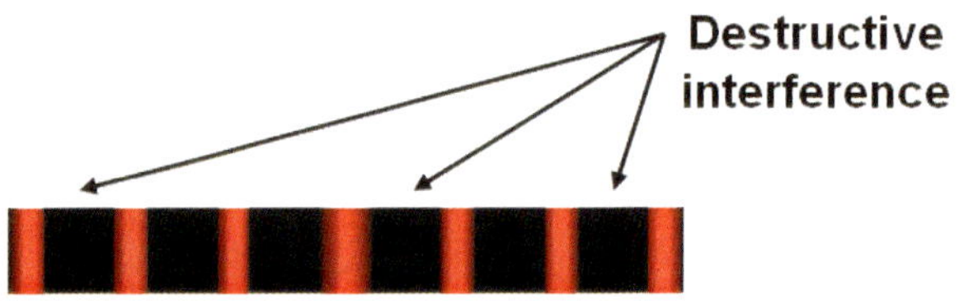

The alternating bright and dark bands resulting from two-slit intererference.

Consider two sound waves traveling in the same medium that are out of phase as shown in the below figure. The sound wave X amplitude is 1 unit and the sound wave Y amplitude is also 1 unit. When these two sound waves having the same frequency meets each other, the amplitudes of the two sound waves cancels each other. Therefore, the resultant sound wave (Z) amplitude is zero or it is lower than that of the original wave.

$$I.e. Z = X + Y = 1 + (-1) = 0 \; units$$

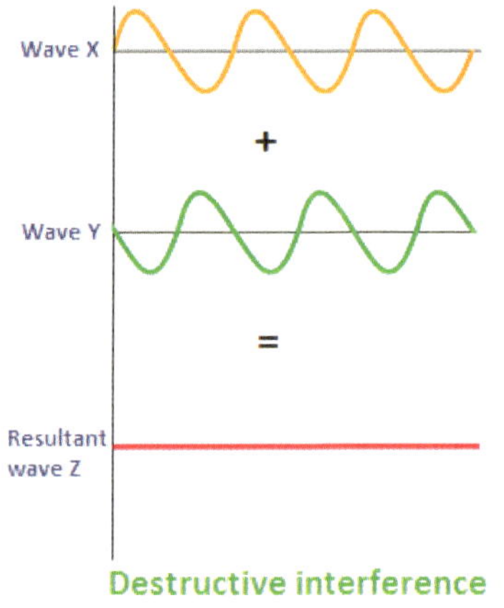

Example:

As we know, if we place two speakers side-by-side, point them in the same direction, and play the same frequency, we get a constructive interference.

Now imagine that we start moving one of the speakers backward, at some point, the two sound waves will become out of phase.

So let's see what will happen if we stand in front of two speakers which are producing the out of phase sound waves.

(Amplitude means the maximum extent of a vibration from the point of equilibrium.)

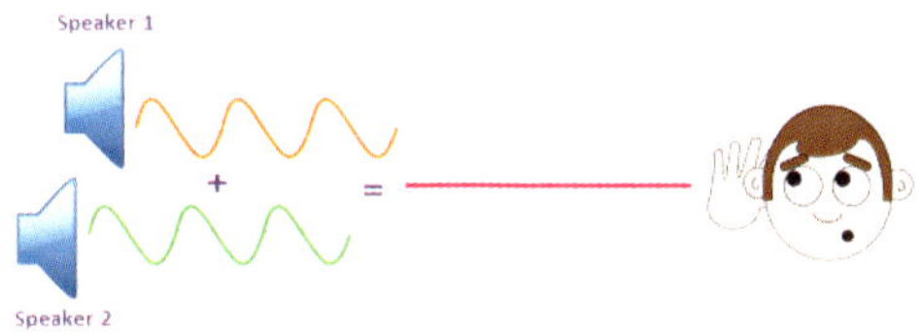

If we stand in front of the two speakers, the speaker 1 one pushes the air in the forward direction and the speaker 2 pulls the air backwards. Hence, the vibrations produced by speaker 1 travels in the forward direction and the vibrations produced by speaker 2 travels in the backward direction. That means the two vibrations produced by speaker 1 and speaker 2 travels in the opposite directions. Hence, the two vibrations cancel out each other. As result, no vibration will reach your ear.

We know that sound means vibrations traveling through the air. Therefore, you will not hear any sound.

This is nothing but a destructive interference.

How far back must we move the speaker to go from constructive to destructive interference?

If the speakers are separated by half a wavelength, then there is a destructive interference. If the speakers are separated by one wavelength, then again there is a constructive interference. That means, for every half-wavelength separation, the constructive and destructive interferences changes alternatively.

Destructive Interference Equation

1. The phase difference between two waves an odd multiple of $\pi\pi$ that is, $(2n-1)\pi$ Here n $= 1, 2, 3$

2. The path difference between two waves is an odd multiple of $\frac{\lambda}{2}$.

$$\Delta = (2n-1)\frac{\lambda}{2} \quad \text{Here n} = 1,\ 2$$

3. The time interval between two waves is odd multiple of $\frac{T}{2}$.

$$\theta = (2n-1)\frac{T}{2} \quad \text{n} = 1,\ 2, 3$$

4. The resultant amplitude of wave is equal to the difference of amplitudes of individual waves

$$\text{Amplitude} = \text{A - B } \textit{if } \phi = \pi, 3\pi, 5......(2n-1)\pi.$$

5. The resultant intensity is less than the sum of intensities of individual waves

$$I_{min} = I1 + I2 - 2\sqrt{I_1 I_2} = \left(\sqrt{I_1} - \sqrt{I_2}\right)^2$$

Destructive Interference Examples

These are the few Examples of Destructive Interference:

- Gravitational waves are the example of Destructive Interference.
- Light beams may show Destructive Interference.
- Acoustics is another example.
- Moving electrons also display this phenomenon.
- Can be observed in radio waves.
- Surface water waves show Destructive Interference.

Complete Destructive Interference

Complete Destructive Interference will never be possible. There would always be some constructive interference associated with it. At some places the energy would be zero (Destructive Interference) but at other places it would appear as the case for constructive interference.

Young's Interference Experiment

Although Christiaan Huygens thought that light was a wave, Isaac Newton did not. Newton felt that there were other explanations for color, and for the interference and diffraction effects that were observable at the time. Owing to Newton's tremendous stature, his view generally prevailed. The fact that Huygens's principle worked was not considered evidence that was direct enough to prove that light is a wave. The acceptance of the wave character of light came many years later when, in 1801, the English physicist and physician Thomas Young (1773–1829) did his now-classic double slit experiment.

Young's double slit experiment. Here pure-wavelength light sent through a pair of vertical slits is diffracted into a pattern on the screen of numerous vertical lines spread out horizontally. Without diffraction and interference, the light would simply make two lines on the screen.

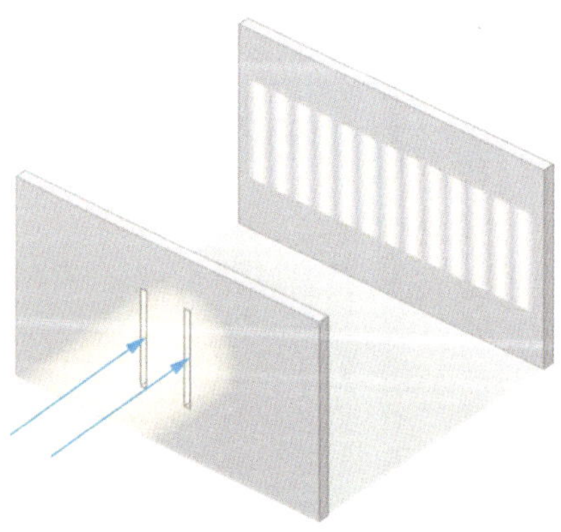

Why do we not ordinarily observe wave behavior for light, such as observed in Young's double slit experiment? First, light must interact with something small, such as the closely spaced slits used by Young, to show pronounced wave effects. Furthermore, Young first passed light from a single source (the Sun) through a single slit to make the light somewhat coherent. By coherent, we mean waves are in phase or have a definite phase relationship. Incoherent means the waves have random phase relationships. Why did Young then pass the light through a double slit? The answer to this question is that two slits provide two coherent light sources that then interfere constructively or destructively. Young used sunlight, where each wavelength forms its own pattern, making the effect more difficult to see. We illustrate the double slit experiment with monochromatic (single λ) light to clarify the effect. shows the pure constructive and destructive interference of two waves having the same wavelength and amplitude.

The amplitudes of waves add. (a) Pure constructive interference is obtained when identical waves are in phase. (b) Pure destructive interference occurs when identical waves are exactly out of phase, or shifted by half a wavelength.

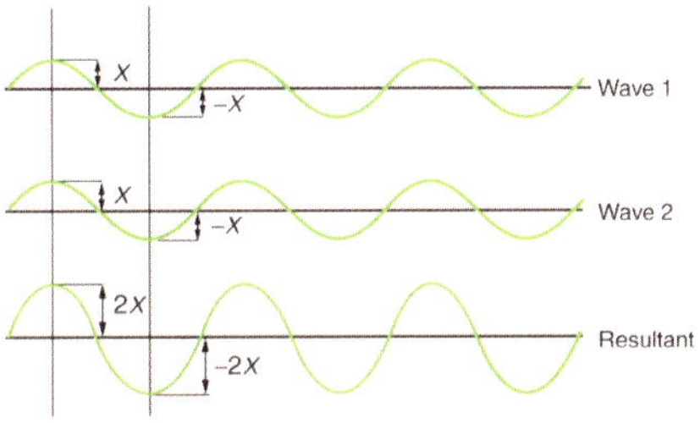

(a)

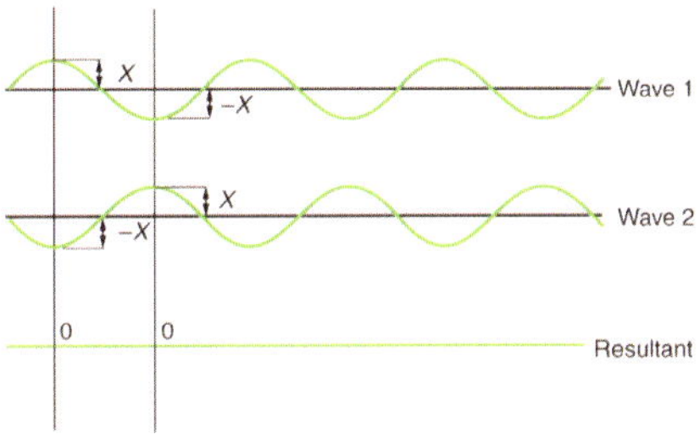

(b)

When light passes through narrow slits, it is diffracted into semicircular waves, as shown in (a). Pure constructive interference occurs where the waves are crest to crest or trough to trough. Pure destructive interference occurs where they are crest to trough. The light must fall on a screen and be scattered into our eyes for us to see the pattern. An analogous pattern for water waves is shown in (b). Note that regions of constructive and destructive interference move out from the slits at well-defined angles to the original beam. These angles depend on wavelength and the distance between the slits.

Double slits produce two coherent sources of waves that interfere. (a) Light spreads out (diffracts) from each slit, because the slits are narrow. These waves overlap and interfere constructively (bright lines) and destructively (dark regions). We can only see this if the light falls onto a screen and is scattered into our eyes. (b) Double slit interference pattern for water waves are nearly identical to that for light. Wave action is greatest in regions of constructive interference and least in regions of destructive interference. (c) When light that has passed through double slits falls on a screen, we see a pattern such as this.

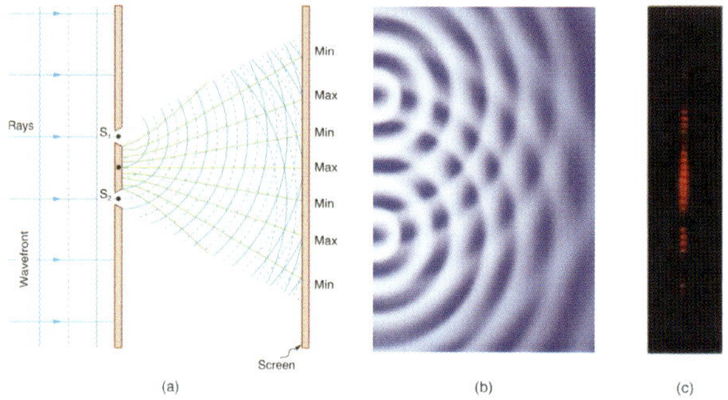

To understand the double slit interference pattern, we consider how two waves travel from the slits to the screen, as illustrated. Each slit is a different distance from a given point on the screen. Thus different numbers of wavelengths fit into each path. Waves start out from the slits in phase (crest to crest), but they may end up out of phase (crest to trough) at the screen if the paths differ in length by half a wavelength, interfering destructively as shown in (a). If the paths differ by a whole wavelength, then the waves arrive in phase (crest to crest) at the screen, interfering constructively as shown in (b). More generally, if the paths taken by the two waves differ by any half-integral number of wavelengths $[(1/2)\lambda,(3/2)\lambda,(5/2)\lambda,\ etc.]$, then destructive interference occurs. Similarly, if the paths taken by the two waves differ by any integral number of wavelengths $(\lambda,2\lambda,3\lambda,\ etc.)$, then constructive interference occurs.

Take-Home Experiment: Using Fingers as Slits

Look at a light, such as a street lamp or incandescent bulb, through the narrow gap between two fingers held close together. What type of pattern do you see? How does it change when you allow the fingers to move a little farther apart? Is it more distinct for a monochromatic source, such as the yellow light from a sodium vapor lamp, than for an incandescent bulb?

Waves follow different paths from the slits to a common point on a screen. (a) Destructive interference occurs here, because one path is a half wavelength longer than the other. The waves start in phase but arrive out of phase. (b) Constructive interference occurs here because one path is a whole wavelength longer than the other. The waves start out and arrive in phase.

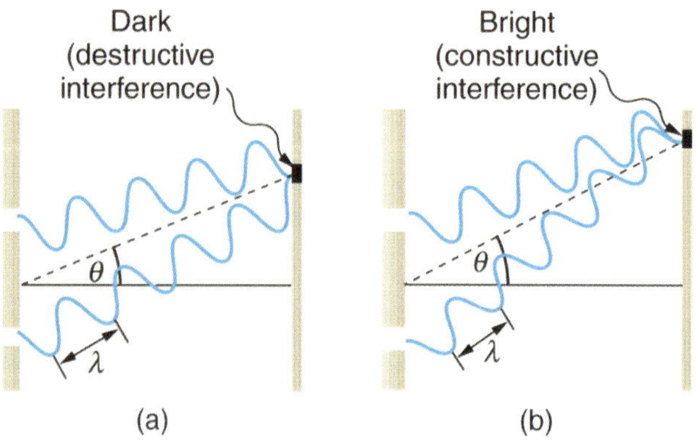

shows how to determine the path length difference for waves traveling from two slits to a common point on a screen. If the screen is a large distance away compared with the distance between the slits, then the angle θ between the path and a line from the slits to the screen is nearly the same for each path. The difference between the paths is shown in the figure; simple trigonometry shows it to be θ $d \sin\theta$, where d is the distance between the slits. To obtain constructive interference for a double slit, the path length difference must be an integral multiple of the wavelength, or

$$d \sin\theta = m\lambda, \textit{ for } m = 0, 1, -1, 2, -2, \ \ldots . \text{ (constructive).}$$

Similarly, to obtain destructive interference for a double slit, the path length difference must be a half-integral multiple of the wavelength, or

$$d \sin\theta = \left(m + \frac{1}{2}\right)\lambda \ \textit{ for } m = 0, 1, -1, \ 2, -2, \ \ldots . \text{ (destructive),}$$

where λ is the wavelength of the light, d is the distance between slits, and θ is the angle from the original direction of the beam as discussed above. We call m the order of the interference. For example, $m = 4$ is fourth-order interference.

The paths from each slit to a common point on the screen differ by an amount $d \sin\theta$, assuming the distance to the screen is much greater than the distance between slits (not to scale here).

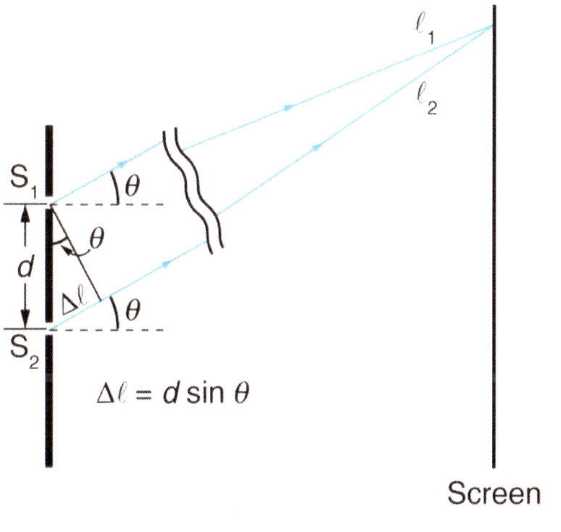

The equations for double slit interference imply that a series of bright and dark lines are formed. For vertical slits, the light spreads out horizontally on either side of the incident beam into a pattern called interference fringes, illustrated in . The intensity of the bright fringes falls off on either side, being brightest at the center. The closer the slits are, the more is the spreading of the bright fringes. We can see this by examining the equation.

$$d \sin\theta = m\lambda, \textit{ for } \ m = 0, 1, -1, \ 2, -2, \ \ldots .$$

For fixed λ and m, the smaller d is, the larger θ must be, since

$$\sin\theta = m\lambda / d.$$

This is consistent with our contention that wave effects are most noticeable when the object the wave encounters (here, slits a distance d apart) is small. Small d gives large θ, hence a large effect.

The interference pattern for a double slit has an intensity that falls off with angle. The photograph shows multiple bright and dark lines, or fringes, formed by light passing through a double slit.

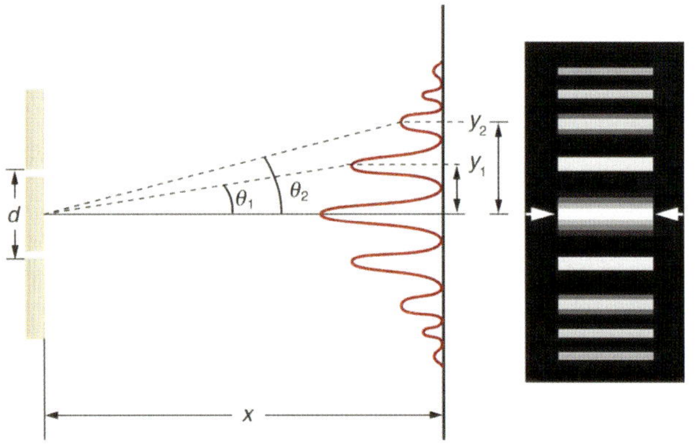

Finding a Wavelength from an Interference Pattern

Suppose you pass light from a He-Ne laser through two slits separated by 0.0100 mm and find that the third bright line on a screen is formed at an angle of 10.95 relative to the incident beam. What is the wavelength of the light?

Strategy

The third bright line is due to third-order constructive interference, which means that $m = 3$. We are given $d = 0.0100$ mm and $\theta = 10.95$. The wavelength can thus be found using the equation $d \sin\theta = m\lambda$ for constructive interference.

Solution

The equation is $d \sin\theta = m\lambda$.. Solving for the wavelength λ gives

$$\lambda = \frac{d \sin \theta}{m}$$

Substituting known values yields

$$\lambda = \frac{(0.0100\,mm)(\sin 10.95)}{3}$$

$$= 6.3310^{-4}\ mm\ = 633\,nm.$$

Discussion

To three digits, this is the wavelength of light emitted by the common He-Ne laser. Not by coincidence, this red color is similar to that emitted by neon lights. More important, however, is the fact that interference patterns can be used to measure wavelength. Young did this for visible wavelengths. This analytical technique is still widely used to measure electromagnetic spectra. For a given order, the angle for constructive interference increases with λ, so that spectra (measurements of intensity versus wavelength) can be obtained.

Calculating Highest Order Possible

Interference patterns do not have an infinite number of lines, since there is a limit to how big m can be. What is the highest-order constructive interference possible with the system described in the preceding example?

Strategy and Concept

The equation $d \sin \theta = m\lambda$ (for $m = 0, 1, -1, 2, -2$............. describes constructive interference. For fixed values of d and λ, the larger m is, the larger θ is. However, the maximum value that $\sin \theta$ can have is 1, for an angle of 90. (Larger angles imply that light goes backward and does not reach the screen at all.) Let us find which m corresponds to this maximum diffraction angle.

Solution

Solving the equation $d \sin \theta = m\lambda$ for m gives

$$m = \frac{d \sin \theta}{\lambda}$$

Taking $\sin \theta = 1$ and substituting the values of d and λ from the preceding example gives

$$m = \frac{(0.0100\ mm\ (1)}{633} \approx 15.8$$

Therefore, the largest integer m can be is 15, or

$$m = 15$$

The number of fringes depends on the wavelength and slit separation. The number of fringes will be very large for large slit separations. However, if the slit separation becomes much greater than the wavelength, the intensity of the interference pattern changes so that the screen has two bright lines cast by the slits, as expected when light behaves like a ray. We also note that the fringes get fainter further away from the center. Consequently, not all 15 fringes may be observable.

Thin-film Interference

Thin films between two media often produce iridescence and other interference patterns: areas of destructive and constructive interference for different wavelengths. When light is involved, the result is often beautifully coloured patterns.

When light hits a surface, it can be absorbed, transmitted or reflected.

Light can reflect from any interface where there is a change in refractive index.

When light hits a material that has multiple layers, each layer can reflect light. The light reflected from any one of these layers might interfere with the light that has reflected off any of the other layers. Each layer creates its own image of a point source. These sources will be mutually coherent because they all arise from a single original source. The basic idea is illustrated in the figure below, which shows a glass surface with a thin coating. There are two main sources of reflection from the surface (ignoring multiple internal reflections) labeled 1 and 2. These two reflections wil linterfere. How they interfere depends on phase difference between them. The phase difference depended on the index of refraction of the coating, the thickness of the coating and what happens to the phase upon reflection.

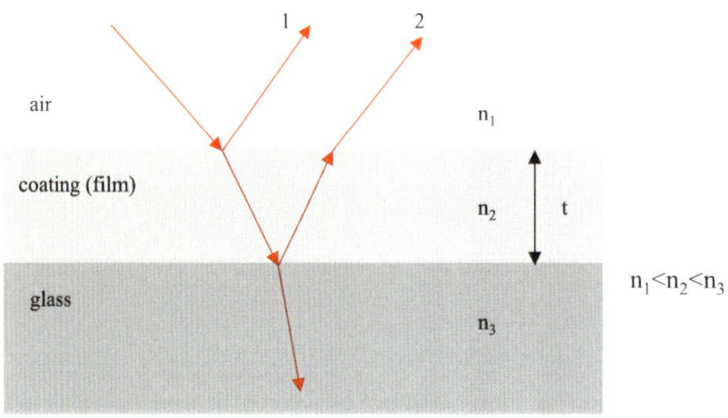

So, what are the rules for how waves reflect from surfaces - what happens to the phase? The rule for reflection is that when the index of refraction after the interface is higher then the index of refraction before the interface then the light, upon reflection, undergoes a 180 degree phase shift. When the index of refraction after the interface is less than the index before the interface, then there is no phase shift in the reflected light.

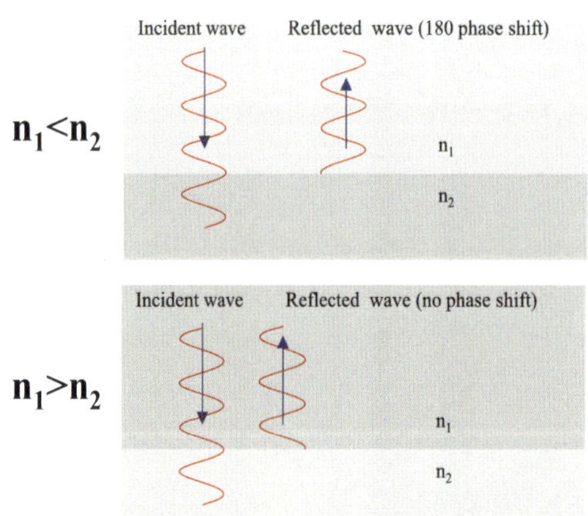

Because each interface is at a transition from a lower to a higher refractive index, there is 180 deg phase shift at each reflection (for 1 and 2). So they undergo the same phase shift and we can therefore ignore it in this example. All we need to do is calculate the phase difference between rays 1 and 2 due to the path length difference. If there is destructive interference, then there will be a minimum in the amount of reflected light (this is the design of an antireflection coating). If light constructively interferes, there will be an increase in the amount of reflection (this is the design of highly reflective mirror). Since light has to go somewhere, the light that is not reflected will have to be transmitted, so both transmission and reflection are affected by thin film coatings.

The difference in actual path length between 1 and 2 is (assuming that the angles are small):

$$\Delta path_{1,\,2} = 2t$$

For destructive interference to occur, we require that rays 1 and 2 be 180 degrees out of phase, or that the number of waves of difference be 1/2.

$$\frac{1}{2} = \frac{2tn_2}{\lambda} \qquad \Rightarrow \qquad t_{dest} = {}^{1}\!/\!_{4}\,\frac{\lambda}{n_2}$$

Similar interference can occur when the difference in the number of waves is 1.5, 2.5 3.5 ..

$$t_{dest} = /_{4}\,\frac{\lambda}{n_2}, {}^{3}\!/\!_{4}\,\frac{\lambda}{n_2}, {}^{5}\!/\!_{4}\,\frac{\lambda}{n_2}, {}^{7}\!/\!_{4}\,\frac{\lambda}{n_2} \ldots\ldots \quad t = \frac{(\qquad /\ )}{\qquad}\frac{\lambda}{n_2}, m = 0, 1, 2\ldots.$$

But recall the requirement that interference can only occur when the difference in path length of two mutually coherent sources of light be within the coherence length. So as the thickness increases, many sources will no longer be mutually coherent.

The condition for interference is that the difference in distance between the sources of light (from the reflections) be within the coherence length of the light.

Diversion: Coherence Length Defined as the length in space over which the light has a predictable phase. A laser, for example, has a long coherence length. Although there are random fluctuations in phase over time they occur after the waves have traveled some meters in distance. A incandescent bulb on the other hand, has tremendously fast variations so the phase is predictable over only a very short period of space. Consider the double slit experiment: The interference arises from the difference in path lengths between the two mutually coherence sources of light. For high angles, the path difference can be come very large (multiple wavelengths). For a laser, which has a long coherence length, the two beams will still retain a fixed phase difference between them. For a normal light source, there will be so much difference in distance between the two waves that there will have been time for random phase jumps to occur. So the two sources are essentially mutually incoherent. The path difference must be within the coherence length for interference to occur. For a laser, the coherence length is on the order of meters. For an incandescent light bulb, the coherence length is only a few micrometers. For the sun, the coherence length is on the order of millimeters. For some lasers, the coherence length can be many kilometers.

The same holds for constructive interference. Constructive interference will occur when the difference in waves between rays 1 and 2 is some integer multiple of wavelengths.

$$1 = \frac{2tn_2}{\lambda} \quad \Rightarrow t_{const} = {}^{1}\!/\!_{2}\,\frac{\lambda}{n_2}, 1\frac{\lambda}{n_2}, {}^{3}\!/\!_{2}\,\frac{\lambda}{n_2}, 2\frac{\lambda}{n_2} \ldots \quad t_{const} = \frac{m}{2}\frac{\lambda}{n_2}, m = 0, 1, 2\ldots$$

So, imagine that you could vary the thickness of a coating on a piece of glass, the amount of reflection would vary as the thickness requirements for constructive and destructive interference were met. Manipulating the coatings on surfaces can be used for many applications, which we will talk about in the next couple of lectures.

Anti Reflection Coatings ARCs

To make an antireflection coating, or ARC, you put a coating on the surface of your glasses with the appropriate thickness that minimizes the reflection. By minimizing the reflection, you automatically increase the transmission.

Lens without antireflection coatings Lens with ARC

4% off the first surface	0.5% off first surface
4% off the second surface	0.5 % off second surface
92% transmittance.	99% transmittance

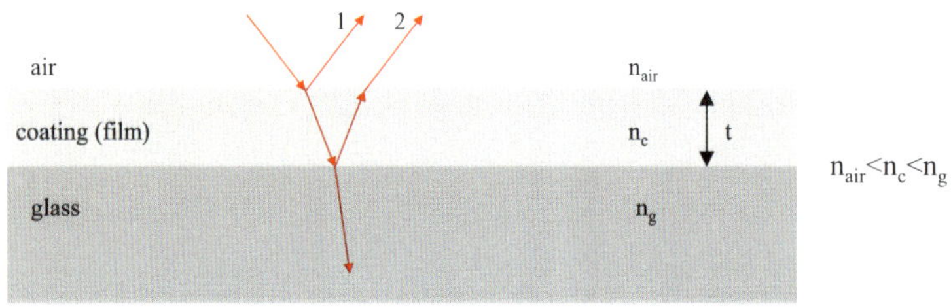

But there is a second condition that needs to be satisfied in order to maximize the amount of destructive interference. For complete destructive interference, we also require that the amplitudes of ray 1 and 2 be the same. To meet this requirement we need to have the condition that the reflectivity of the air coating interface and the coating glass interface be the same.

Recall the reflectance equation for amplitude: $r = \frac{n'-n}{n'+n}$ for intensity: $R = (r)^2 = \left[\frac{n'-n}{n'+n}\right]^2$

For ray 1: $r_1 = \frac{n_c - n_{air}}{n_c + n_{air}}$ For ray 2: $r_2 = \frac{n_g - n_c}{n_g + n_c}$

For light that is out of phase, the resultant amplitude is the difference between the two amplitudes so total destructive interference occurs when

$$r_1 - r_2 = 0$$

$$\therefore \frac{n_c - n_{air}}{n_c + n_{air}} = \frac{n_g - n_c}{n_g + n_c}$$

yields the result:

$$n_c = \sqrt{n_g}$$

For non-ideal refractive indices, the intensity is the square of the resultant nonzero amplitude.

To summarize, an effective ARC must meet the following two conditions:

Path condition (destructive interference): $t_{dest} = \frac{1}{4}\frac{\lambda}{n_c}$

Amplitude condition: $n_c = \sqrt{n_g}$

So, what are the problems with ARCs? They can make coatings of exact thicknesses, but the constraint is on available materials that meet the amplitude condition requirement.

Problem A:

If we start with glass of index 1.5, we are forced to find a suitable coating material that has an index 1.22. Such materials are not readily available so manufacturers simply try to get as close as possible. They typically use a coating made of $MgFl_2$, which has an index of 1.38. The ideal glass for that coating would have an index of 1.9, which is not available. So the higher the index of the glass, the better the ARC, when you are using $MgFl_2$ as a coating.

Problem B:

Wavelength dependency. The path condition can only be exactly met for one wavelength. Ideally, we choose to optimize the coating for wavelengths in the middle of the spectrum. So while 550 does not reflect, the lower and higher wavelengths do, which gives the reflection from ARC glasses a purplish hue.

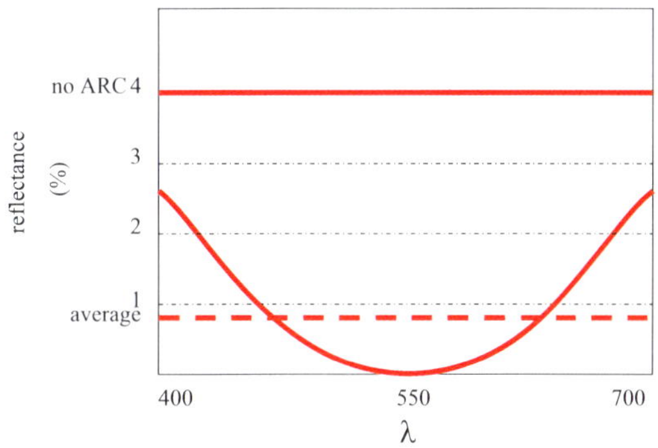

A typical ARC has an average of about 2% reflectance

Problem C:

Only works as designed on axis. When you look at the reflection off axis, it changes color because the path lengths through the coating are increased. It becomes a much more complicated problem

Example:

What is the reflectance of a glass (n=1.5) surface with a MgFl2 coating (n=1.38) optimized for 550 nm light for

550 nm light?

400 nm light?

Solution:

Step 1: What is the thickness of the coating?

$$t_{dest} = \frac{1}{4}\frac{\lambda}{n_c} = \frac{1}{4}\frac{550}{1.38} = 99.64\,\text{nm}$$

Step 2: What is the amplitude of reflectance at the surfaces?

$$r_1 = \frac{n_c - n_{air}}{n_c + n_{air}} = \frac{1.38 - 1}{1.38 + 1} = 0.16 \qquad r_2 = \frac{n_g - n_c}{n_g + n_c} = \frac{1.5 - 1.38}{1.5 + 1.38} = 0.0417$$

Step 3: For 550 nm light....

$$I_{coherent} = (E_1 + E_2)^2 = A_1^2 + A_2^2 + 2 \cdot A_1 A_2 \cos(p_1 - p_2)$$

$p_1 - p_2$ 180 since they are out of phase

$$I_{coherent} = (E_1 + E_2)^2 = A_1^2 + A_2^2 + 2 \cdot A_1 A_2 \times (-1) = 0.159^2 + 0.0417^2 - 2 \times 0.159 \times 0.0417$$

$$= 0.0138 \equiv 1.38\,\%$$

Step 4: For 400 nm light, what is the phase difference?

$$\Delta \text{waves} = \frac{2 \times 99.64}{400/1.38} = 0.687 \text{ waves}$$

$$\Rightarrow \Delta\, phase = 0.687 \times 2 \times \pi = 4.32\, radians$$

Step 5: For 400 nm light...Step 5: For 400 nm light...

$$I_{coherent} = (E_1 + E_2)^2 = A_1^2 + A_2^2 + 2 \cdot A_1 A_2 \cos(p_1 - p_2)$$

$$I_{coherent} = 0.159^2 + 0.0417^2 + 2 \times 0.159 \times 0.0417 \times \cos(4.32)$$

$$= 0.0219 \equiv 2.19\%$$

400 nm light is more reflective than 550 nm light, as expected. Reflectance was not zero for 550 nm light because the amplitude condition was not met.

Anti-Reflection Coating

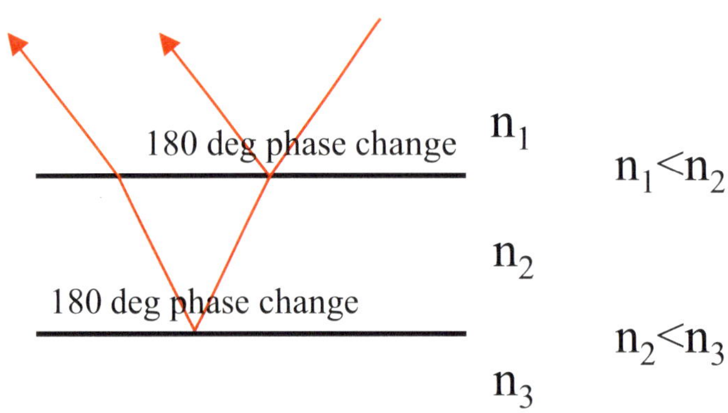

Oil Film (Like a Soap Bubble)

The typical index of refraction of oil is about 1.5, which is similar to glass.

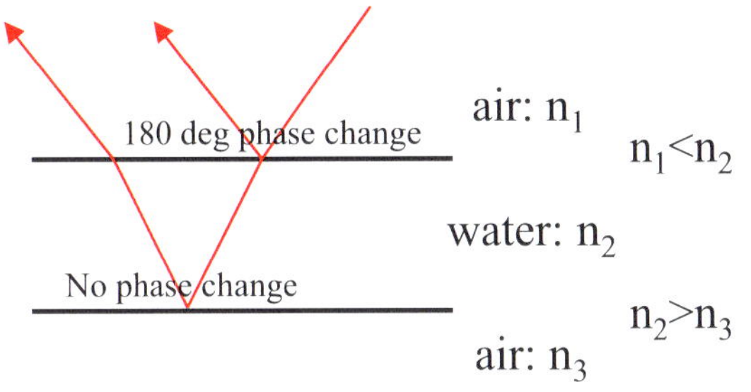

The thickness of the oil layer varies as well as your viewing angle, so you see a spectrum of colors. Like the soap bubble, the oil has low reflectance when the layer is very thin. That is why the reflection from an oily surface on a puddle has a blackish appearance.

If the phase changes are common to both surfaces (eg ARC), then

$$t_{dest} = \frac{\left(m + \frac{1}{2}\right)}{2} \frac{\lambda}{n_2},\ m = 0, 1, 2.... \quad t_{const} = \frac{m}{2} \frac{\lambda}{n_2},\ m = 0, 1, 2...$$

If the phase changes are not common to both surfaces (eg soap bubble, or oil), then

$$t_{dest} = \frac{m}{2} \frac{\lambda}{n_2},\ m = 0, 1, 2... \quad t_{const} = \frac{\left(m + \frac{1}{2}\right)}{2} \frac{\lambda}{n_2},\ m = 0, 1, 2....$$

Fringes of Equal Thickness

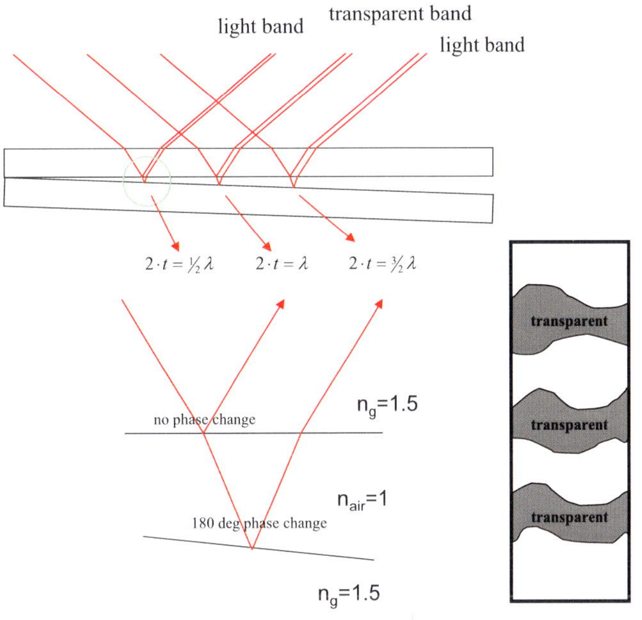

When t = 1/4 λ you get constructive interference, when t = 1/2 = λ you get destructive interference. In this case, the film is air so the wavelength need not be adjusted.

path condition for constructive interference

$$2 \cdot n_c t = \tfrac{1}{2}\lambda, \tfrac{3}{2}\lambda, \ldots \qquad but \;\; n_c = 1$$

$$2 \cdot t = \left(m + \tfrac{1}{2}\right)\lambda, \qquad m = 0, 1, 2 \ldots\ldots\ldots$$

$$\text{or } t_{const} = \frac{\left(m + \tfrac{1}{2}\right)\lambda}{2}, \qquad m = 0, 1, 2 \ldots\ldots\ldots$$

path condition for destructive interference

$$2 \cdot n_c t = 0, \lambda, 2\lambda, \ldots \qquad but \;\; n_c = 1$$

$$2 \cdot t = m\lambda, \qquad m = 0, 1, 2 \ldots\ldots\ldots$$

$$\text{or } t_{const} = \frac{m\lambda}{2}, \qquad m = 0, 1, 2 \ldots\ldots\ldots$$

Special case: Newton's rings

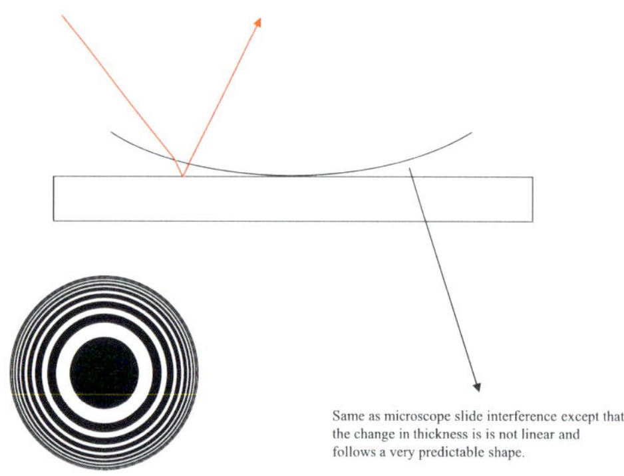

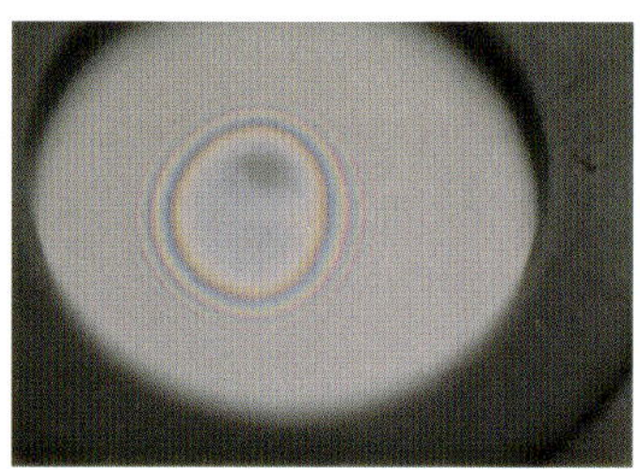

Example

1) Two flat microscope slides, 10 cm long are touching on one side and are separated by 3 microns on the other. How many dark interference bands will appear on the slide if you look at the reflection for 450 nm light?

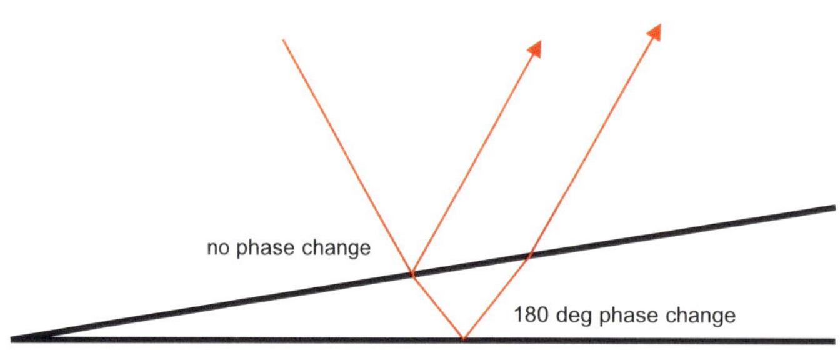

path condition for destructive interference

$$t = \frac{m\lambda}{2}, \qquad m = 0, \pm 1, \pm 2 \ldots\ldots\ldots$$

$$m = 0 \quad \Rightarrow \quad t = 0$$

$$m = 1 \quad \Rightarrow \quad t = 0.225$$

$$m = 2 \quad \Rightarrow \quad t = 0.450$$

$$m = 3 \quad \Rightarrow \quad t = \ldots\ldots$$

a dark band occurs wheneverthe thickness changes by 0.225 microns.

$\frac{3}{0.225} = 13.33$, so the last band occurs when $m = 13$

$\therefore$ there are 14 dark bands (there is one for $m = 0$)

Colours of Thin Films

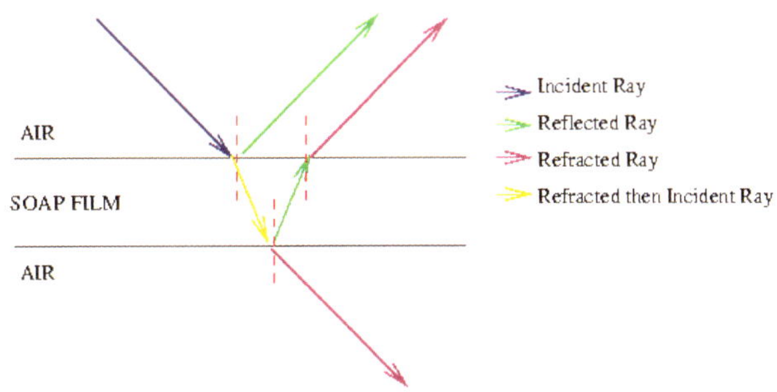

Everyone is familiar with the brilliant colours exhibited by a thin oil film spread on the surface of water and also by a soap bubble. These colours are due to interference between light waves reflected from the top and the bottom surfaces of thin films. When white light is incident on a thin film, the film appears coloured and the colour depends upon the thickness of the film and also the angle of incidence of the light.

Theory

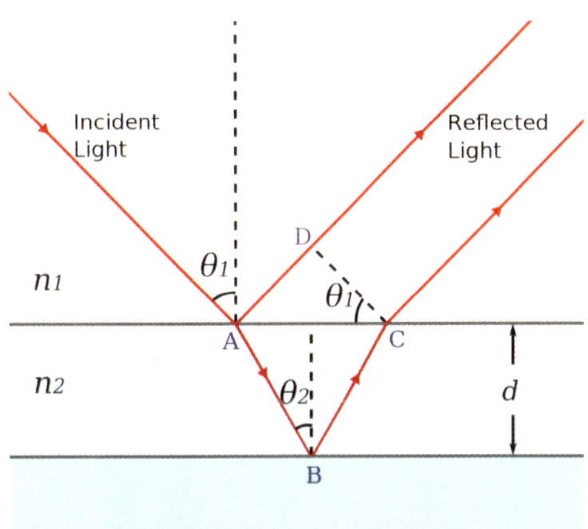

Demonstration of the optical path length difference for light reflected from the upper and lower boundaries of a thin film.

Thin-film interference caused by ITO defrosting coating on an Airbuscockpit window.

A thin film is a layer of material with thickness in the sub-nanometer to micron range. As light strikes the surface of a film it is either transmitted or reflected at the upper surface. Light that is transmitted reaches the bottom surface and may once again be transmitted or reflected. The Fresnel equations provide a quantitative description of how much of the light will be transmitted or reflected at an interface. The light reflected from the upper and lower surfaces will interfere. The degree of constructive or destructive interference between the two light waves depends on the difference in their phase. This difference in turn depends on the thickness of the film layer, the refractive index of the film, and the angle of incidence of the original wave on the film. Additionally, a phase shift of 180° or π radians may be introduced upon reflection at a boundary depending on the refractive indices of the materials on either side of the boundary. This phase shift occurs if the refractive index of the medium the light is travelling through is less than the refractive index of the material it is striking. In other words, if $n_1 < n_2$ and the light is travelling from material 1 to material 2, then a phase shift occurs upon reflection. The pattern of light that results from this interference can appear either as light and dark bands or as colorful bands depending upon the source of the incident light.

Consider light incident on a thin film and reflected by both the upper and lower boundaries. The optical path difference (OPD) of the reflected light must be calculated in order to determine the condition for interference. Referring to the ray diagram above, the OPD between the two waves is the following:

$$OPD = n_2(\overline{AB} + \overline{BC}) - n_1(\overline{AD})$$

Where,

$$\overline{AB} = \overline{BC} = \frac{d}{\cos(\theta_2)}$$

$$\overline{AD} = 2d\tan(\theta_2)\sin(\theta_1)$$

Using Snell's Law, $n_1\sin(\theta_1) = n_2\sin(\theta_2)$

$$OPD = n_2\left(\frac{2d}{\cos(\theta_2)}\right) - 2d\tan(\theta_2)n_2\sin(\theta_2)$$

$$OPD = 2n_2d\left(\frac{1-\sin^2(\theta_2)}{\cos(\theta_2)}\right)$$

$$OPD = 2n_2d\cos(\theta_2)$$

Interference will be constructive if the optical path difference is equal to an integer multiple of the wavelength of light, λ.

$$2n_2 d\cos(\theta_2) = m\lambda$$

This condition may change after considering possible phase shifts that occur upon reflection.

Monochromatic Source

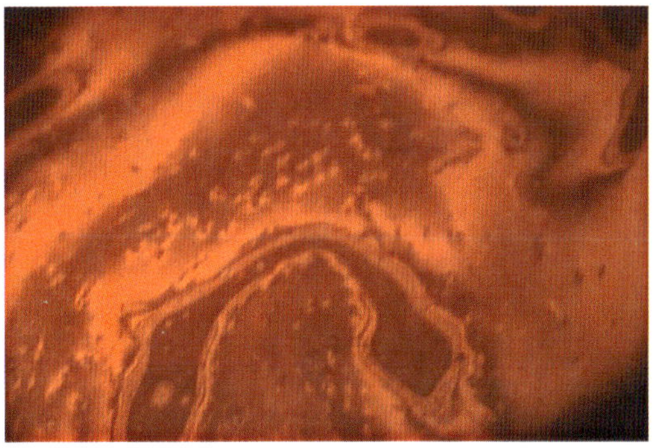

Gasoline on water shows a pattern of bright and dark fringes when illuminated with 589nm laser light.

Where incident light is monochromatic in nature, interference patterns appear as light and dark bands. Light bands correspond to regions at which constructive interference is occurring between the reflected waves and dark bands correspond to destructive interference regions. As the thickness of the film varies from one location to another, the interference may change from constructive to destructive. A good example of this phenomenon, termed "Newton's rings," demonstrates the interference pattern that results when light is reflected from a spherical surface adjacent to a flat surface. Concentric rings are observed when the surface is illuminated with monochromatic light. This phenomenon is used with optical flats to measure the shape and flatness of surfaces.

Broadband Source

If the incident light is broadband, or white, such as light from the sun, interference patterns appear as colorful bands. Different wavelengths of light create constructive interference for different film thicknesses. Different regions of the film appear in different colors depending on the local film thickness.

Phase Interaction

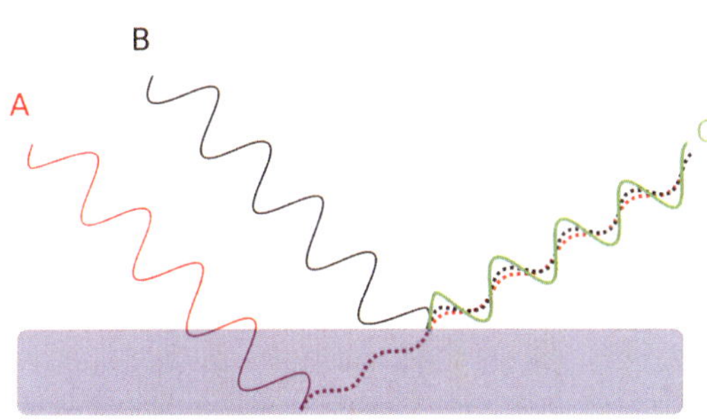

Constructive phase interaction

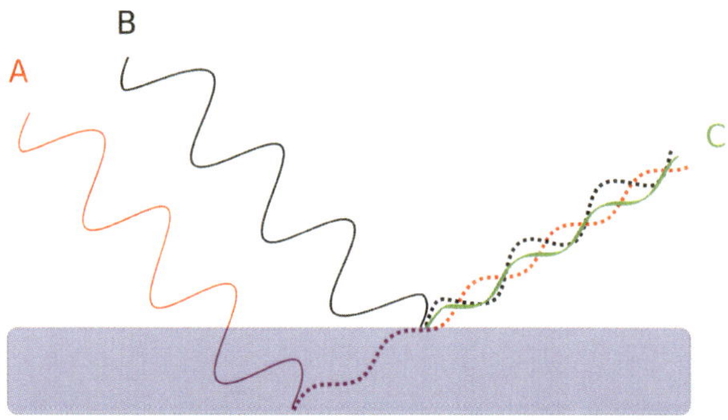

Destructive phase interaction

This section provides a simplified explanation of the phase relationship responsible for most of this phenomenon. The figures show two incident light beams (A and B). Each beam produces a reflected beam (dashed). The reflections of interest are beam A's reflection off of the lower surface and beam B's reflection off of the upper surface. These reflected beams combine to produce a resultant beam (C). If the reflected beams are in phase (as in the first figure) the resultant beam is relatively strong. If, on the other hand, the reflected beams have opposite phase, the resulting beam is attenuated (as in the second figure).

The phase relationship of the two reflected beams depends on the relationship between the wavelength of beam A in the film, and the film's thickness. If the total distance beam A travels in the film is an integer multiple of the wavelength of the beam in the film, then the two reflected beams are in phase and constructively interfere (as depicted in the first figure). If the distance traveled by beam A is an odd integer multiple of the half wavelength of light in the film, the beams destructively interfere (as in the second figure). Thus, the film shown in these figures reflects more strongly at the wavelength of the light beam in the first figure, and less strongly at that of the beam in the second figure.

Colours in a Soap Film

The reflected colours in both the film clip above and the photo below result from white light reflected in a thin film made from water, glycerol (which increases the viscosity) and detergent, which reduces the surface tension (or interfacial free energy) by forming a layer, one molecule thick, on the air-water interface. This reduction in surface tension makes the thin film more stable. The detergent layers at the surface also affect the process of draining, as we explain below.

Once the film is formed, the water-glycerol solution between the two detergent layers at the surface drains downward under gravity, making the film thinnest at the top, as shown in the sketch below right. This leads to the range of colours, as we shall now discuss.

First, consider a region in the top third of the film shown, where it appears black. Here there is no reflected light because of destructive interference. In this region, the thickness is very much less than one wavelength of visible light, as indicated at the top of the sketch. (In fact, the film here is only several molecules thick.) The reflection from the first interface (air towards water) has a phase change of π, while that at the second interface (water towards air) has a phase change of zero. So we have destructive interference and a black film.

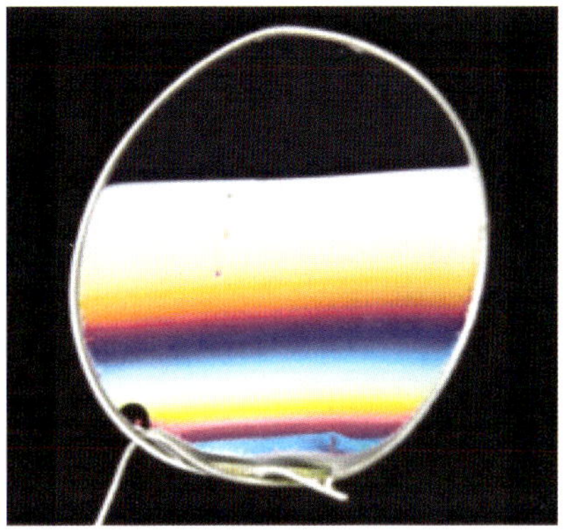

Above: a photograph of a draining soap film and the colours produced by constructive and destructive interference, as we explain with reference to the sketch at right.

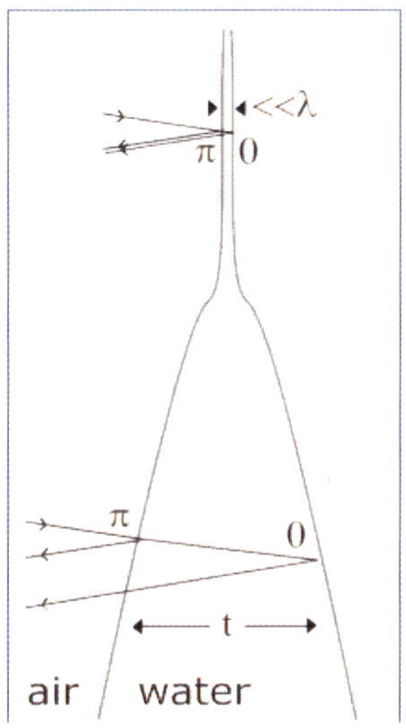

Now look at the yellow band across the middle of the film in the photo. At this point, the thickness t is about half the wavelength of blue light in water (i.e. $t \sim \lambda_{blue}/2n$). For blue light, the phase difference in reflection has a component of one complete cycle or 2π due to path length difference, and π due to the reflections. So there is destructive interference for blue light. Green and red light do not have strong destructive interference, so we see red+green = yellow.

A little below the yellow band (where blue has destructive interference), the film is a little thicker, so there is destructive interference for green light. So we see red+blue = magenta. And somewhat below that, there is a thickness where red light has destructive interference, so we see green+blue = cyan. Then further bands of colours as we reach thicker parts of the film.

Let's repeat those two figures while we discuss a few more details.

Now look at the broad white band, below the black $(t<<\lambda)$ but above the yellow $(t \sim \lambda_{blue}/2n)$. Here, none of the visible wavelengths have destructive interface, so we see red+green+blue = white. It raises the interesting question, however: why is the junction between black and white so sharp? Obviously, the thickness of the film must go rapidly from $<<\lambda$ to a thickness great enough to avoid destructive interference for any wavelengths. What's going on?

In the black film region, there are attractive van der Waals forces between the two detergent layers at the surface. These are opposed by repulsive hydration forces between the hydrophilic sides of the detergent layers, which stop them getting any closer (and thus probably destroying the film). These hydration forces have very short range – a characteristic length of about 0.2 nm. The attractive van der Waals forces have much longer range and their attraction squeezes water out from near the black-white boundary. This is still a fairly thin layer, however, so the last layers of water are slow to leave, so there is a bulge just below the black-white boundary, as shown in the sketch.

References

- Stavenga, D. G. (2014). "Thin Film and Multilayer Optics Cause Structural Colors of Many Insects and Birds" (PDF). Materials Today: Proceedings. 1: 109. doi:10.1016/j.matpr.2014.09.007
- Wave interference, physics: physics-and-radio-electronics.com, Retrieved 19 May 2018
- Thin-film-interference-and-reflections, light: animations.physics.unsw.edu.au, Retrieved 29 March 2018
- Greene, Brian (1999). The Elegant Universe: Superstrings, Hidden Dimensions, and the Quest for the Ultimate Theory. New York: W.W. Norton. pp. 97–109. ISBN 0-393-04688-5
- Constructive-interference, waves: physics.tutorvista.com, Retrieved 29 March 2018
- Interference-by-thin-film, iit-jee-wave-optics: Retrieved 08 May 2018
- Van Der Kooi, C. J.; Elzenga, J.T.M.; Dijksterhuis, J.; Stavenga, D.G. (2017). "Functional optics of glossy buttercup flowers" (PDF). Journal of the Royal Society Interface. 14 (127). doi:10.1098/rsif.2016.0933
- Destructive-interference, waves: physics.tutorvista.com, Retrieved 20 March 2018
- Van Der Kooi, C. J.; Wilts, B. D.; Leertouwer, H. L.; Staal, M.; Elzenga, J. T. M.; Stavenga, D. G. (2014). "Iridescent flowers? Contribution of surface structures to optical signaling" (PDF). New Phytologist. 203 (2): 667. doi:10.1111/nph.12808. PMID 24713039
- Youngs-double-slit-experiment, physics test book2: opentextbc.ca, Retrieved 30 May 2018

Diffraction

When a wave encounters a slit or an obstacle, the bending of light waves occur around the corners of the obstacle or aperture and penetrates into the geometrical shadow behind the obstacle. This phenomenon is called diffraction. This chapter explores the different forms of diffraction such as Fresnel diffraction, Fraunhofer diffraction and particle diffraction as well as the key concepts of diffraction grating, diffraction-limited system, Bragg law and diffraction.

Diffraction refers to various phenomena that occur when a wave encounters an obstacle. In classical physics, the diffraction phenomenon is described as the apparent bending of waves around small obstacles and the spreading out of waves past small openings. Similar effects occur when light waves travel through a medium with a varying refractive index or a sound wave through one with varying acoustic impedance. Diffraction occurs with all waves, including sound waves, water waves, and electromagnetic waves such as visible light, X-rays, and radio waves. As physical objects have wave-like properties (at the atomic level), diffraction also occurs with matter and can be studied according to the principles of quantum mechanics. Italian scientist Francesco Maria Grimaldi coined the word *diffraction* and was the first to record accurate observations of the phenomenon in 1665.

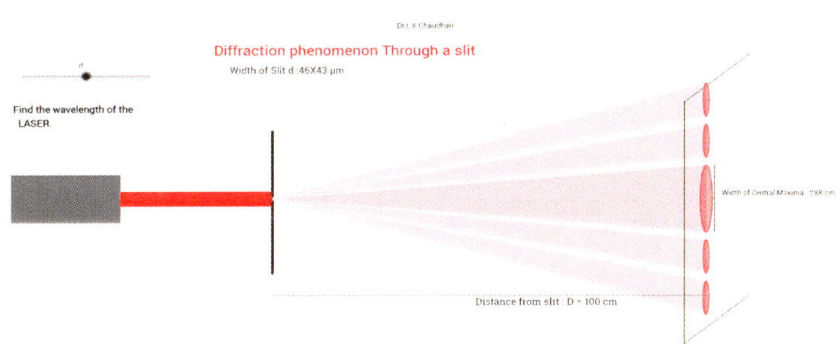

Diffraction In classical physics, the diffraction phenomenon is described as the apparent bending of waves around small obstacles and the spreading out of waves past small openings.

The effects of diffraction are often seen in everyday life. The most striking examples of diffraction are those involving light; for example, the closely spaced tracks on a CD or DVD act as a diffraction grating to form the familiar rainbow pattern seen when looking at a disk. This principle can be extended to engineer a grating with a structure such that it will produce any diffraction pattern desired; the hologram on a credit card is an example. Diffraction in the atmosphere by small particles can cause a bright ring to be visible around a bright light source like the sun or the moon. A shadow of a solid object, using light from a compact source, shows small fringes near its edges. All these effects occur because light propagates as a wave.

Richard Feynman said, "No one has ever been able to define the difference between interference and diffraction satisfactorily. It is just a question of usage, and there is no specific, important physical difference between them."

He suggested that when there are only a few sources, say two, we call it interference (as in Young's slits), but with a large number of sources, the process can be labelled diffraction.

While diffraction occurs whenever propagating waves encounter such changes, its effects are generally most pronounced for waves where the wavelength is roughly similar to the dimensions of the diffracting objects. If the obstructing object provides multiple, closely spaced openings, a complex pattern of varying intensity can result. This is due to the superposition, or interference, of different parts of a wave that traveled to the observer by different paths.

Diffraction effects are classified into either Fresnel or Fraunhofer types.

Fresnel diffraction is concerned mainly with what happens to light in the immediate neighborhood of a diffracting object or aperture, so is only of interest when the illumination source is close by.

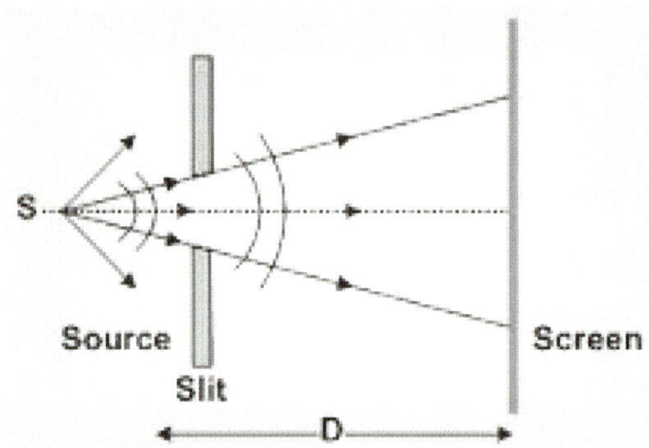

Fraunhofer diffraction is the light-spreading effect of an aperture when the aperture (or object) is lit by plane waves, i.e., waves that effectively come from a source that is infinitely far away.

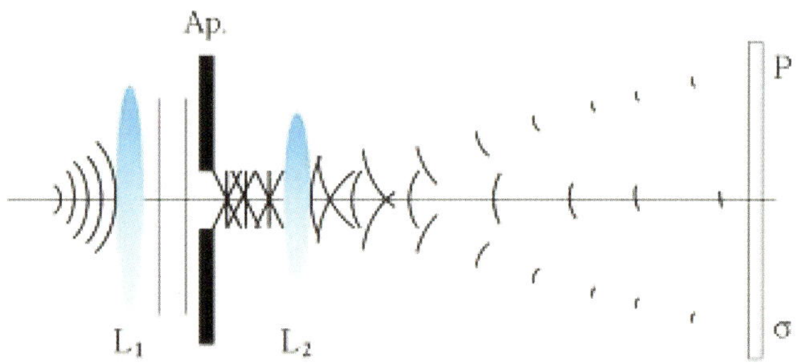

Terms used and their Definitions

- Aperture : Aperture is an optical terminology,it layman's term it is called different names such as an opening, hole, or gap.It often refers to an opening in which light is allowed to pass in optical systems such as cameras and lasers.
- Waves : wave is a disturbance or oscillation that travels through space time, accompanied by a transfer of energy.
- Fringes :the edge and outside boundary of something.
- Interference*:{}interference* is a phenomenon in which two waves superimpose to form a resultant wave of greater or lower amplitude.

- Radiation:Radiation is energy that comes from a source and travels through some material or through space.
- Frequency:is the number of occurrences of a repeating event per unit time.
- Amplitude: amplitude of a wave is a measure of how big its oscillation is.

Types of Diffractions

- Diffraction of light - is the bending of light as it passes the edge of an object.
- water diffraction - is the diffraction of water waves passing through a slit.
- Sound diffraction - This ia a process whereby sound waves can diffract around objects,which is why you can still hear someone calling even when hiding behind a tree.
- Electron diffraction - Electron diffraction refers to the wave nature of electrons.Fast electrons are diffracted from a polycrystalline layer of graphite
- X-Ray Diffraction - X-ray diffraction (XRD) is a tool for characterizing arrangement of atoms in crystals and distances between crystal faces. This can be used to identify atoms and the crystalline form
- Powder Diffraction - Powder diffraction is a scientific technique using X-ray, neutron, or electron diffraction on powder or microcrystalline samples for structural characterization of materials.
- Diffraction from a Single slit - The single slit diffraction is illustrated via the use of finite-difference time-domain (FDTD) simulation in which slits with various widths are illuminated by electromagnetic plane waves at a single frequency
- Selected area diffraction - Selected area (electron) diffraction (abbreviated as SAD or SAED), is a crystallographic experimental technique that can be performed inside a transmission electron microscope (TEM)

Huygens Principle

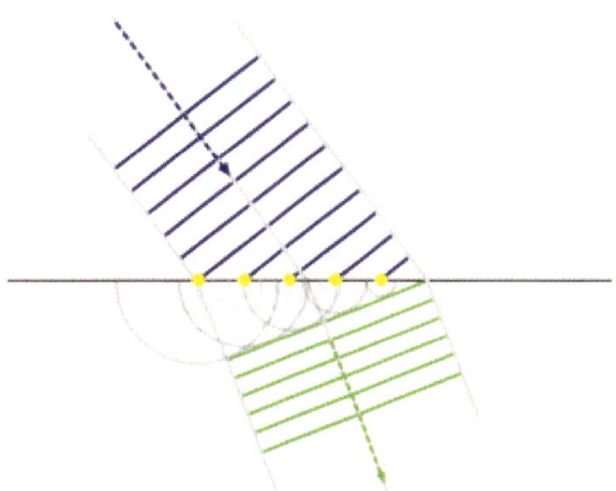

A Dutch physicist Christiaan Huygens earlier proposed in 1678, that “every point in which a luminous disturbance reaches act as a source of a wave spherical in nature”, the summation of this secondary waves determines the form of the wave at any point in time”.

Though his theory adequately explained how linear and spherical wave propagate, it didn't explain how according to his own presumption these secondary wave move only in forward direction.

Huygen-fresnel Principle

However, in 1816, French physicist Augustin-Jean Fresnel would argue that Huygens principle along with his own (principle of interference) can successfully explain not only the rectilinear propagation of light but also the effects of diffraction. To justify his claims he inculcated more arbitrary assumptions of secondary wave's phase and amplitude with and obliquity factor. His predictions were to some extend arrived at some agreement including the Arago spot, though with no physical root.

Though a member of the French Academy, Poisson reviewed Fresnel's theory and suggested that "a bright spot will appear in the center of the shadow of small disc" concluded that theory was insufficient. However in a counter analysis, another member of the same academy concluded after his investigations that Fresnel theory was indeed correct, his investigation was indeed helpful towards justifying the wave theory of light over the corpuscular theory.

The diffraction formula by Kirchhoff based on wave equation, gives a mathematical roots for diffraction. Fresnel assumptions that eventually led to the Huygens-Fresnel equation came from the derivation of Kirchhoff's diffraction formula.

A good instance of this could be explained with a simple showing with two rooms with a common doorway. When sound is produced in one of the rooms, someone in the other room will assume the same produced sound in the next room is emanating from the doorway they share, as s/he view the doorway as the source of the source due to the vibrating of air at the doorway.

Now let's consider a source at the point P_0, with vibrating frequency f. The disturbance we describe as a complex variable U_0 known as the complex amplitude, produces a wave with a wavelength λ, and wave number $k = 2\pi/\lambda$.

The complex amplitude of the primary wave at the point Q located at a distance r_0 from Po can be given as

Diffraction of Light

Some examples of diffraction of light are considered below.

Single-slit Diffraction

Diffraction of a scalar wave passing through a 1-wavelength-wide slit

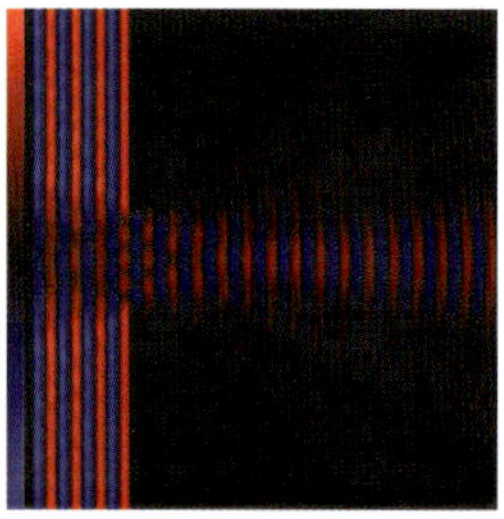

Diffraction of a scalar wave passing through a 4-wavelength-wide slit

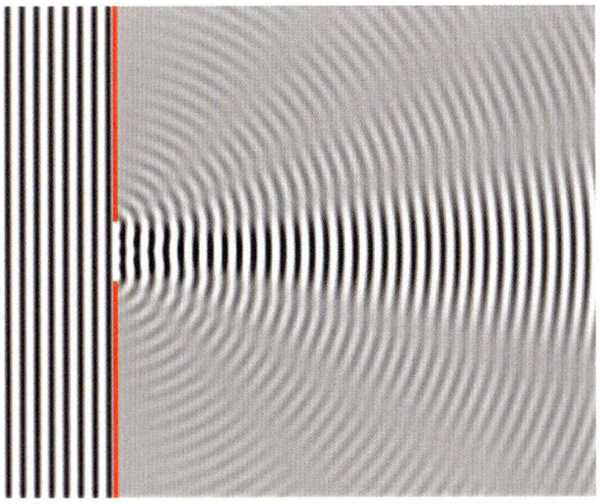

Numerical approximation of diffraction pattern from a slit of width four wavelengths with an incident plane wave. The main central beam, nulls, and phase reversals are apparent.

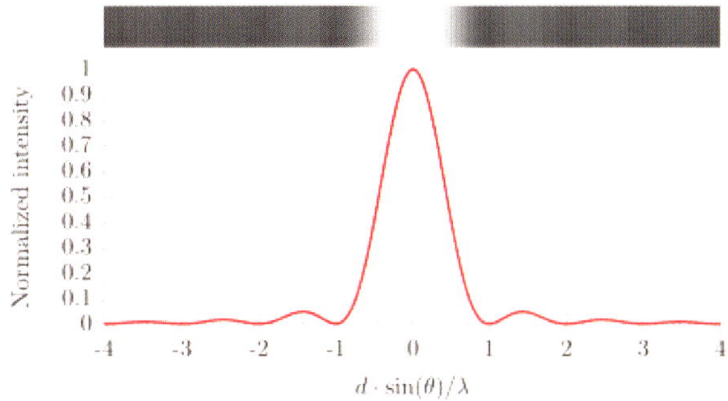

Graph and image of single-slit diffraction.

A long slit of infinitesimal width which is illuminated by light diffracts the light into a series of circular waves and the wavefront which emerges from the slit is a cylindrical wave of uniform intensity.

A slit which is wider than a wavelength produces interference effects in the space downstream of the slit. These can be explained by assuming that the slit behaves as though it has a large number of point sources spaced evenly across the width of the slit. The analysis of this system is simplified if we consider light of a single wavelength. If the incident light is coherent, these sources all have the same phase. Light incident at a given point in the space downstream of the slit is made up of contributions from each of these point sources and if the relative phases of these contributions vary by 2π or more, we may expect to find minima and maxima in the diffracted light. Such phase differences are caused by differences in the path lengths over which contributing rays reach the point from the slit.

We can find the angle at which a first minimum is obtained in the diffracted light by the following reasoning. The light from a source located at the top edge of the slit interferes destructively with a source located at the middle of the slit, when the path difference between them is equal to $\lambda/2$. Similarly, the source just below the top of the slit will interfere destructively with the source located just below the middle of the slit at the same angle. We can continue this reasoning along the entire height of the slit to conclude that the condition for destructive interference for the entire slit is the same as the condition for destructive interference between two narrow slits a distance apart that is half the width of the slit. The path difference is approximately $\frac{d\sin(\theta)}{2}$ so that the minimum intensity occurs at an angle θ_{min} given by

$$d\sin\theta_{min} = \lambda$$

where

- d is the width of the slit,
- θ_{min} is the angle of incidence at which the minimum intensity occurs, and
- λ is the wavelength of the light

A similar argument can be used to show that if we imagine the slit to be divided into four, six, eight parts, etc., minima are obtained at angles θ_n given by

$$d \sin \theta_n = n\lambda$$

where

- n is an integer other than zero.

There is no such simple argument to enable us to find the maxima of the diffraction pattern. The intensity profile can be calculated using the Fraunhofer diffraction equation as

$$I(\theta) = I_0 \operatorname{sinc}^2\left(\frac{d\pi}{\lambda}\sin\theta\right)$$

where

- $I(\theta)$ is the intensity at a given angle,
- I_0 is the original intensity, and
- the unnormalized sinc function above is given by $\operatorname{sinc}(x) = \frac{\sin x}{x}$ if $x \neq 0$, and $\operatorname{sinc}(0) = 1$

This analysis applies only to the far field, that is, at a distance much larger than the width of the slit.

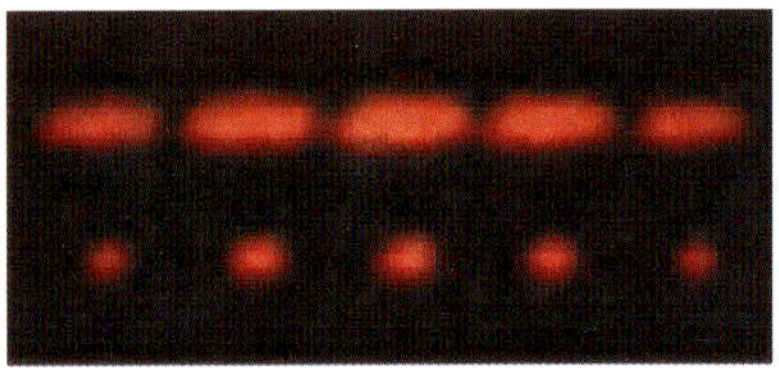

2-slit (top) and 5-slit diffraction of red laser light

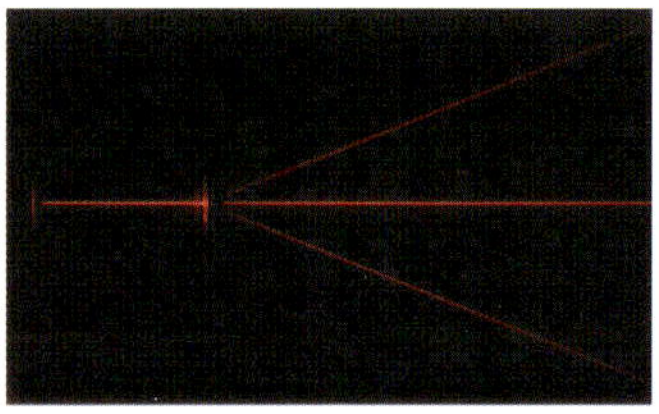

Diffraction of a red laser using a diffraction grating.

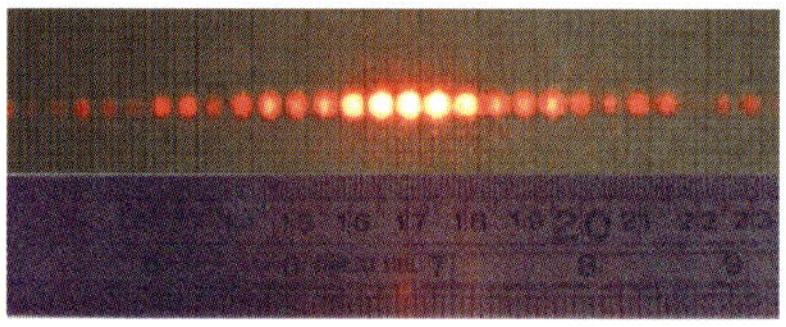

A diffraction pattern of a 633 nm laser through a grid of 150 slits

Diffraction Grating

A diffraction grating is an optical component with a regular pattern. The form of the light diffracted by a grating depends on the structure of the elements and the number of elements present, but all gratings have intensity maxima at angles θ_m which are given by the grating equation

$$d\left(\sin\theta_m + \sin\theta_i\right) = m\lambda.$$

where

- θ_i is the angle at which the light is incident,
- d is the separation of grating elements, and
- m is an integer which can be positive or negative.

The light diffracted by a grating is found by summing the light diffracted from each of the elements, and is essentially a convolution of diffraction and interference patterns.

The figure shows the light diffracted by 2-element and 5-element gratings where the grating spacings are the same; it can be seen that the maxima are in the same position, but the detailed structures of the intensities are different.

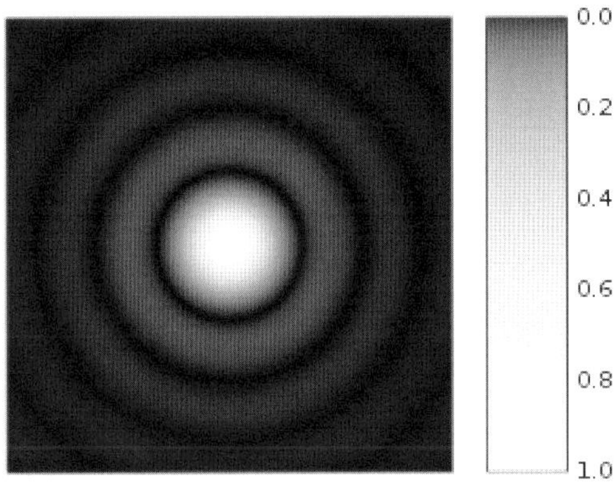

A computer-generated image of an Airy disk.

Computer generated light diffraction pattern from a circular aperture of diameter 0.5 micrometre at a wavelength of 0.6 micrometre (red-light) at distances of 0.1 cm – 1 cm in steps of 0.1 cm. One can see the image moving from the Fresnel region into the Fraunhofer region where the Airy pattern is seen.

Circular Aperture

The far-field diffraction of a plane wave incident on a circular aperture is often referred to as the Airy Disk. The variation in intensity with angle is given by

$$I(\theta) = I_0 \left(\frac{2J_1(ka \sin\theta)}{ka \sin\theta} \right)^2 ,$$

where a is the radius of the circular aperture, k is equal to $2\pi/\lambda$ and J_1 is a Bessel function. The smaller the aperture, the larger the spot size at a given distance, and the greater the divergence of the diffracted beams.

General Aperture

The wave that emerges from a point source has amplitude ψ at location r that is given by the solution of the frequency domain wave equation for a point source (The Helmholtz Equation),

$$\nabla^2\psi + k^2\psi = \delta(\mathbf{r})$$

where $\delta(\mathbf{r})$ is the 3-dimensional delta function. The delta function has only radial dependence, so the Laplace operator (a.k.a. scalar Laplacian) in the spherical coordinate system simplifies to (del in cylindrical and spherical coordinates)

$$\nabla^2\psi = \frac{1}{r}\frac{\partial^2}{\partial r^2}(r\psi)$$

By direct substitution, the solution to this equation can be readily shown to be the scalar Green's function, which in the spherical coordinate system (and using the physics time convention $e^{-i\omega t}$) is:

$$\psi(r) = \frac{e^{ikr}}{4\pi r}$$

This solution assumes that the delta function source is located at the origin. If the source is located at an arbitrary source point, denoted by the vector $\mathbf{r}'$ and the field point is located at the point $\mathbf{r}$, then we may represent the scalar Green's function (for arbitrary source location) as:

$$\psi(\mathbf{r}\,|\,\mathbf{r}') = \frac{e^{ik|\mathbf{r}-\mathbf{r}'|}}{4\pi\,|\mathbf{r}-\mathbf{r}'|}$$

Therefore, if an electric field, $E_{inc}(x,y)$ is incident on the aperture, the field produced by this aperture distribution is given by the surface integral:

$$\Psi(r) \propto \iint_{\text{aperture}} E_{\text{inc}}(x',y')\frac{e^{ik|\mathbf{r}-\mathbf{r}'|}}{4\pi\,|\mathbf{r}-\mathbf{r}'|}dx'dy',$$

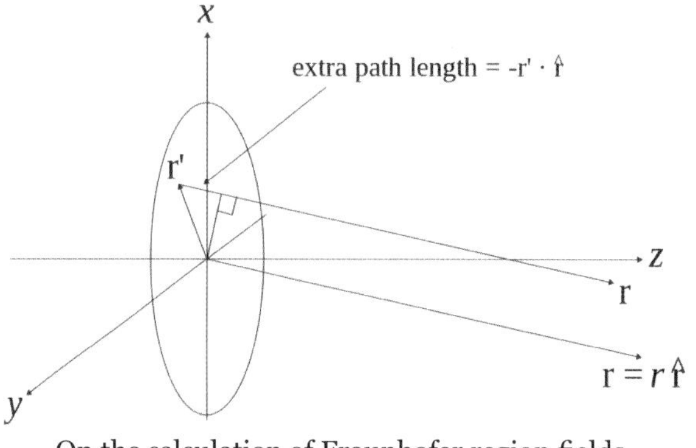

On the calculation of Fraunhofer region fields

where the source point in the aperture is given by the vector

$$\mathbf{r}' = x'\hat{\mathbf{x}} + y'\hat{\mathbf{y}}$$

In the far field, wherein the parallel rays approximation can be employed, the Green's function,

$$\psi(\mathbf{r}\,|\,\mathbf{r}') = \frac{e^{ik|\mathbf{r}-\mathbf{r}'|}}{4\pi\,|\mathbf{r}-\mathbf{r}'|}$$

simplifies to

$$\psi(\mathbf{r}\,|\,\mathbf{r}') = \frac{e^{ikr}}{4\pi r} e^{-ik(\mathbf{r}'\cdot\hat{\mathbf{r}})}$$

The expression for the far-zone (Fraunhofer region) field becomes

$$\Psi(r) \propto \frac{e^{ikr}}{4\pi r} \iint_{\text{aperture}} E_{\text{inc}}(x', y') e^{-ik(\mathbf{r}'\cdot\hat{\mathbf{r}})} dx'dy',$$

Now, since

$$\mathbf{r}' = x'\hat{\mathbf{x}} + y'\hat{\mathbf{y}}$$

and

$$\hat{\mathbf{r}} = \sin\theta\cos\phi\hat{\mathbf{x}} + \sin\theta\,\sin\phi\,\hat{\mathbf{y}} + \cos\theta\hat{\mathbf{z}}$$

the expression for the Fraunhofer region field from a planar aperture now becomes,

$$\Psi(r) \propto \frac{e^{ikr}}{4\pi r} \iint_{\text{aperture}} E_{\text{inc}}(x', y') e^{-ik\sin\theta(\cos\phi x' + \sin\phi y')} dx'dy'$$

Letting,

$$k_x = k\sin\theta\cos\phi$$

and

$$k_y = k\sin\theta\sin\phi$$

the Fraunhofer region field of the planar aperture assumes the form of a Fourier transform

$$\Psi(r) \propto \frac{e^{ikr}}{4\pi r} \iint_{\text{aperture}} E_{\text{inc}}(x', y') e^{-i(k_x x' + k_y y')} dx'dy',$$

In the far-field / Fraunhofer region, this becomes the spatial Fourier transform of the aperture distribution. Huygens' principle when applied to an aperture simply says that the far-field diffraction pattern is the spatial Fourier transform of the aperture shape, and this is a direct by-product of using the parallel-rays approximation, which is identical to doing a plane wave decomposition of the aperture plane fields.

Propagation of a Laser Beam

The way in which the beam profile of a laser beam changes as it propagates is determined by diffraction. When the entire emitted beam has a planar, spatially coherent wave front, it approximates Gaussian beam profile and has the lowest divergence for a given diameter. The smaller the output beam, the quicker it diverges. It is possible to reduce the divergence of a laser beam by first expanding it with one convex lens, and then collimating it with a second convex lens whose focal point is coincident with that of the first lens. The resulting beam has a larger diameter, and hence a lower divergence. Divergence of a laser beam may be reduced below the diffraction of a Gaussian beam or even reversed to convergence if the refractive index of the propagation media increases with the light intensity. This may result in a self-focusing effect.

When the wave front of the emitted beam has perturbations, only the transverse coherence length (where the wave front perturbation is less than 1/4 of the wavelength) should be considered as a Gaussian beam diameter when determining the divergence of the laser beam. If the transverse coherence length in the vertical direction is higher than in horizontal, the laser beam divergence will be lower in the vertical direction than in the horizontal.

Diffraction-limited Imaging

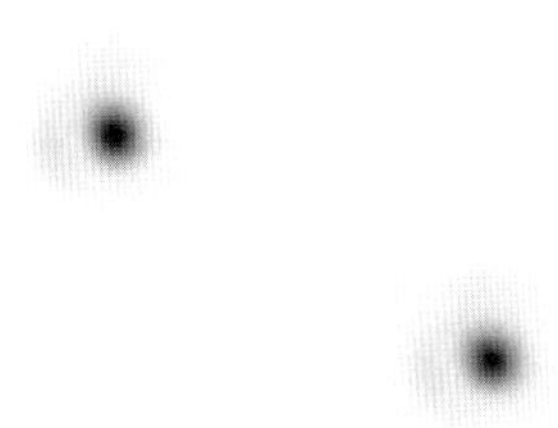

The Airy disk around each of the stars from the 2.56 m telescope aperture can be seen in this *lucky image* of the binary star zeta Boötis.

The ability of an imaging system to resolve detail is ultimately limited by diffraction. This is because a plane wave incident on a circular lens or mirror is diffracted as described above. The light is not focused to a point but forms an Airy disk having a central spot in the focal plane with radius to first null of

$$d = 1.22\lambda N,$$

where λ is the wavelength of the light and N is the f-number (focal length divided by diameter) of the imaging optics. In object space, the corresponding angular resolution is

$$\sin\theta = 1.22\frac{\lambda}{D},$$

where D is the diameter of the entrance pupil of the imaging lens (e.g., of a telescope's main mirror).

Two point sources will each produce an Airy pattern – see the photo of a binary star. As the point sources move closer together, the patterns will start to overlap, and ultimately they will merge to form a single pattern, in which case the two point sources cannot be resolved in the image. The Rayleigh criterion specifies that two point sources can be considered to be resolvable if the separation of the two images is at least the radius of the Airy disk, i.e. if the first minimum of one coincides with the maximum of the other.

Thus, the larger the aperture of the lens, and the smaller the wavelength, the finer the resolution of an imaging system. This is why telescopes have very large lenses or mirrors, and why optical microscopes are limited in the detail which they can see.

Speckle Patterns

The speckle pattern which is seen when using a laser pointer is another diffraction phenomenon. It is a result of the superposition of many waves with different phases, which are produced when a laser beam illuminates a rough surface. They add together to give a resultant wave whose amplitude, and therefore intensity, varies randomly.

Babinet's Principle

Babinet's Principle is a useful theorem stating that the diffraction pattern from an opaque body is identical to that from a hole of the same size and shape, but with differing intensities. This means that the interference conditions of a single obstruction would be the same as that of a single slit.

Patterns

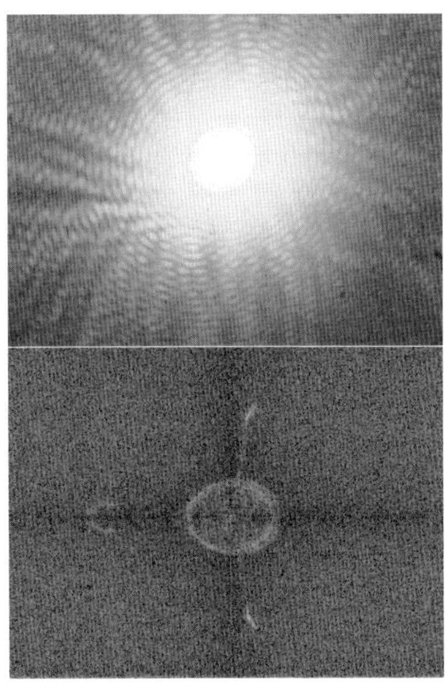

The upper half of this image shows a diffraction pattern of He-Ne laser beam on an elliptic aperture. The lower half is its 2D Fourier transform approximately reconstructing the shape of the aperture.

Several qualitative observations can be made of diffraction in general:

- The angular spacing of the features in the diffraction pattern is inversely proportional to the dimensions of the object causing the diffraction. In other words: The smaller the diffracting object, the 'wider' the resulting diffraction pattern, and vice versa. (More precisely, this is true of the sines of the angles.)

- The diffraction angles are invariant under scaling; that is, they depend only on the ratio of the wavelength to the size of the diffracting object.
- When the diffracting object has a periodic structure, for example in a diffraction grating, the features generally become sharper. The third figure, for example, shows a comparison of a double-slit pattern with a pattern formed by five slits, both sets of slits having the same spacing, between the center of one slit and the next.

Particle Diffraction

Quantum theory tells us that every particle exhibits wave properties. In particular, massive particles can interfere and therefore diffract. Diffraction of electrons and neutrons stood as one of the powerful arguments in favor of quantum mechanics. The wavelength associated with a particle is the de Broglie wavelength

$$\lambda = \frac{h}{p}$$

where h is Planck's constant and p is the momentum of the particle (mass × velocity for slow-moving particles).

For most macroscopic objects, this wavelength is so short that it is not meaningful to assign a wavelength to them. A sodium atom traveling at about 30,000 m/s would have a De Broglie wavelength of about 50 pico meters.

Because the wavelength for even the smallest of macroscopic objects is extremely small, diffraction of matter waves is only visible for small particles, like electrons, neutrons, atoms and small molecules. The short wavelength of these matter waves makes them ideally suited to study the atomic crystal structure of solids and large molecules like proteins.

Relatively larger molecules like buckyballs were also shown to diffract.

Bragg Diffraction

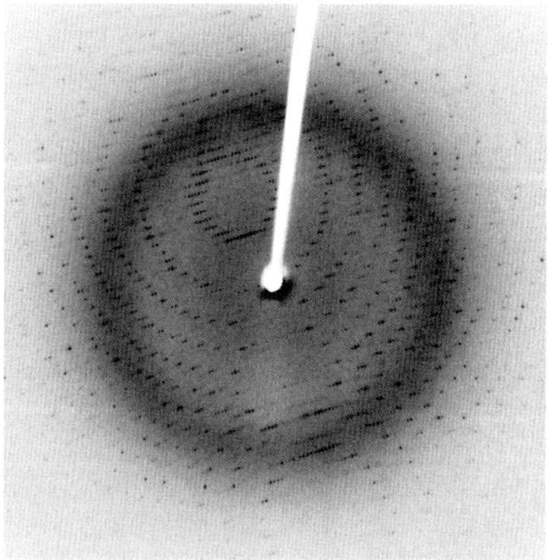

Following Bragg's law, each dot (or *reflection*) in this diffraction pattern forms from the constructive interference of X-rays passing through a crystal. The data can be used to determine the crystal's atomic structure.

Diffraction from a three-dimensional periodic structure such as atoms in a crystal is called Bragg diffraction. It is similar to what occurs when waves are scattered from a diffraction grating. Bragg diffraction is a consequence of interference between waves reflecting from different crystal planes. The condition of constructive interference is given by *Bragg's law*:

$$m\lambda = 2d \sin\theta$$

where

λ is the wavelength,

d is the distance between crystal planes,

θ is the angle of the diffracted wave.

and m is an integer known as the *order* of the diffracted beam.

Bragg diffraction may be carried out using either light of very short wavelength like X-rays or matter waves like neutrons (and electrons) whose wavelength is on the order of (or much smaller than) the atomic spacing. The pattern produced gives information of the separations of crystallographic planes d, allowing one to deduce the crystal structure. Diffraction contrast, in electron microscopes and x-topography devices in particular, is also a powerful tool for examining individual defects and local strain fields in crystals.

Coherence

The description of diffraction relies on the interference of waves emanating from the same source taking different paths to the same point on a screen. In this description, the difference in phase between waves that took different paths is only dependent on the effective path length. This does not take into account the fact that waves that arrive at the screen at the same time were emitted by the source at different times. The initial phase with which the source emits waves can change over time in an unpredictable way. This means that waves emitted by the source at times that are too far apart can no longer form a constant interference pattern since the relation between their phases is no longer time independent.

The length over which the phase in a beam of light is correlated, is called the coherence length. In order for interference to occur, the path length difference must be smaller than the coherence length. This is sometimes referred to as spectral coherence, as it is related to the presence of different frequency components in the wave. In the case of light emitted by an atomic transition, the coherence length is related to the lifetime of the excited state from which the atom made its transition.

If waves are emitted from an extended source, this can lead to incoherence in the transversal direction. When looking at a cross section of a beam of light, the length over which the phase is correlated is called the transverse coherence length. In the case of Young's double slit experiment, this would mean that if the transverse coherence length is smaller than the spacing between the two slits, the resulting pattern on a screen would look like two single slit diffraction patterns.

In the case of particles like electrons, neutrons and atoms, the coherence length is related to the spatial extent of the wave function that describes the particle.

Application of Diffraction

There are several applications of diffraction and these are somewhat evident in our daily contact with objects exhibiting diffraction or diffraction properities, without we noticing,or on the other hand noticing the object but not knowing the principle behind it or what made the object look or act like that.

Here are some of the applications of diffraction

Scientific Application

- Diffraction is used a lot in discovering the structures of materials and atoms.
- It has been used a lot in discovering medicines and drugs.
- Diffraction is also fundamental in other applications such as x-ray diffraction studies of crystals and holography

Gadget that uses or Work using Diffraction Principles

The effects of diffraction are usually seen in everyday life. One of the most evident examples of diffraction are those involving light; for example,when you take a keen look at a CD or DVD the closely spaced tracks on a CD or DVD act as a diffraction grating to form the familiar rainbow pattern.

The hologram on a credit card,work with the principle of diffraction,as you can see we pay our bills through a credit card without noticing or caring to know how it works or on what principle is the credit card working on.Our cameras use the principle of diffraction,diffrication in the apertures makes our pictures comes out sharp and clear.All these gadget we use every day or once since our existence and we dkont know how they operate or work.

Fresnel Diffraction

Fresnel diffraction, or near-field diffraction, occurs when a wave passes via a small hole and diffracts, creating a diffraction pattern. The size of the pattern depends on the distance between a projection and an aperture.

This type of diffraction occurs when the distance of propagation of diffracted waves is short, resulting in a Fresnel number greater than . Diffracted waves become planar when the distance of propagation is increased.

Fresnel Zones

In the study of Fresnel diffraction it is convenient to divide the aperture into regions called Fresnel zones. Figure below shows a point source, S, illuminating an aperture a distance z1away. The observation point, P, is a distance to the right of the aperture. Let the line SP be normal to the plane containing the aperture. Then we can write

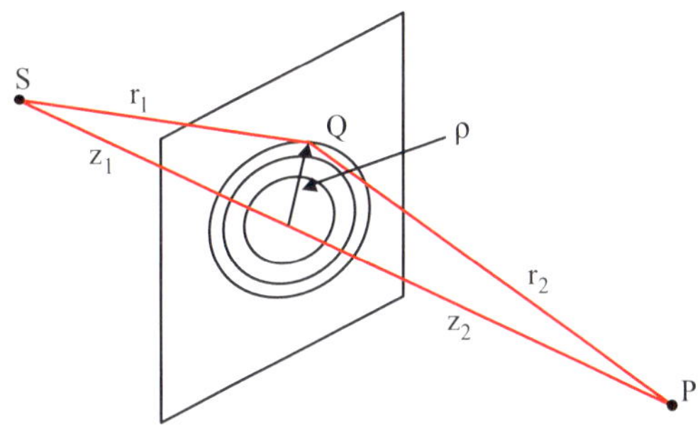

Figure: Spherical wave illuminating aperture.

$$\mathrm{SQP} = r_1 + r_2 = \sqrt{z_1^2 + \rho^2} + \sqrt{z_2^2 + \rho^2}$$

$$= z_1 + z_2 + \frac{1}{2}\rho^2\left(\frac{1}{z_1} + \frac{1}{z_2}\right) + \dots$$

The aperture can be divided into regions bounded by concentric circles $\rho =$ constant defined such that $r_1 + r_2$ differ by $\lambda / 2$ in going from one boundary to the next. These regions are called Fresnel zones or half-period zones. If z_1 and z_2 are sufficiently large compared to the size of the aperture the higher order terms of the expansion can be neglected to yield the following result.

$$\mathrm{n}\frac{\lambda}{2} = \frac{1}{2}\rho_{\mathrm{n}}^2\left(\frac{1}{z_1} + \frac{1}{z_2}\right)$$

Solving for ρ_{n}, the radius of the nth Fresnel zone, yields

$$\rho_{\mathrm{n}} = \sqrt{\mathrm{n}\,\lambda\,\mathrm{L}} \text{ or } \rho_1 = \sqrt{\lambda\,\mathrm{L}},\ \rho_2 = \sqrt{2\lambda\,\mathrm{L}},\ \dots, \text{ where}$$

$$\mathrm{L} = \frac{1}{\frac{1}{z_1} + \frac{1}{z_2}}$$

Figure shows a drawing of Fresnel zones where every other zone is made dark. Note that in the center the zones are widely spaced and the spacing decreases as the radius increases. As shown above, the radius of the zones increase as the square root of integers.

Figure: Fresnel zones

If ρ_n and ρ_{n+1} are inner and outer radii of the nth zone then the area of the nth zone is given by

$$\text{Area of nth Fresnel zone} = \pi\,\rho_{n+1}^{2} - \pi\,\rho_n^{2}$$
$$= \pi(n+1)\lambda L - \pi(n)\lambda L$$
$$= n\lambda L = \pi\,\rho_1^{2}\text{ , independent of n.}$$

That is, the area of all zones are equal. If the higher order terms in the expansion for SRQ are maintained the area of the zones would slightly increase with increasing ρ. Generally, it is assumed that z_1 and z_2 are sufficiently large compared to ρ that the higher order terms can be neglected and the area of all zones are equal.

Circular Aperture

On-Axis Irradiance

In discussing the Fresnel diffraction of a circular aperture it is convenient to first look at the on-axis irradiance. Consider a circular aperture of radius R, and let S and P lie on the normal through the center of the circular aperture as shown in Figure.

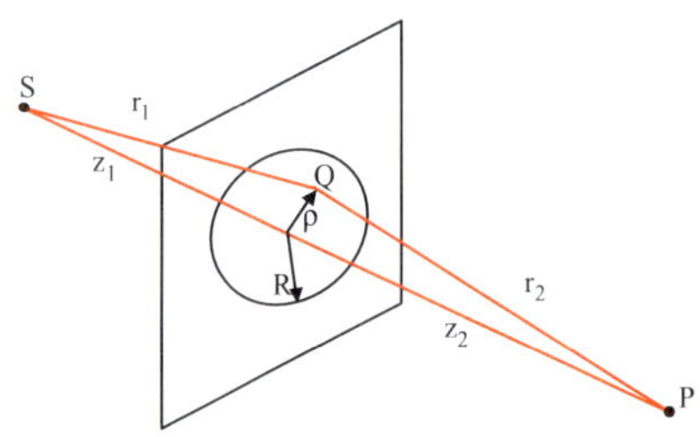

Figure: Circular aperture of radius R

If R is small compared with z_1 and z_2 the usual approximations can be made in the Kirchhoff integral that the obliquity factor is equal to 1 and the $1/r_1 r_2$ in the denominator can be replaced with $1/z_1 z_2$. The cylindrical symmetry about the axis SP suggests using polar coordinates in the integration. Then

$$u_o[P] = \frac{A}{i\lambda z_1 z_2}\int_0^R e^{ik(r_1+r_2)}\,2\pi\rho\,d\rho$$

Since $r_1^2 = z_1^2 + \rho^2$ and $r_2^2 = z_2^2 + \rho^2$,

$$\rho\,d\rho = r_1\,dr_1 = r_2\,dr_2$$

$$d(r_1 + r_2) = \left(\frac{1}{r_1} + \frac{1}{r_2}\right)\rho\,d\rho$$

Therefore,

$$\rho\,d\rho = \frac{r_1 r_2}{r_1 + r_2}\,d(r_1 + r_2) \approx \frac{z_1 z_2}{z_1 + z_2}\,d(r_1 + r_2)$$

$$u_o[P] = \frac{2\pi A}{i\lambda(z_1 + z_2)} \int_{l[0]}^{l[R]} e^{ik(r_1 + r_2)} \, d(r_1 + r_2)$$

where $l[0] = z_1 + z_2$ and $l[R]$ is equal to $r_1 + r_2$ when , that is when point Q is on the rim of the aperture.

It is convenient to introduce a variable q defined by

$$r_1 + r_2 = z_1 + z_2 + q\frac{\lambda}{2}$$

Thus q measures the difference between $r_1 + r_2$ and $z_1 + z_2$ in units of half-wavelengths and is equal to the number of Fresnel zones within the circular aperture. $u_o[P]$ can now be written as

$$u_o[P] = \frac{\pi u_{oo}[P]}{i} \int_0^{\zeta} e^{i\pi q} \, dq$$

where $\zeta = q[R]$, that is the value of q associated with a point on the rim of the aperture, and $u_{oo}[p]$ is the unobstructed value of the amplitude at point P.

$$u_o[P] = \frac{A}{z_1 + z_2} e^{ik(z_1 + z_2)}$$

Solving the integral yields

$$u_o[P] = u_{oo}[P]\left(1 - e^{i\pi\zeta}\right)$$

The on-axis irradiance is given by

$$i[P] = 2i_{oo}[P]\left(1 - \cos[\pi\zeta]\right) = 4\, i_{oo}[p] \; \mathrm{Sin}\left[\frac{\zeta\pi}{2}\right]^2$$

As the radius of the aperture increases, the irradiance at point P oscillates between zero and 4 times the unobstructed irradiance. Thus, the irradiance produced by the first Fresnel zone is equal to 4 times the irradiance obtained with a clear aperture. As the equation shows, for odd numbers of Fresnel zones present

$$i[P] = 4i_{oo}[P].$$

For even numbers of Fresnel zones present the Fresnel zones cancel one another and the on-axis irradiance is zero.

The relationship between ζ and R is found as follows: At the edge of the aperture $\rho = R$ and

$$r_1 = \left(z_1^2 + R^2\right)^{1/2} \approx z_1 + \frac{R^2}{2z_1} \text{ and}$$

$$r_2 = \left(z_2^2 + R^2\right)^{1/2} \approx z_2 + \frac{R^2}{2z_2}$$

But

$$r_1 + r_2 = z_1 + z_2 + \zeta \frac{\lambda}{2}$$

Solving for ζ, the number of Fresnel zones present, yields

$$\zeta = \frac{R^2 (z_1 + z_2)}{\lambda z_1 z_2}$$

Off-Axis Irradiance Distribution

Now we want to calculate the irradiance of the Fresnel diffraction of a circular aperture at off-axis positions. Figure shows the schematic setup.

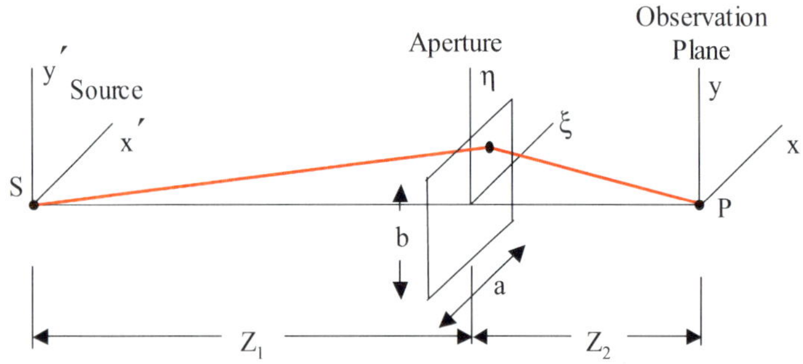

Figure: Fresnel diffraction of circular aperture.

The general equation for the amplitude of the Fresnel diffraction pattern is given by

$$u[x, y] = \frac{e^{i k z_2}}{i \lambda z_2} e^{i \frac{\pi}{\lambda z_2}(x^2 + y^2)} \int_{-\infty}^{\infty} \int_{-\infty}^{\infty} \left(u[\xi, \eta] e^{i \frac{\pi}{\lambda z_2}(\xi^2 + \eta^2)} \right) e^{i \frac{2\pi}{\lambda z_2}(x\xi + y\eta)} d\xi d\eta$$

One important fact to note about this equation is that we have the Fourier transform of $u[\xi, \eta] e^{i \frac{\pi}{\lambda z_2}(\xi^2 + \eta^2)}$ Where $u[\xi, \eta]$ is the amplitude of the wave front illuminating the aperture which we will take as the amplitude of a spherical wave. For an on-axis spherical wave illuminating the aperture $u[\xi, \eta]$ can be written as $\frac{A}{z_1} e^{i k z_1} e^{i \frac{k}{2 z_1}(\xi^2 + \eta^2)}$ If Pupil $[\xi, \eta]$ describes the aperture amplitude transmission function we can write

$$u[x, y] = A \frac{e^{i k (z_1 + z_2)}}{i \lambda z_1 z_2} e^{i \frac{\pi}{\lambda z_2}(x^2 + y^2)}$$

$$\int_{-\infty}^{\infty} \int_{-\infty}^{\infty} \left(\text{pupil}[\xi, \eta] e^{i \frac{\pi (z_1 + z_2)}{\lambda z_1 z_2}(\xi^2 + \eta^2)} \right) e^{-i \frac{2\pi}{\lambda z_2}(x\xi + y\eta)} d\xi d\eta .$$

Let r_{max} be the maximum value of $\sqrt{\left(\xi^2+\eta^2\right)}$ and nFZones be the number of Fresnel zones in the aperture. Then

$$\text{nFZones} = \frac{(z_1+z_2)}{l\, z_1+z_2}\, r_{max^2} \text{ and we can write}$$

$$u[x,\ y] = A\frac{e^{i k (z_1+z_2)}}{i\,\lambda\, z_1\, z_2}\, e^{i\frac{\pi}{\lambda z_2}\left(x^2+y^2\right)}$$

$$\int_{-\infty}^{\infty}\int_{-\infty}^{\infty}\left(\text{pupil}[\xi,\eta]\, e^{i(\text{nFZones}\,\pi)\left(\frac{\xi^2+\eta^2}{r\max^2}\right)}\right) e^{-i\frac{2\pi}{\lambda z_2}(x\xi+y\eta)}\, d\xi d\eta$$

$$\text{But } \frac{r_{max^2}}{\lambda\, z_1+z_2} = \frac{\text{nFZones}}{(z_1+z_2)},\ \text{so}$$

$$u[x,\ y] = A\frac{e^{i k (z_1+z_2)}}{i\,(z_1+z_2)}\,\frac{\text{nFZones}}{r_{max^2}}\, e^{i\frac{\pi}{\lambda z_2}\left(x^2+y^2\right)}$$

$$\int_{-\infty}^{\infty}\int_{-\infty}^{\infty}\left(\text{pupil}[\xi,\ \eta]\, e^{i(\text{nFZones}\,\pi)\left(\frac{\xi^2+\eta^2}{r\max^2}\right)}\right) e^{-i\frac{2\pi}{\lambda z_2}(x\xi+y\eta)}\, d\xi d\eta$$

$$u[x,\ y] = U_{OO}\frac{\text{nFZones}}{i\, r_{max^2}}\, e^{i\frac{\pi}{\lambda z_2}\left(x^2+y^2\right)}$$

$$\int_{-\infty}^{\infty}\int_{-\infty}^{\infty}\left(\text{pupil}[\xi,\ \eta]\, e^{i(\text{nFZones}\,\pi)\left(\frac{\xi^2+\eta^2}{r\max^2}\right)}\right) e^{-i\frac{2\pi}{\lambda z_2}(x\xi+y\eta)}\, d\xi d\eta$$

Where $U_{OO} = A\frac{e^{i k (z_1+z_2)}}{(z_1+z_2)}$ is the unobstructed on-axis irradiance.

Fresnel Half Period Zones

In the following we first learn a qualitative treatment of Fresnel diffraction due to a circular aperture. Later we go for a quantitative treatment.

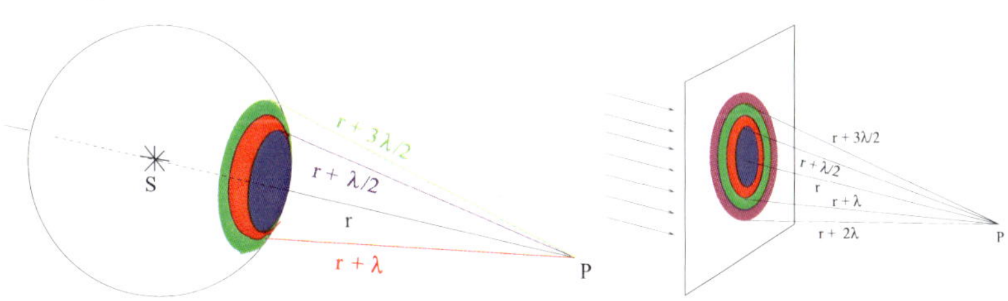

Figure: Fresnel zone construction for a point source and a plane wave (source at infinity)

We refer to the fig., where Fresnel half period zones (shown with different colours) are created on an arbitrary plane where we intend to put the diffracting element. The 'half period' refers to the fact that the path difference between two rays coming form the two consecutive zones and reaching the point of observation, P, is λ/2 (i.e. half the wavelength of the monochromatic source used). The zones are created concentrically around the central one, which is created by the straight line joining the source and the point P. Now the rays from one zone and reaching the observation point are (almost) travelling the same distance and hence reach the point P with the same phase. Small amplitudes of individual rays from the same zone will then add up to produce a large amplitude at P. Whereas two rays coming from two consecutive zones have a path difference of λ/2 (which corresponds to a phase difference of π) and hence cancel each other. We will see in the following that each half period zone has almost the same area so the amplitude of light coming from the different zones are almost the same only their phases are changing alternatively. So the contribution from two consecutive zones would produce a null amplitude. Now we put a circular aperture on the plane where we constructed the zones. The aperture would allow up to certain number of zones and the rest would be cut off. We count the number of zones allowed by the circular aperture and if that is even then the intensity at P will be almost zero due to the mutual cancellations of the consecutive zones. And if the number of zones is odd we have some intensity at the point P and which is equal to the intensity produced by a single Fresnel zone.

Let   be the radius of the mth Fresnel half period zone, then from fig.

$$r^2 + \rho_m^2 = \left(r + m\frac{\ddot{e}}{2} \right)^2$$

In usual circumstances λ << r (r is order of meters and λ is order of microns) and hence the λ^2 term can be neglected from the right hand side of the above equation. So we have,

$$\rho_m = \sqrt{mr\lambda}$$

The area, a_m, of the mth zone is given by

$$a_m = \pi\left(\rho_m^2 - \rho_{m-1}^2\right) = \pi r\lambda$$

which is constant and independent of zone number as mentioned earlier. So the amplitudes reaching from the different zones are almost the same. If one sees more carefully one finds that it decreases slightly with the increase of zone number due to an 'obliquity factor'. The situation is depicted in fig, where the contribution from the first three Fresnel zones are shown in different colours. The second diagram shows the situation if one considers semicircular aperture and the observation point is situated above the centre of the circle and the it is covering first three Fresnel zones.

In the fig. the angle between the ray direction from the source on a certain half period zone (which is also the normal direction of the zone) and the direction of the point of observation from the same zone is θ. The obliquity factor, f(θ), by Kirchoff is given by (1 + cos θ)/2. This modifies Huygens secondary wavelets. Now with the obliquity factor the Huygens secondary wavelets are not isotropic. They will have reduced amplitude as θ increases. So, in the backward direction (θ = π) the factor becomes zero and hence the amplitude of the secondary wavelets in the backward direction is zero.

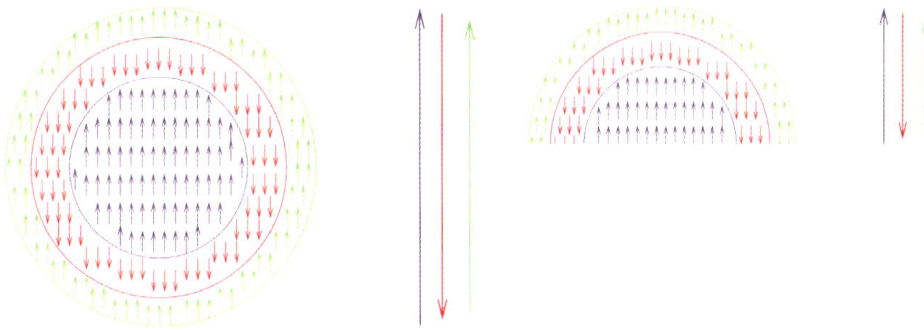

Figure: Contribution from the first three Fresnel zones for circular and semi-circular apertures

This results in a forward direction propagation of wavefronts and restricts the propagation of wavefronts in the backward direction from a source and which matches with the common observation.

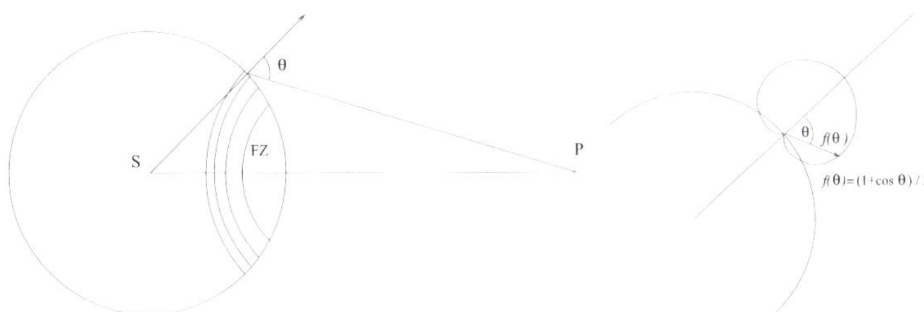

Figure: Obliquity factor and Huygens-Kirchoff secondary wavelet

For simplicity we consider the second diagram of fig where the source is at infinity and we have plane waves. Now suppose the amplitudes contributing from the consecutive Fresnel zones are A_1, A_2, $A_3 \cdots$ etc. We would now like to sum these amplitudes for apertures allowing arbitrary number of zones, say first odd m zones. The resultant contribution, A, would be,

$$A = A_1 - A_2 + A_3 - A_4 + A_5 \ldots + A_m$$

The alternate positive and negative signs are due to the π phase difference between the two consecutive zones. The first one is arbitrarily chosen as positive. The above series can be arranged as the following,

$$A = \frac{A_1}{2} + \left(\frac{A_1}{2} - A_2 + \frac{A_3}{2}\right) + \left(\frac{A_3}{2} - A_4 + \frac{A_5}{2}\right) + \ldots\ldots + \left(\frac{A_{m-2}}{2} - A_{m-1} + \frac{A_m}{2}\right) + \frac{A_m}{2}.$$

If the terms in the brackets of the equation (24.5) are positive (the amplitude of any zone is less than the average of its adjacent zones, i.e. $\left(A_{i-1} + A_{i+1}\right)/2 > A_i\right)$ then $A > \left(A_1 + A_m\right)/2$ Again we can arrange the same sum in the following way,

$$A = A_1 - \frac{A_2}{2} - \left(\frac{A_2}{2} - A_3 + \frac{A_4}{2}\right) - \left(\frac{A_4}{2} - A_5 + \frac{A_6}{2}\right)$$

$$- \ldots - \left(\frac{A_{m-3}}{2} - A_{m-2} + \frac{A_{m-1}}{2}\right) - \frac{A_{m-1}}{2} + A_{m.}$$

The above equation says, $A < A_1 - \left(A_2 + A_{m-1}\right)/2 < +A_{m.}$ Since, $A_1 \approx A_2$ and $A_{m-1} \approx A_{m.}$ the above argument shows $\left(A_1 + A_m\right)/2 < A < \left(A_1 + A_m\right)/2,$

which means

$$A = (A_1 + A_m)/2$$

for odd m. The same argument for even m would conclude

$$A = (A_1 - A_m)/2$$

If the terms of the brackets of the equation (rearrange) are negative (the amplitude of any zone is more than the average of its adjacent zones, i.e. $(A_{i-1} + A_{i+1})/2 > A_i)$ then by the same type of arguments as the above, we have $(A_1 + A_m)/2 > A > A_1 - (A_2 + A_{m-1})/2 + A_m$ and the results of the previous paragraph stay unaltered. When one sums very large number of zones $(m \to \infty)$ as if there were no aperture, then the last zone amplitude, A_m is negligibly small due to the obliquity factor and the total amplitude at the observation point is nothing but the original incident plane wave amplitude A. So we obtain from equation (24.7). So we find, $A_1 = 2A$, that is the amplitude from the first Fresnel zone is twice the incident amplitude and has the same phase as the incident one. Now if there is a circular aperture capturing only the first Fresnel zone for the observation point P, the intensity at P will be four times the incident amplitude.

Fresnel Zone Plate

In discussing the Fresnel diffraction of a circular aperture we utilized the fact that successive Fresnel zones nullify each other. This suggests that we will observe a tremendous increase in irradiance at the on-axis point P if we remove all the odd or all the even zones. A screen that alters the light, either in amplitude or phase, coming from every other half-period zone is called a zone plate.

Binary Amplitude Fresnel Zone Plate

Figure is an example of a Fresnel zone plate where every other Fresnel zone is blocked.

From Eq. the radius of the nth Fresnel zone is given by

$$\rho_n = \left(\frac{\lambda\, z_1\, z_2\, n}{z_1 + z_2} \right)^{1/2}$$

The amplitude at on-axis point P can be written as

$$\begin{aligned} U[P] &= U_1 + U_3 + U_5 + \ldots + U_N \\ &= \frac{\pi\, u_{oo}[P]}{i} \left(\int_0^1 e^{i\pi q}\, dq + \int_2^3 e^{i\pi q}\, dq + \int_4^5 e^{i\pi q}\, dq + \ldots + \int_{N-1}^{N} e^{i\pi q}\, dq \right) \\ &= u_{oo}[p](N+1) \\ I[P] &= (N+1)^2\, I_{oo}[P] \end{aligned}$$

Thus we get a very intense spot of light.

If we keep λ, z_1, and and the size of the zone plate constant, but vary z_2, the limits on the integrals will no longer be integers. The integrals will give complex numbers whose magnitudes are less than $2/\pi$. Therefore, the zone plate has focal properties like for a lens and in this respect differs from a simple opaque disk. The focal length of the zone plate follows from Eq.

$$\frac{1}{z_1}+\frac{1}{z_2}=\frac{n\lambda}{\rho_{1^2}}$$

Letting n =1,

$$\frac{1}{z_1}+\frac{1}{z_2}=\frac{\lambda}{\rho_{1^2}}$$

The zone plate behaves like a lens where the focal length is given by $f=\frac{\rho_{1^2}}{\lambda}$ This is a focal length that is inversely proportional to the wavelength. Likewise, n could be -1 and the focal length would be $-\frac{\rho_{1^2}}{\lambda}$.

z_2 could be changed so each clear portion of the zone plate contains a multiple number of Fresnel zones. If the clear portion contains an even number of Fresnel zones the Fresnel zones will cancel one another. However, if an odd number of Fresnel zones are contained one of the zones will not be cancelled and a fainter focus will be obtained. For example if 3 zones are present, the effects of two of the Fresnel zones will cancel, but that of the third is left over. In this case the amplitude for the third-order will be 1/3 that of the first order and the irradiance will be 1/9 that of the first order. Thus we have fainter images for focal lengths of

$$f=\pm\frac{\rho_{1^2}}{3\lambda},\ \pm\frac{\rho_{1^2}}{5\lambda},\ \pm\frac{\rho_{1^2}}{7\lambda},$$

The focal lengths and image irradiance are the same for transparent odd Fresnel zones as for transparent even Fresnel zones, but the phase of the image will differ by 180° for the two cases.

Binary Phase Fresnel Zone Plate

Instead of having every other Fresnel zone opaque a Fresnel zone plate can be made by having all Fresnel zones transparent, but having a π phase step on every other Fresnel zones. In this way all the Fresnel zones contribute in phase and the resulting amplitude is twice that of a binary amplitude (every other zone blocked) Fresnel zone plate. Since the amplitude increases by a factor of 2, the irradiance increases by a factor of 4.

Polarization Fresnel Zone Plate

An interesting Fresnel zone plate is one where vertical polarizers cover the odd zones and horizontal polarizers cover the even zones. In this situation the odd and even zones operate independently since the polarization of the light transmitted through them is orthogonal. There are three interesting situations to look at.

- Incident radiation is horizontally or vertically polarized

  Only even or odd Fresnel zones transmit radiation and image irradiance and behavior is identical to that of regular binary amplitude zone plate.

- Incident radiation unpolarized

  Even and odd Fresnel zones each transmit half of the incident radiation and the image irradiance and behavior is identical to that of regular binary amplitude zone plate.

- Incident radiation linearly polarized at angle θ relative to horizontal

  Let the incident radiation have amplitude E then $E\,\mathrm{Cos}[\theta]$ is transmitted through even zones and $E\,\mathrm{Sin}[\theta]$ is transmitted through odd zones. The resultant irradiance goes as $E^2\,\mathrm{Cos}[\theta]^2 + E^2\,\mathrm{Sin}[\theta]^2 = E^2$ Thus, the image irradiance and behavior is identical to that of regular binary amplitude zone plate.

  The interesting result is that independent of the polarization of the incident light the amount of light in the orders is the same as for a conventional binary amplitude Fresnel zone plate.

Rectangular Aperture

Next we will look at the Fresnel diffraction of a rectangular aperture as shown in Figure. It follows from Eq. that if the source is a point source of amplitude A at coordinates (x', y') a distance z1 to the left of the aperture and Pupil $[\xi,\eta]$ is the aperture amplitude transmittance function

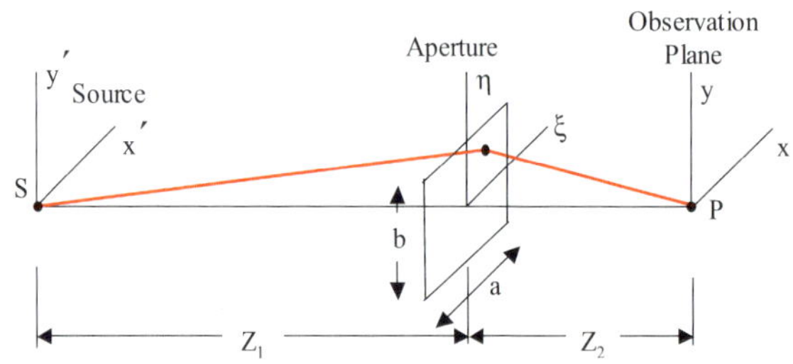

Figure: Fresnel diffraction of rectangular aperture

$$u[x,y] = \frac{A\,e^{i\,k(z_1+z_2)}}{i\,\lambda\,z_1\,z_2} \int_{-\infty}^{\infty} \text{pupil}\,[\xi,\eta]\, e^{i\frac{k}{2z_2}\left((x-\xi)^2+(y-\eta)^2\right)}\, e^{i\frac{k}{2z_1}\left((x'-\xi)^2+(y'-\eta)^2\right)}\, d\xi\, d\eta$$

If the illuminating source is on axis so x' = y' = 0, then we get Eqn. 9 which is repeated below.

$$u[x,y] = \frac{A\,e^{i\,k(z_1+z_2)}}{i\,\lambda\,z_1\,z_2}\, e^{i\frac{\pi}{\lambda z_2}(x^2+y^2)} \int_{-\infty}^{\infty} \text{pupil}\,[\xi,\eta]\, e^{i\frac{\pi}{\lambda}\left(\frac{1}{z_1}+\frac{1}{z_2}\right)(\xi^2+\eta^2)}\, e^{-i\frac{2\pi}{\lambda z_2}(x\xi+y\eta)}\, d\xi\, d\eta$$

Rather than finding the irradiance at various points of the observing screen with the aperture fixed, the normal way of solving for the Fresnel diffraction pattern of a rectangular aperture is to fix the observation point, P, and cause the pattern to sweep over it by translating the aperture in its own plane. The two approaches are not strictly equivalent since the pattern on the screen changes both in size and distribution of irradiance as the aperture is moved. These changes are negligible if the aperture motion is small compared to z_1. The following three cases can be studied without loss of generality.

1. If a and b are small compared to z1 the entire pattern can be examined.
2. If a is small and b is large, the pattern can be investigated as the aperture is translated parallel to the ξ axis.
3. If a and b are both large the pattern can be investigated near the geometrical shadow. (This is the case of the semi-infinite screen with a straight edge.)

Letting x and y equal zero yields

$$u[P] = \frac{A\, e^{i k (z_1 + z_2)}}{i \lambda z_1 z_2} \int_{-\xi 1}^{\xi_2} \int_{-\eta_1}^{\eta 2} e^{i \frac{\pi}{\lambda} \left(\frac{1}{z_1} + \frac{1}{z_2} \right) (\xi^2 + \eta^2)} d\xi\, d\eta$$

$$= \frac{A\, e^{i k (z_1 + z_2)}}{i \lambda z_1 z_2} \int_{-\xi 1}^{\xi_2} \int_{-\eta_1}^{\eta_2} e^{i \frac{\pi}{\lambda} \left(\frac{z_1 + z_2}{z_1 z_2} \right) (\xi^2 + \eta^2)} d\xi\, d\eta$$

Making a substitution of variables of

$$u = \eta \sqrt{\frac{2\,(z_1 + z_2)}{\lambda z_1 z_2}} \quad \text{and} \quad v = \xi \sqrt{\frac{2\,(z_1 + z_2)}{\lambda z_1 z_2}} \quad \text{yields}$$

$$u[P] = \frac{A\, e^{i k (z_1 + z_2)}}{2 i \lambda z_1 z_2} \int_{u_1}^{u_2} e^{i \pi u^2 / 2}\, du \int_{v_1}^{v_2} e^{i \pi v^2 / 2}\, dv$$

$$\text{But} \quad \frac{A\, e^{i k (z_1 + z_2)}}{2 i\, (z_1 + z_2)} = \frac{u_{oo}[P]}{2 i}$$

where $u_{oo}[P]$ is the unobstructed amplitude at point P. Therefore, we can write

$$u[P] = \frac{u_{oo}[P]}{2 i} \int_{u_1}^{u_2} e^{i \pi u^2 / 2}\, du \int_{v_1}^{v_2} e^{i \pi v^2 / 2}\, dv$$

These integrals are not elementary but they can be written in terms of the Fresnel integrals

$$C[u] = \int_0^u \mathrm{Cos}\left[\frac{\pi t^2}{2}\right] dt \quad \text{and} \quad S[u] = \int_0^u \mathrm{Sin}\left[\frac{\pi t^2}{2}\right] dt$$

$$u[P] = \frac{u_{oo}[P]}{2i}\left(C[u_2]-C[u_1]+i\left(S[u_2]-S[u_1]\right)\right)\left(C[v_2]-C[v_1]+i\left(S[v_2]-S[v_1]\right)\right)$$
$$= \frac{u_{oo}[P]}{2i} a\, e^{i\psi}\, b e^{i\chi}$$

Then

$$i[P] = \frac{1}{4} i_{oo}[P] a^2 b^2$$

Effect of Translating Aperture Sideways

In calculating the effects of translating the aperture sideways it is convenient to introduce some new parameters. Let b be the aperture width and η_0 be the coordinate of the center of the aperture. Then

$$\Delta u = u_2 - u_1 = b\left(\frac{2\left(z_1 + z_2\right)}{\lambda z_1 z_2}\right)^{1/2} \quad \text{and}$$

$$u_o = \frac{1}{2}\left(u_2 + u_1\right) = \eta_0\left(\frac{2\left(z_1 + z_2\right)}{\lambda z_1 z_2}\right)^{1/2}$$

The integral will now have the limits $u_o - \frac{\Delta u}{2}$ and $u_o + \frac{\Delta u}{2}$. By similar triangles we see the if the aperture is displaced a distance η_0 sideways the corresponding point in the observation plane is

$$y_o = \eta_0\left(\frac{z_1 + z_2}{z_1}\right)$$

If the aperture is illuminated with collimated light $y_o = \eta_0$

Fresnel Number for Rectangular Aperture

It can be instructive to determine a relationship between the Fresnel number of a rectangular aperture and Δu. If the aperture width is b the Fresnel number, nFZones, is given by

$$\left(\frac{b}{2}\right)^2\left(\frac{1}{2z_1} + \frac{1}{2z_2}\right) = \text{nFZones}\ \frac{\lambda}{2}$$

$$\text{nFZones} = \frac{\frac{b^2}{8}\left(\frac{z_1 + z_2}{z_1 z_2}\right)}{\lambda/2} = \frac{1}{8} b^2\left(\frac{2\left(z_1 + z_2\right)}{\lambda z_1 z_2}\right) = \frac{\left(\Delta u\right)^2}{8}$$

$$\text{nFZones} = \frac{\left(\Delta u\right)^2}{8}$$

Plots of Fresnel diffraction of rectangular aperture of width b

We will now make some plots of the Fresnel diffraction of a rectangular aperture of width b.

From above, the width of the geometrical shadow will be $w = \Delta u = b\left(\frac{2(z_1 + z_2)}{\lambda z_1 z_2}\right)^{1/2}$. As will be seen below, when w is greater than 10 $(\text{nfzones} \approx 12)$, the pattern is roughly that predicted by geometrical optics, the principal difference being the prominent fringes near the edge of the pattern. As the aperture shrinks, the pattern shrinks roughly in proportion, but the system of fringes becomes more complicated. When w passes a value of about 2 $(\text{nfzones} \approx 1/2)$, the pattern ceases to shrink and begins to expand. When w is less than unity (1/8 Fresnel zones), the pattern shows roughly the profile characteristic of Fraunhofer diffraction. Note that when w is greater than unity the irradiance in the bright region hovers about the unobstructed irradiance, but after w becomes less than unity the maximum irradiance falls rapidly because a decreasing amount of power is being spread over an increasing area.

The Mathematical equations to be used in the calculations are

$$\text{amplitude}[u1\_,\ u2\_] :=$$

$$\frac{1}{\sqrt{2}}\left(\text{Fresnel C}[u2] - \text{Fresnel C}[u1] + i\left(\left(\text{Fresnel S}[u2] - \text{Fresnel S}[u1]\right)\right)\right)$$

$$\text{irradiance}[u1\_,\ u2\_] := \left(\text{Abs}\left[\text{amplitude}[u1,\ u2]\right]\right)^2$$

The dashed line in the following plots shows the geometrical shadow.

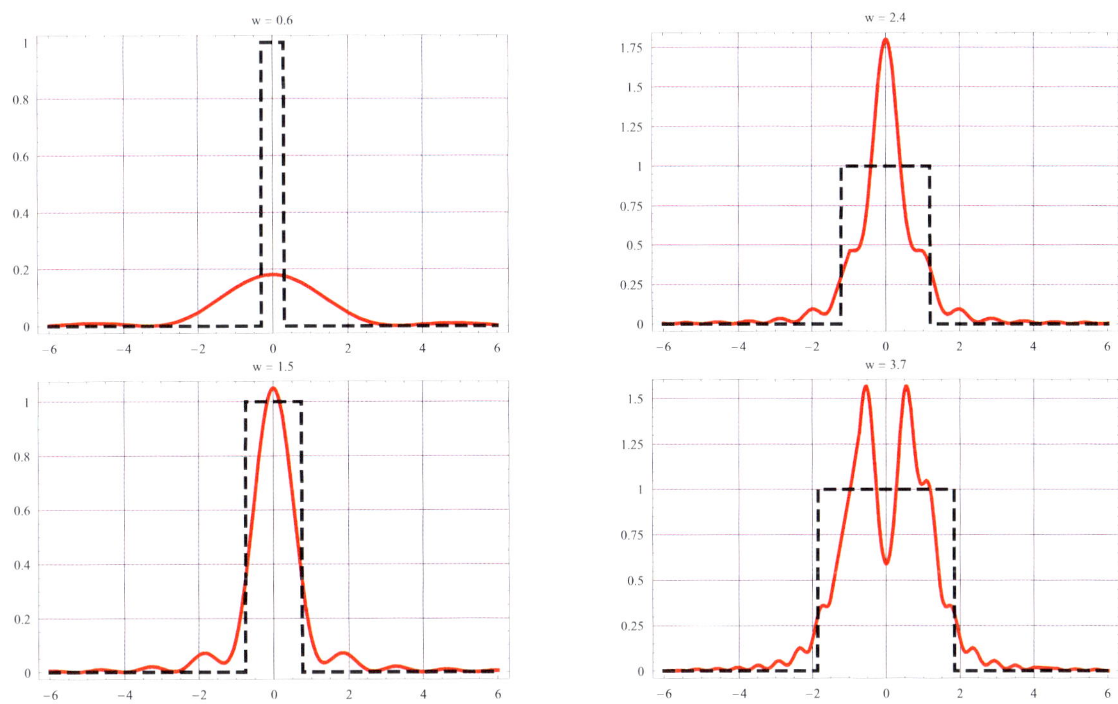

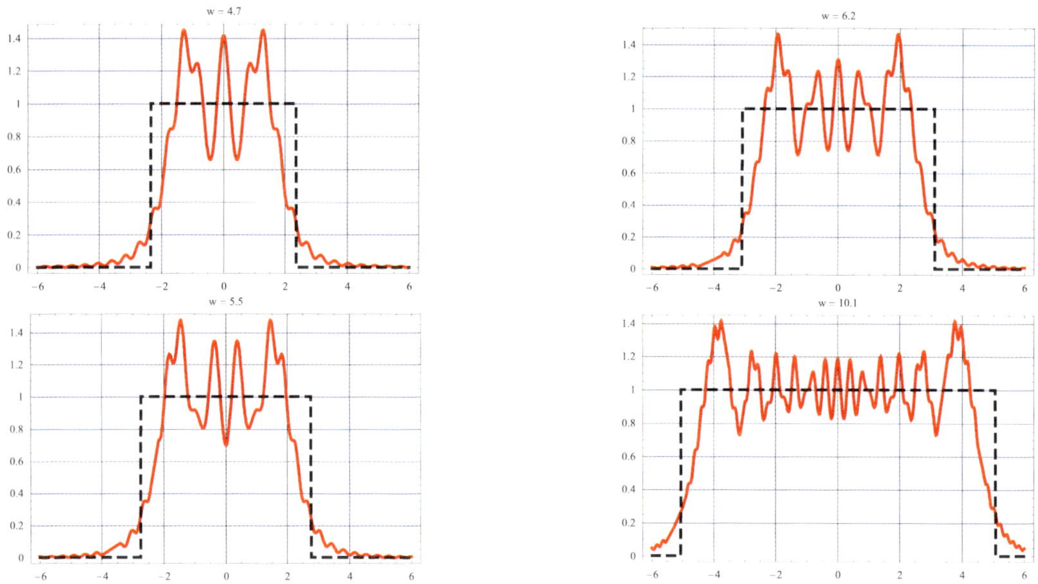

Fresnel Diffraction for Double Slits

The same approach described above for finding the Fresnel diffraction of a single slit can be used to find the Fresnel diffraction of a double slit. As an example we will calculate the Fresnel diffraction pattern of the double slits shown in figure below. Let the slits have a width D and a separation L. The source S is at infinity and the observation plane is at a distance F=1,000,000 wavelengths. D=1000 wavelengths and L=3000 wavelengths. To complicate the situation let the lower slit have a retardation of 1/2 wavelength. The diffraction pattern is calculated as follows.

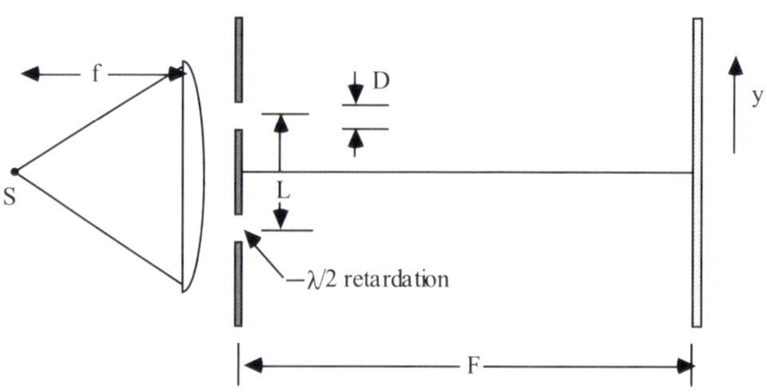

Figure: Double slit configuration.

Remember from above that

$$u = \eta \sqrt{\frac{2(z_1 + z_2)}{\lambda z_1 z_2}}$$

Since the light illuminating the double slits is collimated we can write

$$u = y\left(\frac{2}{\lambda F}\right)^{1/2} = y\left(\frac{2}{\lambda(1{,}000{,}000\lambda)}\right)^{1/2} = \frac{y\sqrt{2}}{1000\lambda}$$

Top Slit

$$u_1 = \frac{(y - 2000\,\lambda)\sqrt{2}}{1000\,\lambda}, \quad u_2 = \frac{(y - 1000\,\lambda)\sqrt{2}}{1000\,\lambda}$$

Bottom Slit

$$u_1 = \frac{(y + 1000\,\lambda)\sqrt{2}}{1000\,\lambda}, \quad u_2 = \frac{(y + 2000\,\lambda)\sqrt{2}}{1000\,\lambda}$$

The dashed line in the plot shows the geometrical shadow.

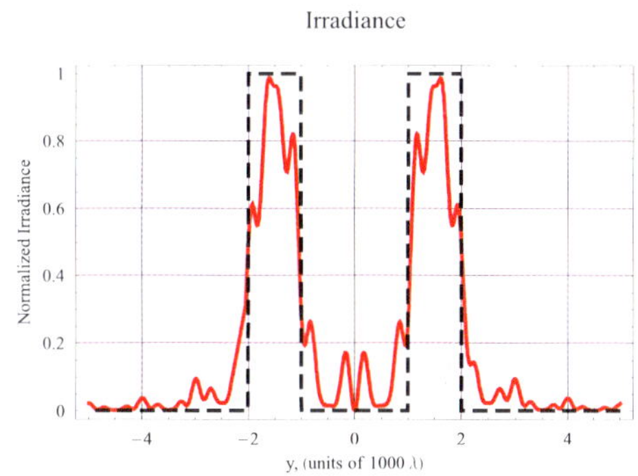

If the $\lambda/2$ retardation plate is not present the amplitudes due to the two slits add and we get

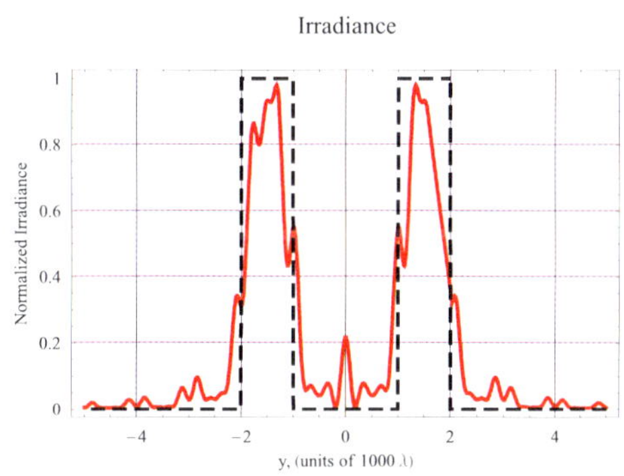

$y, (\text{units of } 1000\,\lambda)$

Large Aperture - Long Slit and Straight Edge

If the dimensions of the aperture, a and b, both become large, approximations concerning the obliquity factor, $\frac{1}{r_1 r_2}$, and $e^{i(r_1 + r_2)}$ breakdown. A more exact solution is rather difficult, one problem being that the double integral does not break down into a product of simple integrals. However, if we ignore this problem we end up with

$$a\,e^{i\psi} = C[\infty] - C[-\infty] + i\,(S[\infty] - S[-\infty])4$$
$$= \sqrt{2}e^{i\pi 4}$$

Likewise $b\ e^{ix} = \sqrt{2}e^{i\pi 4}$

Therefore,

$$u[P] = \frac{u_{oo}[P]}{2\,i}\, a\, e^{i\psi}\, b\, e^{ix} \text{ and } u[P] = u_{oo}[P]$$

It is not surprising we have the correct result since the fundamental restriction in the Kirchhoff theory was that z_1 and

W will now look at a semi-infinite diffracting screen with a straight edge as shown in Fig. Points in the $\xi - \eta$ plane for $\eta > \eta_1$ will be covered and we will assume $|\eta_1| \ll z_1$ so the observation point is always near the edge of the geometrical shadow.

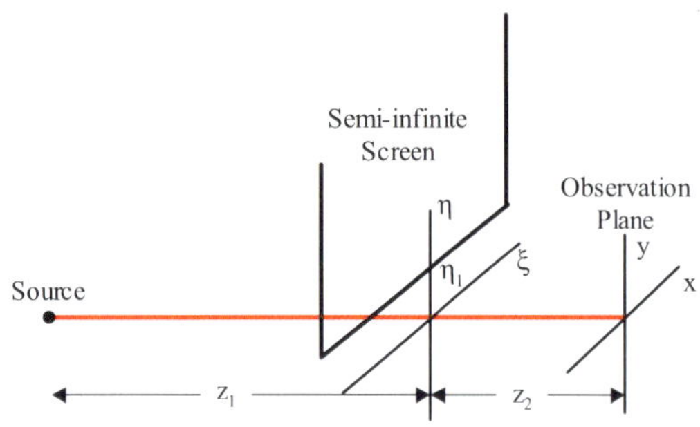

Figure: Semi-infinite diffracting screen.

$$\eta_2 = -\infty, \xi_1 = \infty, \xi = -\infty$$
$$e = c[\eta_1] - c[-\infty] + i(s[\eta_1] - S - \infty])$$

$\eta_1 = 0$ corresponds to the observation point being at the edge of the geometrical shadow. As shown below, at this point the amplitude is $\frac{1}{2}$ and the irradiance is $\frac{1}{4}$ the unobstructed irradiance.

Examples are shown below for the diffraction pattern for a semi-infinite plane. The dashed line shows the geometrical shadow

Fresnel Diffraction Pattern for Semi-infinite Plane

- Calculate amplitude pattern for semi-infinite plane

  Amplitude distribution for Fresnel diffraction of semi-infinite plane

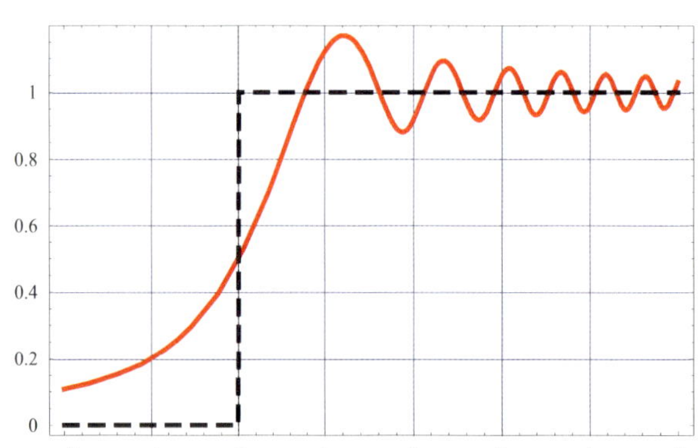

- Calculate irradiance pattern for semi-infinite plane

  Irradiance distribution for Fresnel diffractionof semi-infinite plane

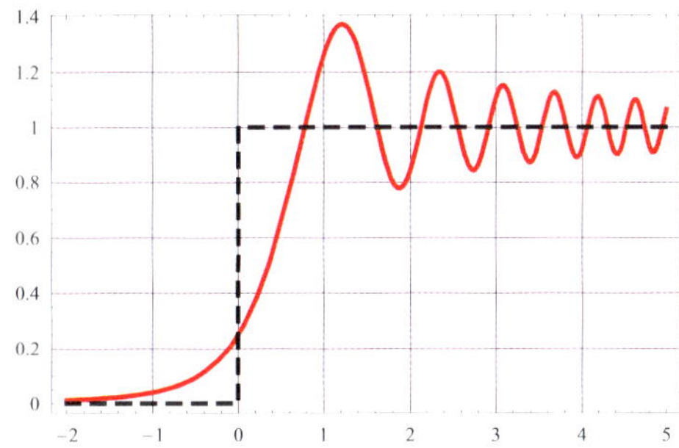

- Grey scale irradiance plot

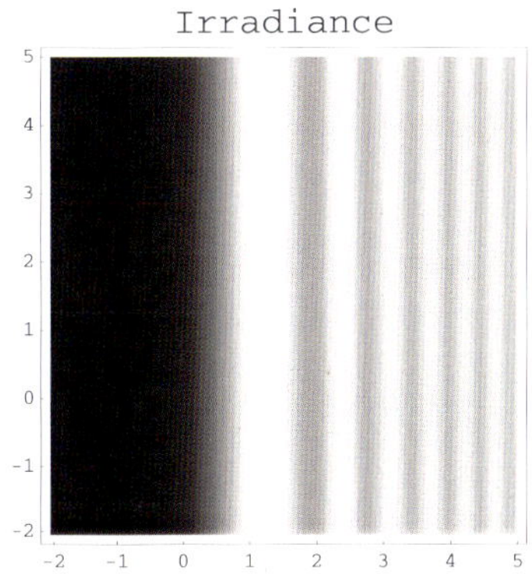

Rectangular Obstacle

The Fresnel diffraction pattern for a rectangular obstacle can be calculated using Babinet's principle. Let

u_{slit} = amplitude for slit

$u_{obstacle}$ = amplitude for obstacle

u_{oo} = amplitude for no obscuration, then

$u_{obstacle} = u_{oo} - u_{slit}$

The Fresnel Diffraction Integral

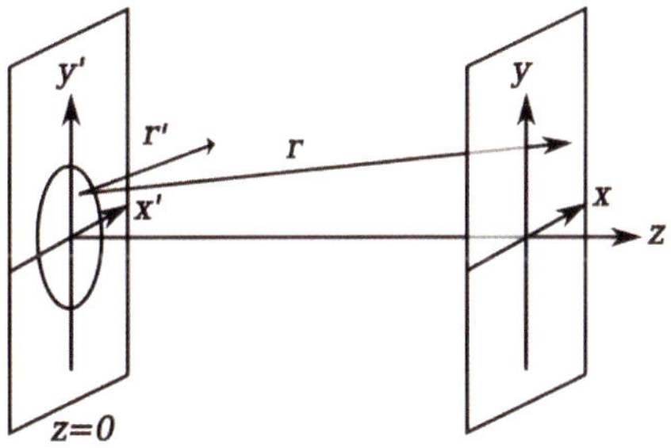

Diffraction geometry, showing aperture (or diffracting object) plane and image plane, with coordinate system.

The electric field diffraction pattern at a point (x, y, z) is given by:

$$E(x,y,z)=\frac{1}{i\lambda}\iint_{-\infty}^{+\infty} E(x',y',0)\frac{e^{ikr}}{r}dx'dy'$$

where

$E(x',y',0)$ is the aperture,

$r=\sqrt{(x-x')^2+(y-y')^2+z^2}$,

k is the wavenumber $2\pi/\lambda$

i is the imaginary unit.

Analytical solution of this integral is impossible for all but the simplest diffraction geometries. Therefore, it is usually calculated numerically.

The Fresnel Approximation

The main problem for solving the integral is the expression of r. First, we can simplify the algebra by introducing the substitution:

$$\rho^2=(x-x')^2+(y-y')^2$$

Substituting into the expression for r, we find:

$$r=\sqrt{\rho^2+z^2}=z\sqrt{1+\frac{\rho^2}{z^2}}$$

Next, using the Taylor series expansion

$$\sqrt{1+u}=(1+u)^{\frac{1}{2}}=1+\frac{u}{2}-\frac{u^2}{8}+\cdots$$

we can express r as

$$=z\sqrt{1+\frac{\rho^2}{z^2}}$$

$$=z\left[1+\frac{\rho^2}{2z^2}-\frac{1}{8}\left(\frac{\rho^2}{z^2}\right)^2+\cdots\right]$$

$$=z+\frac{\rho^2}{2z}-\frac{\rho^4}{8z^3}+\cdots$$

If we consider all the terms of Taylor series, then there is no approximation. Let us substitute this expression in the argument of the exponential within the integral; the key to the Fresnel approximation is to assume that the third term is very small and can be ignored. In order to make this possible, it has to contribute to the variation of the exponential for an almost null term. In other words, it has to be much smaller than the period of the complex exponential; i.e., 2π :

$$k\frac{\rho^4}{8z^3} \ll 2\pi$$

expressing k in terms of the wavelength,

$$k = \frac{2\pi}{\lambda}$$

we get the following relationship:

$$\frac{\rho^4}{z^3\lambda} \ll 8$$

Multiplying both sides by z^3 / λ^3, we have

$$\frac{\rho^4}{\lambda^4} \ll 8\frac{z^3}{\lambda^3}$$

or, substituting the earlier expression for ρ^2,

$$\frac{1}{\lambda^4}\left[(x-x')^2+(y-y')^2\right]^2 \ll 8\frac{z^3}{\lambda^3}$$

If this condition holds true for all values of x, x', y and y', then we can ignore the third term in the Taylor expression. Furthermore, if the third term is negligible, then all terms of higher order will be even smaller, so we can ignore them as well.

For applications involving optical wavelengths, the wavelength λ is typically many orders of magnitude smaller than the relevant physical dimensions. In particular:

$$\lambda \ll z$$

and

$$\lambda \ll \rho$$

Thus, as a practical matter, the required inequality will always hold true as long as

$$\rho \ll z$$

We can then approximate the expression with only the first two terms:

$$r \approx z + \frac{\rho^2}{2z} = z + \frac{(x-x')^2+(y-y')^2}{2z}$$

This equation, then, is the Fresnel approximation, and the inequality stated above is a condition for the approximation's validity.

Fresnel Diffraction

The condition for validity is fairly weak, and it allows all length parameters to take comparable values, provided the aperture is small compared to the path length. For the r in the denominator we go one step further, and approximate it with only the first term, $r \approx z$. This is valid in particular if we are interested in the behaviour of the field only in a small area close to the origin, where the values of x and y are much smaller than z. In general, Fresnel diffraction is valid if the Fresnel number is approximately.

For Fresnel diffraction the electric field at point (x, y, z) is then given by :

$$E(x,y,z) = \frac{e^{ikz}}{i\lambda z} \iint_{-\infty}^{+\infty} E(x',y',0) e^{\frac{ik}{2z}\left[(x-x')^2+(y-y')^2\right]} dx'dy'$$

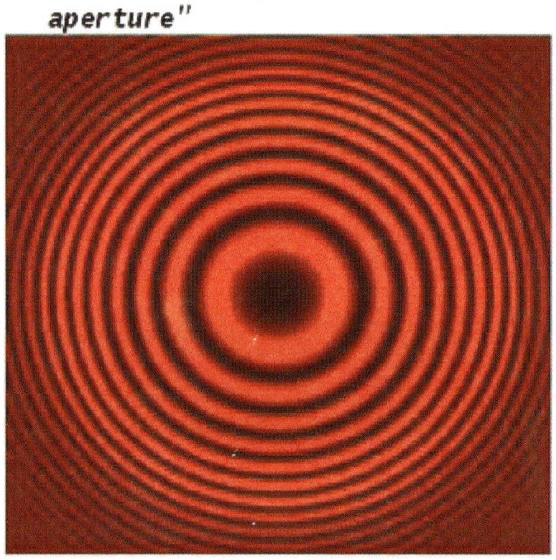

Fresnel diffraction of circular aperture, plotted with Lommel functions

This is the Fresnel diffraction integral; it means that, if the Fresnel approximation is valid, the propagating field is a spherical wave, originating at the aperture and moving along z. The integral modulates the amplitude and phase of the spherical wave. Analytical solution of this expression is still only possible in rare cases. For a further simplified case, valid only for much larger distances from the diffraction source, see Fraunhofer diffraction. Unlike Fraunhofer diffraction, Fresnel diffraction accounts for the curvature of the wavefront, in order to correctly calculate the relative phase of interfering waves.

Alternative Forms

Convolution

The integral can be expressed in other ways in order to calculate it using some mathematical properties. If we define the following function:

$$h(x,y,z) = \frac{e^{ikz}}{i\lambda z} e^{i\frac{k}{2z}\left(x^2+y^2\right)}$$

then the integral can be expressed in terms of a convolution:

$$E(x,y,z)=E(x,y,0)*h(x,y,z)$$

in other words we are representing the propagation using a linear-filter modeling. That is why we might call the function $h(x,\ y,\ z)$ the impulse response of free space propagation.

Fourier Transform

Another possible way is through the Fourier transform. If in the integral we express k in terms of the wavelength:

$$k=\frac{2\pi}{\lambda}$$

and expand each component of the transverse displacement:

$$(x-x')^2=x^2+x'^2-2xx',$$
$$(y-y')^2=y^2+y'^2-2yy',$$

then we can express the integral in terms of the two-dimensional Fourier transform. Let us use the following definition:

$$G(p,q)=\mathcal{F}\{g(x,y)\}\equiv\iint_{-\infty}^{\infty}g(x,y)e^{-i2\pi(px+qy)}\,dx\,dy,$$

where p and q are spatial frequencies (wavenumbers). The Fresnel integral can be expressed as

$$E(x,y,z)=\frac{e^{ikz}}{i\lambda z}e^{i\frac{\pi}{\lambda z}(x^2+y^2)}\mathcal{F}\left\{E(x',y',0)e^{i\frac{\pi}{\lambda z}(x'^2+y'^2)}\right\}\Bigg|_{p=\frac{x}{\lambda z},q=\frac{y}{\lambda z}}$$
$$=H(p,q)\cdot G(p,q)\Big|_{p=\frac{x}{\lambda z},q=\frac{y}{\lambda z}},$$

where

$$H(p,q)=\mathcal{F}\{h(x,y)\}.$$

That is, first multiply the field to be propagated by a complex exponential, calculate its two-dimensional Fourier transform, replace (p,q) with $\left(\frac{x}{\lambda z},\frac{y}{\lambda z}\right)$ and multiply it by another factor. This expression is better than the others when the process leads to a known Fourier transform, and the connection with the Fourier transform is tightened in the linear canonical transformation, discussed below.

Linear Canonical Transformation

From the point of view of the linear canonical transformation, Fresnel diffraction can be seen as a shear in the time-frequency domain, corresponding to how the Fourier transform is a rotation in the time-frequency domain.

Fraunhofer Diffraction

Fraunhofer diffraction, or far-field diffraction, is a type of wave diffraction which occurs when field waves pass via a slit or an aperture, thereby changing the size of an observed aperture image with respect to the increasingly planar nature of diffracted waves and far-field location of observation.

Theory

Suppose, as depicted in figure, that a laser is shone upon a small slit. The light passing through the slit is then allowed to fall on a screen, positioned at some distance, L, from the slit, with its plane perpendicular to the path of the light's propagation. If light traveled in straight lines one would expect the screen to display a single image of the slit with the rest of the screen in shadow. For a sufficiently small slit, however, it is found that this is not the case. Instead, a diffraction pattern is observed, consisting of a central bright fringe along the slitscreen axis with alternating dark and bright fringes on either side of the central fringe. It is evident, based on this observation, that light does not travel in straight lines since bright fringes are seen where shadow would be expected. It follows then, that light has the property of being able to "bend" around corners, a property called diffraction.

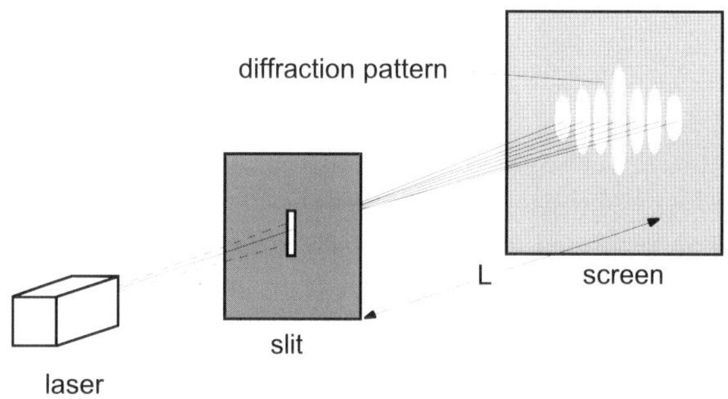

In order to examine diffraction we will assume that light behaves like a wave. With this assumption one can imagine a beam of light to consist of a series of wave fronts aligned perpendicular to the direction of light propagation. As shown in figure below, a ray can be drawn perpendicular to the wave fronts, indicating the direction of light propagation. This wave model of light allows us to make use of Huygens' principle which claims that any point on a wave front can itself be regarded as a point source emitting circular wave fronts. Applying this principle to the situation under consideration, every point located in the slit acts as a point source. Consequently, the diffraction pattern must be the result of the constructive and destructive interference of the various waves generated by these point sources.

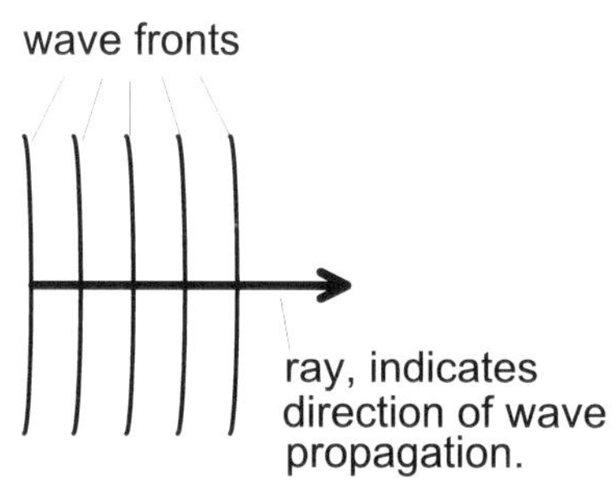

Diffraction phenomena are usually separated into two different types called Fraunhofer and Fresnel diffraction. The latter describes the near-field pattern observed when the screen is placed close to the slit (i.e. at small values of L), so the wave effects make only a small correction to the geometric shadow of the aperture. Fraunhofer diffraction, on the contrary, describes the diffraction pattern observed in the far field (i.e. at large values of L) where geometric optics is completely inapplicable. We will postpone Fresnel diffraction until a later date, and concentrate here on the far-field case.

Consider a general aperture illuminated by light as in figure. In the plane of the aperture, perpendicular to the plane-wave propagation axis $\hat{z}$, the electric field is given by:

$$E\,(x\;y)\;E_0\,f\;(x\;y)\;e^{-i\omega t}$$

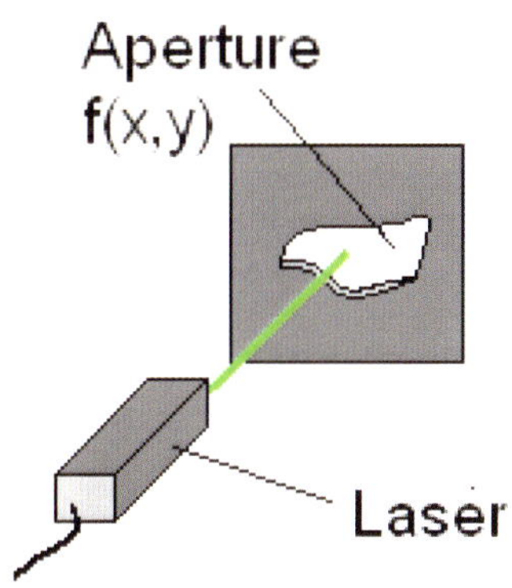

where $f = (x\;y)$ reflects the geometry of the aperture. We treat this field according to Huygens' principle as an antenna (extended oscillator) that generates secondary waves that form the diffraction pattern. To calculate this pattern, we re-express the oscillator as a linear combination of infinite plane antennas:

$$E\,(x\;y)\;E_0 \iint\limits_{K_x, K_z} f\left(K_x, K_y\right) e^{i(K_x x + K_y y) - iwt}\; dk_x\; dk_y\;,$$

where each antenna is characterized by a pair of wave vectors $\left(K_x, K_y\right)$ and $f\left(K_x, K_y\right)$ is the inverse Fourier transform of the aperture function:

$$f\left(K_x, K_y\right) = \Im^{-1}\{f = (x\;y)\} = \frac{1}{2\pi}\int_{-\infty}^{\infty}\int_{-\infty}^{\infty} e^{-ik_x x + ik_y y} f\;(x,\,y)dxdy$$

Each antenna in Eq. (2) generates a plane wave

$$e^{i(k_x x + k_y y) - iwt} \;\rightarrow\; e^{i(k_x x + k_y y + ik_z z) - iwt}$$

with $k_y^2 + k_x^2 + k_z^2 = (2\pi / \pi)^2 = \omega^2 / c^2$, so the field behind the aperture can be written as

$$(E\;x\;y\;z)\;E_0 \iint\limits_{K_x, K_y} f\;(k_x, k_y)\; e^{i(k_x x + k_y y + k_z z) - iwt}\; dk_x\; dk_y.$$

The field produced by each plane wave antenna propagates in the directions defined by angles $\left(\theta_x, \theta_y\right)$ with respect to the z axis, such that

$$\sin\theta_x = \frac{k_x}{k}$$

$$\sin\theta_y = \frac{k_y}{k}.$$

We continue our treatment within the so-called paraxial approximation, where we assume that $k_z \gg k_x, k_y$, so $k_z \approx k = \frac{2\pi}{\lambda}$, independent of k_x and k_y. The paraxial approximation also implies that $\theta_x \approx \sin\theta_x$ and $\theta_y \approx \sin\theta_y$. If the diffraction pattern is observed on a screen very far away from the aperture (i.e. in the far-field approximation relevant to the Fraunhofer diffraction), we can assume that the irradiance at some point (x', y') on the screen will be determined by the electric field of the plane wave propagating at angles

$$\theta_x = \frac{x'}{L}$$

$$\theta_y = \frac{y'}{L}$$

with respect to the z axis. Combining Eqs. $\begin{matrix}\theta_x = \frac{x'}{L} \\ \theta_y = \frac{y'}{L}\end{matrix}$ and $E(x', y') = f(k\frac{x'}{L}, k\frac{y'}{L})$ we find for the far field amplitude

$$E(x', y') = f(k\frac{x'}{L}, k\frac{y'}{L})$$

In other words, in the paraxial approximation the far-field diffraction pattern is just a scaled inverse Fourier transform of the aperture. The intensity of the diffraction pattern is found according to

$$I(x', y') = \left\langle \left| E(x', y') \right| \right\rangle^2,$$

Consider now a single slit, with the aperture function as described in below figure. The vertical dimension of the slit is sufficiently large so that we may treat the problem as one-dimensional. The aperture function here is the “top-hat” function:

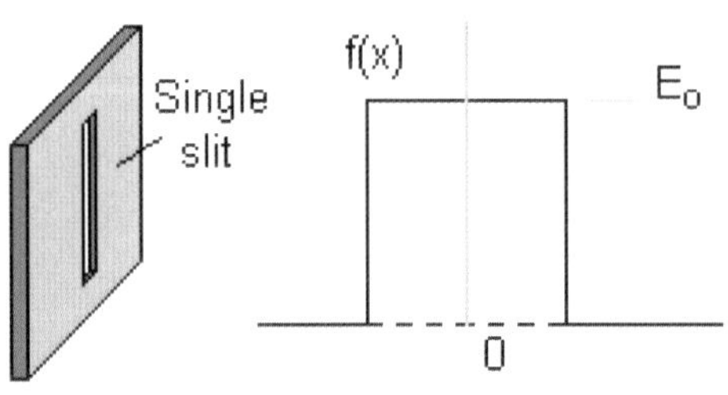

$$f(x) = E_0, x \in [-\frac{a}{2}, \frac{a}{2}]$$

$$f(x) = 0, x \notin [-\frac{a}{2}, \frac{a}{2}]$$

We can then evaluate the Fourier transformation of the aperture function to find the electric field at a point on a screen in the far-field regime

$$E(x') = \frac{1}{2\pi}\int_{-\infty}^{\infty} f(k_x)e^{ik_x x}\,dx = \frac{1}{2\pi}\int_{-a/2}^{a/2} E_0\,e^{ik_x x}\,dx = E_0\,\frac{e^{ik_x a/2} - e^{-ik_x a/2}}{2\pi ik} = \frac{2aE_0}{2\pi ak}\sin(k_x a/2)$$

or, using equation (4) for k_x and defining

$$ak_x/2 = ak\sin\theta/2 \equiv \beta$$

we have:

$$E(\beta) = \frac{aE_0}{2\pi}\frac{\sin\beta}{\beta}.$$

Recalling that the irradiance at x' is just the time-averaged electric field at that point, we can write:

$$I(\theta) = I_0\,[\sin c(\beta)]^2$$

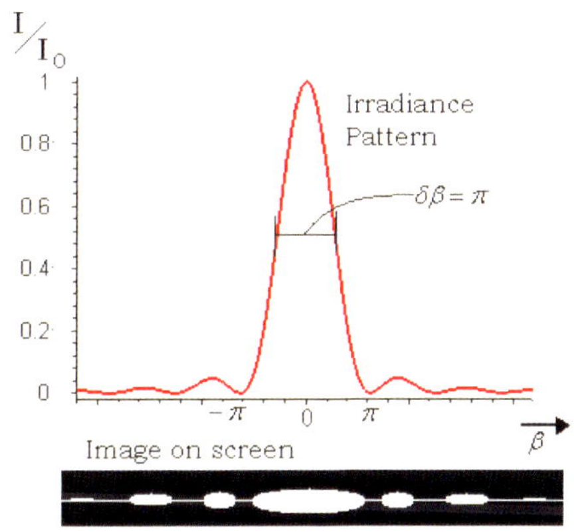

Roots of sin (β) - β

n	β
1	1.43 π
2	2.46 π
3	3.47 π
4	4.48 π
5	5.48 π
6	6.48 π
7	7.49 π

Table

where we've defined I_0 to be $\left(\frac{aE_0}{2\pi}\right)^2$ and used $\sin(x) \equiv \frac{\sin x}{x}$. displays the resultant irradiance pattern. Note that equation (12) will have minima whenever $\beta = n\pi$, for all non-zero integers n (recall that $\lim_{n \to 0} \frac{\sin n}{n} = 1$). Taking the derivative and setting it equal to zero, we see that sinc β reaches maxima whenever $\beta = \tan\beta$. This is a transcendental equation, for which some roots are listed in table. These values may be used, to find the angular positions of the maxima of the refracted beams.

Fraunhofer Diffraction from a more complicated apparatus can be calculated by using the fact that the Fourier transform of the convolution two functions $f(x)$ and $g(x)$ is the product of the individual Fourier transforms:

$$\Im^{-1}\{f(x) * g(x)\} = \frac{1}{2\pi}\Im^{-1}\{f(x)\} \bullet \Im^{-1}\{g(x)\},$$

where the convolution of functions $f(x)$ and $g(x)$ is given by

$$h(x) = f(x) * g(x) = \int_{-\infty}^{\infty} f(t)\, x(x-t)\, dt.$$

For example, the irradiance pattern for a two slits separated by a distance b, with each slit having width a may be evaluated using as the convolution of two Dirac delta functions and a single slit (Figure)

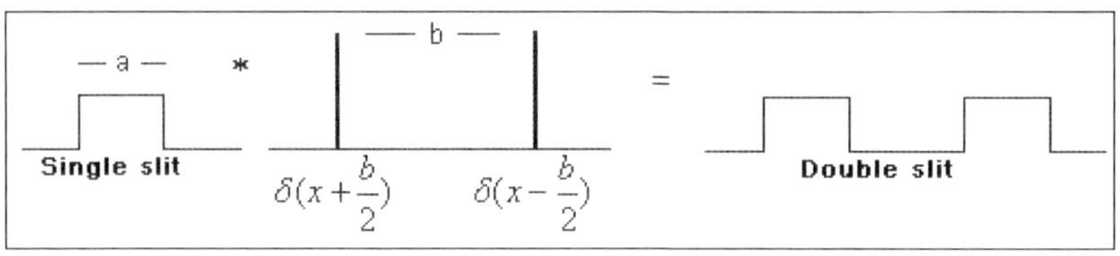

Using equation:

$$\Im^{-1}\left\{f_{slit}(x) * \left[\delta(x-\frac{b}{2}) + \delta(x+\frac{b}{2})\right]\right\} = \Im^{-1}\ (f_{slit})\, \Im^{-1}\left[\delta(x-\frac{b}{2}) + (x+\frac{b}{2})\right]$$

With

$$\Im^{-1}\,(\delta(x-\frac{b}{2}) + \delta(x+\frac{b}{2})) = \frac{1}{2\pi}(e^{-ik_x b/2} + e^{ik_x b/2}) = \frac{1}{\pi}\cos(k_x b/2).'$$

Using our previous result for the single slit and equation, we conclude that

$$I(\theta) \propto \operatorname{sinc}^2\left(\frac{k_x a}{2}\right)\cos^2\left(\frac{k_x a}{2}\right)$$

where the constant of proportionality is unimportant. We see that the single-slit diffraction pattern $\operatorname{sinc}^2(k_x a/2)$ is now modulated by a finer pattern $\cos^2(k_x b/2)$ due to the two-slit diffraction. The distance between the maxima of this pattern is given by

$$\frac{k_x a}{2} = m\pi \text{ for } m = 0, \pm 1, \pm 2, ...$$

or, according to Eq.

$$\frac{kb\sin\theta}{2} = m\pi$$

or simply

$$b\sin\theta = m\lambda$$

The same analysis applies to many slit apertures and in more generality, to an arbitrary collection of similar apertures.

When the number of equidistant slits further increases, the maxima of the diffraction pattern remain at the same positions defined by Eq. (21), but become sharper. An extreme case of such an arrangement is a diffraction grating – a repetitive array of slits (or obstacles). Because of their dispersive properties, gratings are commonly used in monochromators and spectrometers. The first artificial diffraction grating was made around 1785 by Philadelphia inventor David Rittenhouse, who strung hairs between two finely threaded screws.

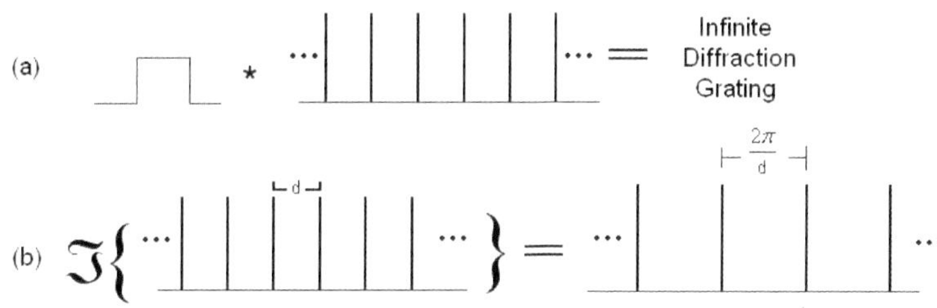

As mentioned, the diffraction grating will create a number of angularly separated maxima, with the angle of diffraction determined by the wavelength of light. The angular position of the mth maximum of the diffracted beam can be obtained by performing the inverse Fourier transform of the grating as illustrated in figure. An infinite periodic grating can be considered to be the convolution of a Dirac comb function (an infinite periodic array of delta functions) and a single slit. The Fourier transform of the Dirac comb function of period d is also a Dirac comb function of period $2\pi/d$. The mth maximum is thus given by the equation

$$k_x = m\frac{2\pi}{d},$$

from which we find the condition for the maximum

$$d\sin\theta = m\lambda.$$

in analogy to Eq.

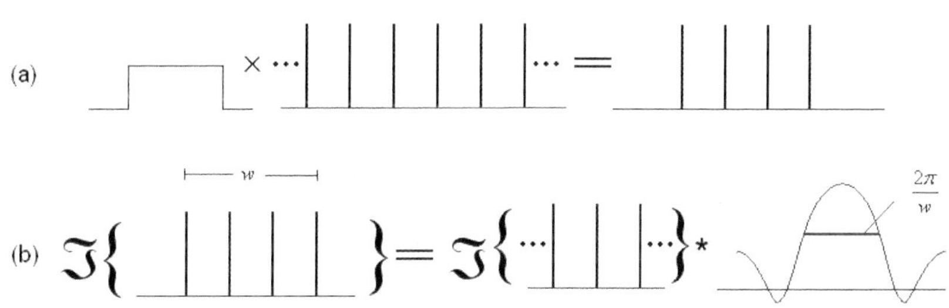

The above treatment is valid for an infinite grating, which produces infinitely sharp diffraction maxima. In a realistic experiment, the grating (more precisely, the illuminated area of the grating) is finite. To find the Fraunhofer diffraction pattern from such a grating, we notice that its aperture function equals to that of an infinite grating times that of a single slit (figure). The resultant far field image is then the convolution of the irradiance pattern for a single slit and that of an infinite grating, and the width $\delta \sin\theta$ of a given peak is found according to Eqs.:

$$wk\delta \sin\theta / 2 = \pi$$

Or

$$\delta \sin\theta = \frac{\lambda}{w}$$

w being the grating width

Illuminating a greater portion of the grating will generate sharper peaks. This is important since gratings are typically used to separate light of different wavelengths. As mentioned, a property of the diffraction grating is that the angle of deviation depends on the wavelength of the incident light. However if two wavelengths differ by very little, so will their angular separation. If the peaks of each diffracted beam overlap too much, the resultant image will appear as a single, brighter peak (see figure). If $\Delta\lambda_{min}$ is the smallest possible difference in wavelength which allows for the identification of individual peaks (known as the limit of resolution), we define the chromatic resolving power of the diffraction grating as:

$$\Re \equiv \frac{\lambda}{\Delta\lambda_{min}}$$

In order to have a high resolving power, we wish to increase the effective width of the grating, thus making the diffraction peaks sharper. Therefore, for a diffraction grating, it is advantageous to illuminate as large an area as possible.

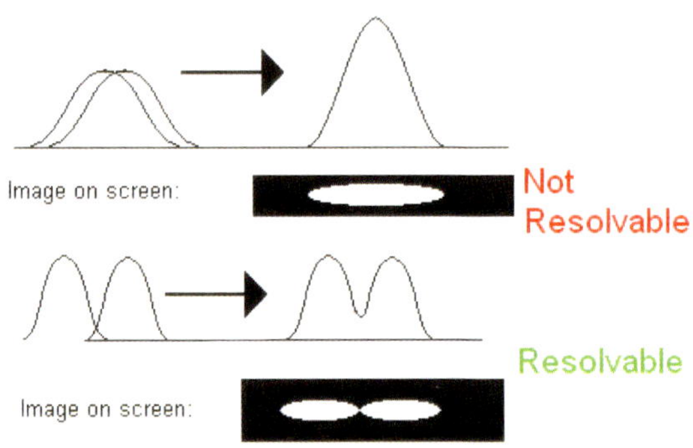

Two spectral lines separated by an interval $\Delta\lambda$ produce two diffraction peaks whose separation $\Delta \sin\theta$ can be found according to Eq.:

$$\Delta\lambda = \frac{d\Delta \sin\theta}{m}$$

In order to resolve these lines, their separation $\Delta \sin\theta$ should not exceed the individual peak width given by Eq. We find

$$\Delta\lambda = \frac{d\Delta \sin\theta}{m} = \frac{d}{mw}\lambda$$

so the resolving power of the mth maximum of the grating is

$$\Re = \frac{\lambda}{\Delta\lambda_{\min}} = \frac{wm}{d} = mN.$$

The resolution of the grating in the first diffraction order equals the number of lines illuminated.

Figure shows the experimental arrangement that will be used to study diffraction patterns. The laser, mounted on the optical bench by means of the two-axis translation stage, shines on the slide holder which holds one of the various diffraction slides or gratings. The translation stage is used to accurately adjust the x and y position of the laser. The resulting diffraction pattern is observed on the screen.

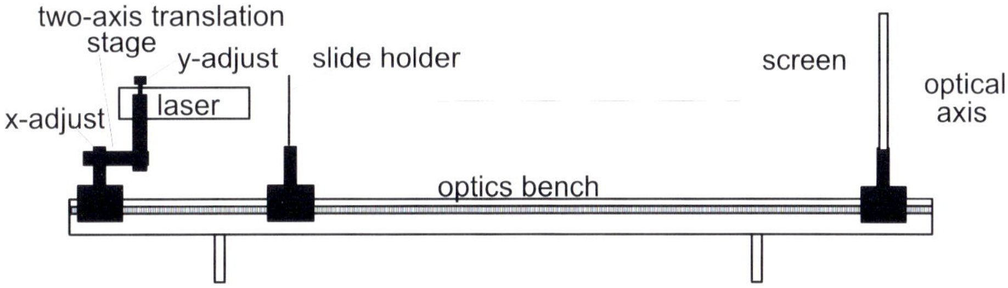

In this experiment, it is frequently required to measure the angle of a diffracted beam by measuring the distance from the aperture to the slide. From basic trigonometry (figure) the sine of the diffracted angle may be calculated from the displacement x of the pattern from the primary (undiffracted) beam as follows:

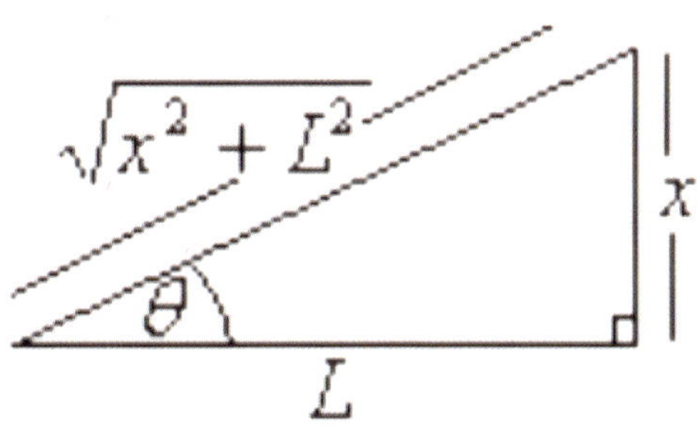

$$\sin\theta = \frac{x}{\sqrt{x^2 + L^2}}$$

Diffraction Grating

Diffraction grating, component of optical devices consisting of a surface ruled with close, equidistant, and parallel lines for the purpose of resolving light into spectra. A grating is said to be a transmission or reflection grating according to whether it is transparent or mirrored—that is, whether it is ruled on glass or on a thin metal film deposited on a glass blank. Reflection gratings are further classified as plane or concave, the latter being a spherical surface ruled with lines that are the projection of equidistant and parallel lines on an imaginary plane surface. The advantage of a concave grating over a plane grating is its ability to produce sharp spectral lines without the aid of lenses or additional mirrors. This makes it useful in the infrared and ultraviolet regions in which these radiations would otherwise be absorbed upon passage through a lens.

The lines on gratings are made by an extremely precise machine called a ruling engine, which uses a diamond-tipped tool to press thousands of very fine, shallow lines onto a highly polished surface. Newer techniques rule the lines photographically, using laser interferometry.

A diffraction grating is able to disperse a beam of various wavelengths into a spectrum of associated lines because of the principle of diffraction: in any particular direction, only those waves of a given wavelength will be conserved, all the rest being destroyed because of interference with one another. Gratings give exceptionally high resolutions of spectral lines. The resolving power (R) of an optical instrument represents the ability to separate closely spaced lines in a spectrum and is equal to the wavelength λ divided by the smallest difference $(\Delta\lambda)$ in two wavelengths that can be detected; *i.e.*, $R = \lambda / \Delta\lambda$. Thus, for a grating 10 centimetres wide and ruled with 10,000 lines per centimetre, the resolution in the first diffraction order would be 100,000. For a wavelength emission in the ultra-violet, say $\lambda = 300$ nanometres (3×10^{-7} metre), a wavelength difference of $\Delta\lambda = 3 \times 10^{-12}$ metre (about $^{1}/_{100}$ the diameter of an atom) should be theoretically possible.

Theory of Operation

The relationship between the grating spacing and the angles of the incident and diffracted beams of light is known as the grating equation.

When a wave propagates, each point on the wavefront can be considered to act as a point source, and the wavefront at any subsequent point can be found by adding together the contributions from each of these individual point sources.

An idealized grating is considered here which is made up of a set of long and infinitely narrow slits of spacing d. When a plane wave of wavelength λ, is incident normally on the grating, each of the point slits in the grating acts as a set of point sources which propagate in all directions. The light in a particular direction, θ, is made up of the interfering components from each slit. Generally, the phases of the waves from different slits will vary from one another, and will cancel one another out partially or wholly. However, when the path difference between the light from adjacent slits is equal to the wavelength, λ, the waves will all be in phase. This occurs at angles θ_m which satisfy the relationship $d sin\theta_m / \lambda = |m|$ where d is the separation of the slits and m is an integer. Thus, the diffracted light will have maxima at angles θ_m given by

$$d sin\theta_m = m\lambda$$

It is straightforward to show that if a plane wave is incident at an angle θ_i, the grating equation becomes

$$d\,(\sin\theta_m + \sin\theta_i) = m\lambda$$

The light that corresponds to direct transmission (or specular reflection in the case of a reflection grating) is called the zero order, and is denoted $m = 0$. The other maxima occur at angles which are represented by non-zero integers m. Note that m can be positive or negative, resulting in diffracted orders on both sides of the zero order beam.

This derivation of the grating equation has used an idealized grating. However, the relationship between the angles of the diffracted beams, the grating spacing and the wavelength of the light apply to any regular structure of the same spacing, because the phase relationship between light scattered from adjacent elements of the grating remains the same. The detailed distribution of the diffracted light depends on the detailed structure of the grating elements as well as on the number of elements in the grating, but it will always give maxima in the directions given by the grating equation.

Gratings can be made in which various properties of the incident light are modulated in a regular pattern; these include

- transparency (transmission amplitude gratings)
- reflectance (reflection amplitude gratings)
- refractive index (phase gratings)
- direction of optical axis (optical axis gratings)

The grating equation applies in all these cases.

The Diffraction Grating Equation in Practice

When a parallel beam of monochromatic light is incident on a grating, the light is diffracted from the grating in directions corresponding to m = -2, -1, 0, 1, 2, 3, etc. This is shown i. Figure, and discussed further under "Grating Order".

When a parallel beam of polychromatic light is incident on a grating then the light is dispersed so that each wavelength satisfies the diffraction grating equation. This is shown in Figure.

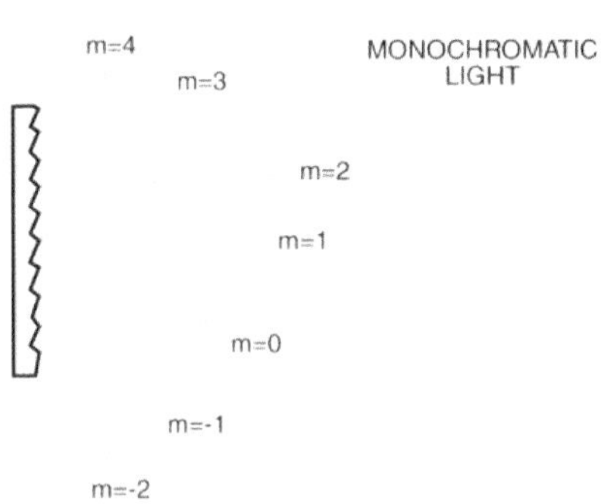

The "Grating Equation" satisfied for a parallel beam of monochromatic light.

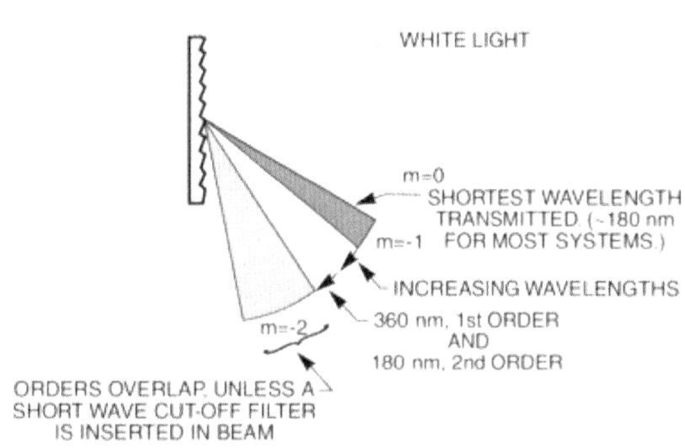

Polychromatic light diffracted from a grating. Positive orders have been omitted for clarity.

In most monochromators, the input slit and collimating mirror fix the direction of the input beam that strikes the grating. The focusing mirror and exit slit fix the output direction. Only wavelengths that satisfy the grating equation pass through the exit slit. The remainder of the light is scattered and absorbed inside the monochromator. As the grating is rotated, the angles I and D change, although the difference between them remains constant and is fixed by the geometry of the monochromator. A more convenient form of the grating equation for use with monochromators is:

$$m\lambda = 2x \,.\, x \cos\varphi \, x \sin\theta$$

Where:

φ = Half the included angle between the incident ray and the diffracted ray at the grating

θ = Grating angle relative to the zero order position

These terms are related to the incident angle I and diffracted angle D by:

$$I = \theta + \kappa \; and \; D = \theta - \varphi$$

Diffraction Grating Order

It is important to note the sign of "m" is given by either form of the grating equation and can be positive or negative. In a monochromator, the angles I and D are determined by the rotational position of the grating. We use the sign convention that all angles that are counter clockwise from the grating normal are positive, and all angles that are clockwise to the grating are negative. The incident light, diffracted light and grating rotation can be at positive or negative angles depending on which side of the grating normal they are. The half angle is always regarded as positive.

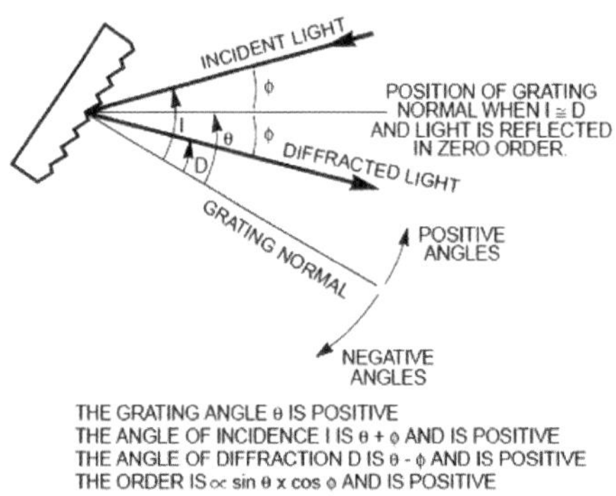

The sign convention for the angle of incidence, angle of diffraction and grating angle.

If the angle D is equal to I and of opposite sign, then the grating angle and order are zero, and the light is simply being reflected. If the grating angle is positive then the order is positive (m=1), if the grating angle is negative, then the order is negative (m = -1). Table lists the half angles and the sign of the orders passed by our monochromators.

The grating equation is also satisfied for wavelengths in higher orders, when $|m| \; is > 1$. Therefore $\lambda_2 = \lambda_1 / 2 \; for \; m = \pm 2$, $\lambda_3 = \lambda_1 / 3 \; for \; m = \pm 3$, etc. The wavelength λ_2 is in the second order and λ_3 is in the third order, etc. Again, this concept is illustrated in figure above.

Usually only the first order, positive or negative, is desired. The other wavelengths in higher orders may need to be blocked. The input spectrum and detector sensitivity determine whether order sorting or blocking filters are needed.

Blaze Wavelength

If monochromatic light strikes a grating then a fraction of it is diffracted into each order. The fraction diffracted into any order can be termed the efficiency of the grating in that order. All Oriel Gratings are designed for efficient diffraction into the first order. Gratings are not equally "efficient" at all wavelengths for a variety of detailed reasons. The efficiency can be "tuned" by changing the groove facet angles, or shape and depth. The optimization of efficiency by appropriate groove shaping has become known as blazing. The "Blaze Wavelength" is the wavelength for which the blazed grating is most efficient. Figure shows a typical efficiency vs. wavelength curve for a ruled and a blazed holographic grating.

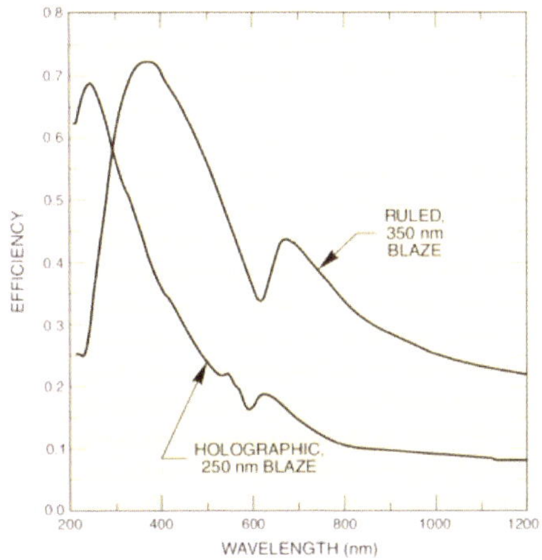

The efficiency of a ruled and a blazed holographic grating, both have 1200 l/mm.

Holographic Gratings

Holographic gratings are created using a sinusoidal interferometric pattern and sometimes an etching process. Sinusoidally grooved gratings produce very little scattered light but have low, flat efficiency curves, although they are generally quite broad. Blazed holographic gratings use etching during the interferometric process, or an ion gun to form a blaze angle in a secondary process. The former does not produce strong blazing, and while the latter produces high efficiencies at the blaze wavelengths, light scatter is increased due to the formation of micro structure along the edges of the grooves. Newport offers both ruled and holographic gratings to provide the best combination of blaze, efficiency, and low light scatter for different wavelength ranges.

Table Half Angles and Orders for Oriel Monochromators

Model	Half Angle (degree)	Orders
77250	5.1	Positive
77400	12,5	Positive
77200	13.04	Negative
77700	11.83	Positive
74000	5.1	Positive
74100	11,83	Positive

Dispersion

If we fix the angle I in the grating equation and differentiate with respect to wavelength we get:

$$a \times \cos D \times \delta D = m \times \delta\lambda$$

. thus:

$$\frac{\delta D}{\delta\lambda} = \frac{m}{a \times \cos D}$$

$\delta D/\delta\lambda$ is the angular dispersion or change of diffraction angle corresponding to a small change in wavelength. It is greater for smaller groove spacings, a (greater number of lines per millimeter); larger orders, m; and larger diffraction angles, D. The linear dispersion, $\delta L/\delta\lambda$, at a monochromator's exit slit will vary with output focal length, f, and angle D and is the product of the focal length and angular dispersion:

$$\frac{\delta L}{\delta\lambda} = f \times \frac{\delta D}{\delta\lambda} = \frac{f \times m}{a \times \cos D}$$

For our 77250 Monochromator and MS257 Monochromato. and Spectrograph each with 1200 l/mm gratings, the linear dispersion is:

77250

$$\frac{125 \times m \times 1200}{10^6 \times \cos D} mm/nm$$

77000

$$\frac{257 \times m \times 1200}{10^6 \times \cos D} mm/nm$$

Normally, we use the "reciprocal linear dispersion", which gives the wavelength dispersion in nm/mm of slit width.

Bandpass and Resolution

The bandpass is the width of the spectrum passed by a monochromator or spectrograph when illuminated by a light source with a continuous spectrum.

In a monochromator, the bandpass may be decreased by reducing the width of the slits until a limiting bandpass is reached. The limiting bandpass is called the resolution of the instrument. In spectral analysis, the resolution is a measure of the ability of the instrument to separate two spectral lines that are close together. The resolution of the 77250 is about 0.5 nm and that of the 77700 is 0.1 nm with 1200 l/mm gratings in the visible. For a properly illuminated grating, see below, the aberrations of the optical system determine the resolution. For any instrument, the optical aberrations are minimized by illuminating only the central zone of the input slit.

For large bandpasses, compared with the resolution, the bandpass is given by the product of the reciprocal linear dispersion and the slit width.

Diffraction Grating Illumination and Resolution

The resolving power R of a grating is defined in terms of wavelengths λ and $\lambda + \delta\lambda$, where $\lambda + \delta\lambda$ is the closest wavelength to λ that can be resolved. Theoretically:

$$R = \frac{\lambda}{\delta\lambda} = \frac{W(\sin I + \sin D)}{\lambda}$$

Where:

$\delta\lambda$ = Resolution

W = Illuminated width of the grating.

Normally the smallest slit available and optical aberrations, rather than the grating, determine the attainable resolution. For the 77700 at λ = 500 nm, with a 1200 l/mm grating R is 60000 with a fully illuminated grating. Based on this, $\delta\lambda$ is 0.008 nm. With 25 μm slits the product of slitwidth and reciprocal linear dispersion is 0.09 nm, which is close to the measured value of 0.1 nm and R = 5000 instead of 60000. If however, only a few mm of the grating are illuminated, as is sometimes the case with laser sources incorrectly coupled to the monochromator, the grating resolution can broaden the measured bandwidths. For example, if a laser illuminates only 2 mm of the grating width, the measured bandwidth will be 0.2 nm, even for a very narrowband laser.

Tech Note

Very short pulse (ps) lasers will only illuminate a small portion of the grating at any instant. The resolution is degraded in agreement with Heisenberg's Uncertainty Principle.

Double Dispersion

If two monochromators are used in tandem then the reciprocal linear dispersion of the combination is half that of a single monochromator. For two 77700 Monochromators in series, the reciprocal linear dispersion is 1.6 nm/mm at 500 nm for a 1200 l/mm grating. The bandpass of the combination is half that of the single monochromators for the same slits, except at very narrow slit widths. Practically, for two 77700 Monochromators arranged as a double monochromator, the bandpass is around 0.07 nm with 10 μm slits (1200 l/mm gratings).

Stray Light

The exit beam from a typical monochromator should be monochromatic but usually contains undesired light of other wavelengths. Filters can be used to block higher order diffracted wavelengths. Other unwanted light, called stray light, has several origins and should be minimized. Usually, part of this unwanted light, called re-entrant spectra, is due to diffracted light being directed onto the entrance, collimating or focusing mirror, and back towards the grating. Light can also be reflected off a focal plane detector such as a diode array back into a spectrograph and similarly cause re-entrant spectra. The Ebert-Fastie Out of Plane design used in the 77250 Monochromator has no re-entrant spectra, neither does the 77700 Monochromator. Re-entrant spectra are absorbed on baffles and by tilting the detector in the 77400 Spectrograph. For a major improvement in signal/stray light ratio, use a double monochromator.

Gratings as Dispersive Elements

The wavelength dependence in the grating equation shows that the grating separates an incident polychromatic beam into its constituent wavelength components, i.e., it is dispersive. Each wavelength of input beam spectrum is sent into a different direction, producing a rainbow of colors under white light illumination. This is visually similar to the operation of a prism, although the mechanism is very different.

A light bulb of a flashlight seen through a transmissive grating, showing three diffracted orders. The order $m = 0$ corresponds to a direct transmission of light through the grating. In the first positive order ($m = +1$), colors with increasing wavelengths (from blue to red) are diffracted at increasing angles.

The diffracted beams corresponding to consecutive orders may overlap, depending on the spectral content of the incident beam and the grating density. The higher the spectral order, the greater the overlap into the next order.

The grating equation shows that the angles of the diffracted orders only depend on the grooves' period, and not on their shape. By controlling the cross-sectional profile of the grooves, it is possible to concentrate most of the diffracted energy in a particular order for a given wavelength. A triangular profile is commonly used. This technique is called *blazing*.The incident angle and wavelength for which the diffraction is most efficient are often called *blazing angle* and *blazing wavelength*. The efficiency of a grating may also depend on the polarization of the incident light. Gratings are usually designated by their *groove density*, the number of grooves per unit length, usually expressed in grooves per millimeter (g/mm), also equal to the inverse of the groove period. The groove period must be on the order of the wavelength of interest; the spectral range covered by a grating is dependent on groove spacing and is the same for ruled and holographic gratings with the same grating constant. The maximum wavelength that a grating can diffract is equal to twice the grating period, in which case the incident and diffracted light will be at ninety degrees to the grating normal. To obtain frequency dispersion over a wider frequency one must use a prism. In the optical regime, in which the use of gratings is most common, this corresponds to wavelengths between 100 nm and 10 μm. In that case, the groove density can vary from a few tens of grooves per millimeter, as in *echelle gratings*, to a few thousands of grooves per millimeter.

When groove spacing is less than half the wavelength of light, the only present order is the $m = 0$ order. Gratings with such small periodicity are called subwavelength gratings and exhibit special optical properties. Made on an isotropic material the subwavelength gratings give rise to form birefringence, in which the material behaves as if it were birefringent.

Fabrication

Originally, high-resolution gratings were ruled using high-quality *ruling engines* whose construction was a large undertaking. Henry Joseph Grayson designed a machine to make diffraction gratings, succeeding with one of 120,000 lines to the inch (approx. 47 000 per cm) in 1899. Later, photolithographic techniques allowed gratings to be created from a holographic interference pattern. Holographic gratings have sinusoidal grooves and may not be as efficient as ruled gratings, but are often preferred in monochromators because they lead to much less stray light. A copying technique allows high quality replicas to be made from master gratings, thereby lowering fabrication costs.

Another method for manufacturing diffraction gratings uses a photosensitive gel sandwiched between two substrates. A holographic interference pattern exposes the gel which is later developed. These gratings, called *volume phase holography diffraction gratings* (or VPH diffraction gratings) have no physical grooves, but instead a periodic modulation of the refractive index within the gel. This removes much of the surface scattering effects typically seen in other types of gratings. These gratings also tend to have higher efficiencies, and allow for the inclusion of complicated patterns into a single grating. In older versions of such gratings, environmental susceptibility was a trade-off, as the gel had to be contained at low temperature and humidity. Typically, the photosensitive substances are sealed between two substrates which make them resistant to humidity, thermal and mechanical stresses. VPH diffraction gratings are not destroyed by accidental touches and are more scratch resistant than typical relief gratings.

Semiconductor technology today is also utilized to etch holographically patterned gratings into robust materials as fused silica. In this way, low stray-light holography is combined with the high efficiency of deep, etched transmission gratings, and can be incorporated into high volume, low cost semiconductor manufacturing technology.

A new technology for grating insertion into integrated photonic lightwave circuits is digital planar holography (DPH). DPH gratings are generated in computer and fabricated on one or several interfaces of an optical waveguide planar with standard micro-lithography or nano-imprinting methods, compatible with mass-production. Light propagates inside the DPH gratings, confined by the refractive index gradient, which provides longer interaction path and greater flexibility in light steering.

Examples

The grooves of a compact disc can act as a grating and produce iridescent reflections.

Diffraction gratings are often used in monochromators, spectrometers, wavelength division multiplexing devices, optical pulse compressing devices, and many other optical instruments.

Ordinary pressed CD and DVD media are every-day examples of diffraction gratings and can be used to demonstrate the effect by reflecting sunlight off them onto a white wall. This is a side effect of their manufacture, as one surface of a CD has many small pits in the plastic, arranged within concentric rings; that surface has a thin layer of metal applied to make the pits more visible. The structure of a DVD is optically similar, although it may have more than one pitted surface, and all pitted surfaces are inside the disc.

In a standard pressed vinyl record when viewed from a low angle perpendicular to the grooves, a similar but less defined effect to that seen in a CD/DVD. This is due to viewing angle (less than the critical angle of reflection of the black vinyl) and the path of the light being reflected due to this being changed by the grooves, leaving a rainbow relief pattern behind.

The image sensor of a digital camera has a fine pattern which can produce a diffraction artifact on the image.

Diffraction gratings are also present in nature. For example, the iridescent colors of peacock feathers, mother-of-pearl, butterfly wings, and some other insects are caused by very fine regular structures that diffract light, splitting it into its component colors.

Diffraction-limited System

Every lens has an absolute upper performance limit dictated by the laws of physics. This limitation is controlled by the working f/# of the lens and the wavelength(s) of light that pass through the lens. Known as the Diffraction Limit, this limitation is given in line pairs/mm and determines the theoretical maximum resolving power of the lens. Even a perfect lens that is not limited by design will be diffraction limited. This limit is the point where two Airy patterns are no longer distinguishable from each other. To calculate the diffraction limit, a simple formula that relates it to the f/# of the lens and the wavelength of light can be used.

$$Diffraction\ \ Limit\,(lp\,/\,mm) = \frac{1000\,\mu m\,/\,mm}{f\,/\,\#\,x\,wavelength\,(in\,\mu m)}$$

After the diffraction limit is reached, the lens becomes incapable of resolving greater frequencies. The diffraction limit detailed in table shows contrast at 0% for given frequencies. These numbers may appear rather high, but are strictly theoretical – a number of other practical factors must also be considered. First, as a general rule, imaging sensors cannot reproduce information at or near 0% contrast. Due to inherent noise, contrast generally needs to be above 10% to be reliably detected on standard imaging sensors. To avoid imaging complications, it is recommended to target 20% contrast or higher at the application's critical lp/mm resolution. Additionally, lens aberrations and variations associated with manufacturing tolerances also reduce performance. Modulation Transfer Function (MTF) curves are used to determine whether a lens will effectively utilize a sensors capabilities and fulfill the desired application's requirements.

f/#	0% Contrast Limit (lp/mm) @0.520μm
1.4	1374
2	962
2.8	687
4	481
5.6	343
8	240
11	175
16	120

Table: The diffraction limit calculated at different f/#s for 0.520μm light (green light).

The Abbe Diffraction Limit for a Microscope

The observation of sub-wavelength structures with microscopes is difficult because of the Abbe diffraction limit. Ernst Abbe found in 1873 that light with wavelength λ, traveling in a medium with refractive index n and converging to a spot with half-angle θ will make a spot with diameter

$$d = \frac{\lambda}{2n\sin\theta} = \frac{\lambda}{2\text{NA}}$$

The portion of the denominator $n\sin\theta$ is called the numerical aperture (NA) and can reach about 1.4–1.6 in modern optics, hence the Abbe limit is $d = \lambda / 2.8$. Considering green light around 500 nm and a NA of 1, the Abbe limit is roughly $d = \lambda / 2 = 250\,nm$ (0.25 μm), which is small compared to most biological cells (1 μm to 100 μm), but large compared to viruses (100 nm), proteins (10 nm) and less complex molecules (1 nm). To increase the resolution, shorter wavelengths can be used such as UV and X-ray microscopes. These techniques offer better resolution but are expensive, suffer from lack of contrast in biological samples and may damage the sample.

Implications for Digital Photography

In a digital camera, diffraction effects interact with the effects of the regular pixel grid. The combined effect of the different parts of an optical system is determined by the convolution of the point spread functions (PSF). The point spread function of a diffraction limited lens is simply the Airy disk. The point spread function of the camera, otherwise called the instrument response function (IRF) can be approximated by a rectangle function, with a width equivalent to the pixel pitch. A more complete derivation of the modulation transfer function (derived from the PSF) of image sensors is given by Fliegel. Whatever the exact instrument response function we may note that it is largely independent of the f-number of the lens. Thus at different f-numbers a camera may operate in three different regimes, as follows:

1. in the case where the spread of the IRF is small with respect to the spread of the diffraction PSF, in which case the system may be said to be essentially diffraction limited (so long as the lens itself is diffraction limited).
2. in the case where the spread of the diffraction PSF is small with respect to the IRF, in which case the system is instrument limited.
3. in the case where the spread of the PSF and IRF are of the same order of magnitude, in which case both impact the available resolution of the system.

The spread of the diffraction-limited PSF is approximated by the diameter of the first null of the Airy disk,

$$d/2 = 1.22\lambda N,$$

where λ is the wavelength of the light and N is the f-number of the imaging optics. For f/8 and green (0.5 μm wavelength) light, d = 9.76 μm. This is of the same order of magnitude as the pixel size for the majority of commercially available 'full frame' (43mm sensor diagonal) cameras and so these will operate in regime 3 for f-numbers around 8 (few lenses are close to diffraction limited at f-numbers smaller than 8). Cameras with smaller sensors will tend to have smaller pixels, but their lenses will be designed for use at smaller f-numbers and it is likely that they will also operate in regime 3 for those f-numbers for which their lenses are diffraction limited.

Obtaining Higher Resolution

There are techniques for producing images that appear to have higher resolution than allowed by simple use of diffraction-limited optics. Although these techniques improve some aspect of resolution, they generally come at an enormous increase in cost and complexity. Usually the technique is only appropriate for a small subset of imaging problems, with several general approaches outlined below.

Extending Numerical Aperture

For a given numerical aperture (NA), the resolution of microscopy for flat objects under coherent illumination can be improved using interferometric microscopy. Using the partial images from a holographic recording of the distribution of the complex optical field, the large aperture image can be reconstructed numerically. Another technique, 4 Pi microscopy uses two opposing objectives to double the effective numerical aperture, effectively halving the diffraction limit.

Among sub-diffraction limited techniques, structured illumination holds the distinction of being one of the only methods that can work with simple reflectance without the need for special dyes or fluorescence and at very long working distances. In this method, multiple spatially modulated illumination patterns are used to double the effective numerical aperture. In principle, the technique can be used at any range and on any target provided that illumination can be controlled. Additionally, if exogenous contrast agents are used, the technique can also achieve more than a two-fold increase in resolution.

Near-field Techniques

The diffraction limit is only valid in the far field. Various near-field techniques that operate less than 1 wavelength of light away from the image plane can obtain substantially higher resolution. These techniques exploit the fact that the evanescent field contains information beyond the diffraction limit which can be used to construct very high resolution images, in principle beating the diffraction limit by a factor proportional to how far into the near field an imaging system extends. Techniques such as total internal reflectance microscopy and metamaterials-based superlens can image with resolution better than the diffraction limit by locating the objective lens extremely close (typically hundreds of nanometers) to the object. However, because these techniques cannot image beyond 1 wavelength, they cannot be used to image into objects thicker than 1 wavelength which limits their applicability.

Far-field Techniques

Far-field imaging techniques are most desirable for imaging objects that are large compared to the illumination wavelength but that contain fine structure. This includes nearly all biological applications in which cells span multiple wavelengths but contain structure down to molecular scales. In recent years several techniques have shown that sub-diffraction limited imaging is possible over macroscopic distances. These techniques usually exploit optical nonlinearity in a material's reflected light to generate resolution beyond the diffraction limit.

Among these techniques, the STED microscope has been one of the most successful. In STED, multiple laser beams are used to first excite, and then quench fluorescent dyes. The nonlinear response to illumination caused by the quenching process in which adding more light causes the image to become less bright generates sub-diffraction limited information about the location of dye molecules, allowing resolution far beyond the diffraction limit provided high illumination intensities are used.

Laser Beams

The limits on focusing or collimating a laser beam are very similar to the limits on imaging with a microscope or telescope. The only difference is that laser beams are typically soft-edged beams. This non-uniformity in light distribution leads to a coefficient slightly different from the 1.22 value familiar in imaging. But the scaling is exactly the same.

The beam quality of a laser beam is characterized by how well its propagation matches an ideal Gaussian beam at the same wavelength. The beam quality factor M squared (M^2) is found by measuring the size of the beam at its waist, and its divergence far from the waist, and taking the product of the two, known as the beam parameter product. The ratio of this measured beam parameter product to that of the ideal is defined as M^2, so that $M^2=1$ describes an ideal beam. The M^2 value of a beam is conserved when it is transformed by diffraction-limited optics.

The outputs of many low and moderately powered lasers have M^2 values of 1.2 or less, and are essentially diffraction-limited.

Other Waves

The same equations apply to other wave-based sensors, such as radar and the human ear.

As opposed to light waves (i.e., photons), massive particles have a different relationship between their quantum mechanical wavelength and their energy. This relationship indicates that the effective "de Broglie" wavelength is inversely proportional to the momentum of the particle. For example, an electron at an energy of 10 keV has a wavelength of 0.01 nm, allowing the electron microscope (SEM or TEM) to achieve high resolution images. Other massive particles such as helium, neon, and gallium ions have been used to produce images at resolutions beyond what can be attained with visible light. Such instruments provide nanometer scale imaging, analysis and fabrication capabilities at the expense of system complexity.

Particle Diffraction

Diffraction of Particles is the scattering of microparticles (for example, electrons, neutrons, and atoms) by the crystals or molecules of liquids and gases, during which there arise from the initial beam of particles additional deflected beams of these particles; the direction and intensity of such deflected beams depend on the structure of the scattering object.

The diffraction of particles can be understood only on the basis of quantum theory. Diffraction is a wave phenomenon; it isobserved in the propagation of a variety of waves: the diffraction of light, sound waves, waves on the surface of a liquid, andso on. From the standpoint of classical physics, diffraction is impossible in the scattering of particles. Quantum mechanicshas eliminated the absolute boundary between the wave and the particle. Particle-dualism, that is, the dual nature ofmicroparticles, is the principal proposition of quantum mechanics, which describes the behavior of microentities. Thus, thebehavior of electrons in some phenomena (such as in the observation of the motion of electrons in a Wilson cloud chamberor in the measurement of electric charge in the photoelectric effect) can be described on the basis of the particle concept,while in other phenomena (especially in diffraction phenomena) it can be described only on the basis of the wave concept.The idea of "waves of matter" was advanced by the French physicist L. de Broglie in 1924 and soon received brilliantconfirmation in experiments on the diffraction of particles.

According to quantum mechanics, the free motion of a particle with mass m and momentum p = mv (where v is the velocity ofthe particle) can be represented as a planar monochromatic wave ψ_0 (a de Broglie wave) with a wavelength

$$\lambda = h / p$$

that propagates in the same direction (for example, in the direction of the x = axis) as that in which the particle is moving(Figure). Here h is Planck's constant. The relationship of ψ_0 to the coordinate x is given by the formula

$$\psi \sim cos(k_0 x)$$

where $k_0 = | k_0 | = 2\pi/\lambda$, the so-called wave number, while the wave vector

$$k_0 = \frac{2\pi}{h} p$$

is directed toward the propagation side of the wave or along the direction of motion of the particle.

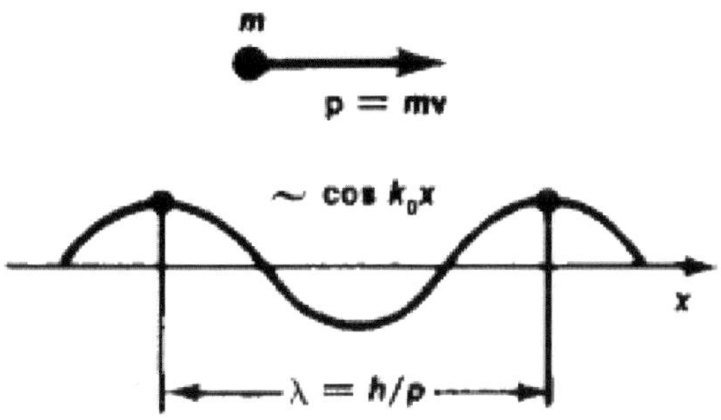

Comparison of a wave and a freelymoving particle. Shown at top is the linear motion ofa particle with mass m and momentum p = mv,where v is the velocity of the particle; below, thepropagation of the corresponding matter wave $\psi \sim \cos k_0 x$ with wavelength " = h/p.

Thus, the wave vector of a monochromatic wave connected with a freely moving microparticle is proportional to themomentum and inversely proportional to the wavelength.

Since the kinetic energy of a particle that is moving comparatively slowly is S = mv², the wave-length can also be expressedin terms of the energy:

$$\lambda = \frac{h}{\sqrt{2m\delta}}$$

When a particle interacts with some object, such as a crystal or molecule, its energy changes: the potential energy of thisinteraction is added to it, leading to a change in the particle's motion. The character of the propagation of the waveassociated with the particle changes accordingly; this takes place according to the principles common to all wavephenomena. Therefore, the main geometric principles of the diffraction of particles differ in no way from the principlesgoverning the diffraction of any waves. A general condition for the diffraction of waves of any nature is the commensurabilityof the length of the incident wave λ with the distance d between the scattering centers: $\lambda \leq d$.

The first experiment on the diffraction of particles, which brilliantly confirmed the original concept of quantum mechanics—that is, particle-wave dualism—was that of the American physicists K. Davisson and L. Germer on the diffraction ofelectrons in monocrystals of nickel (figure). If electrons are accelerated by an electrical field of intensity V, then they willacquire the kinetic energy $S = eV$ (e being the electron charge). After the numerical values are inserted in equation.

this gives

$$\lambda = 12.26 / \sqrt{V}$$

Here V is expressed in volts (V) and λ is expressed in angstroms (1 υ = 10^{-8} cm). At voltages V on the order of 100 V (thequantity used in these experiments), so-called slow electrons with a λ on the order of 1 υ are obtained. This quantity is closeto the interatomic distances d in crystals (several angstroms or less), and the ratio $\lambda \leq d$, which is necessary for diffraction to occur, is satisfied.

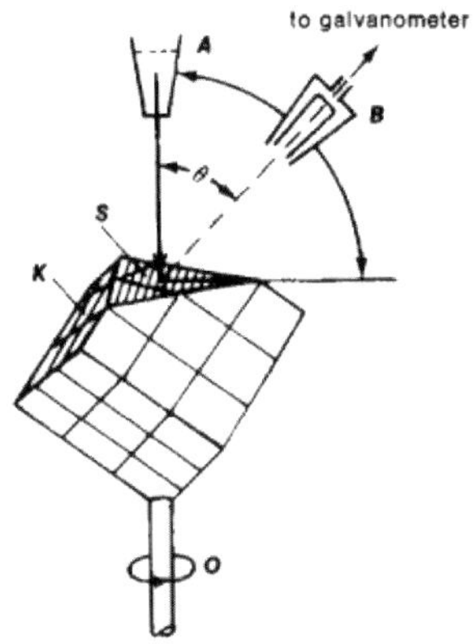

Diagram of the Davisson-Germerexperiment. (*K*) monocrystal of nickel, (*A*)electron source, (*B*) electron detector, (*θ*) angleof deflection of the electron beams. Theelectron beam is perpendicularly incident to thepolished crystal plane *S*. When the crystal isrotated around the axis *O*, a galvanometerattached to detector *B* shows periodic maxima.

Crystals are highly regular. The atoms within them are arranged in a three-dimensional periodic crystal lattice, that is, theyform a spatial diffraction grating for the appropriate wavelengths. The diffraction of waves in such a grating takes place as aresult of scattering in systems of parallel crystal planes, on which the scattering centers are arranged in strict order. TheBragg-Vul'f condition is the necessary condition for observation of the diffraction maximum on reflection from a crystal:

$$2d \sin\theta = n\lambda$$

where θ is the angle at which the electron beam is incident to the given crystal plane (the slip angle) and d is the distancebetween the corresponding crystal planes.

In the Davisson and Germer experiment, maxima occurred when the electrons were "reflected" from the surface of the nickelcrystal at specific angles of reflection (Figure). These maxima of the reflected electron beams corresponded to formula and their appearance could be explained in no way other than on the basis of the concepts of waves and wave diffraction;thus, the wave properties of particles—electrons—were proved by experiment.

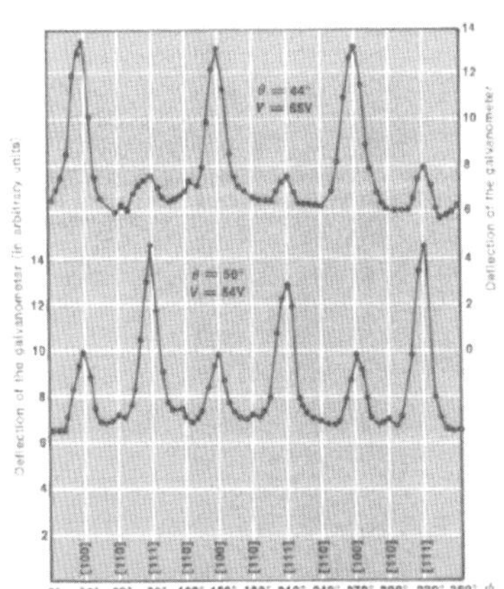

Diffraction maxima in the Davisson-Germer experiment on electrondiffraction at various angles of rotation of a crystal, φ, for two values of the angle ofdeflection of the electrons, *θ*, and two accelerating voltages *V*. The maximacorrespond to reflection from the various crystal planes, whose indexes areindicated in brackets.

At higher accelerating voltages (dozens of kilovolts) electrons acquire sufficient kinetic energy to penetrate fine films ofmatter (with a thickness on the order of 10^{-5} cm, that is, thousands of υ). There then occurs the so-called transmissiondiffraction of fast electrons which the English scientist J. J. Thomson and the Soviet physicist P. S. Tartakovskii firstinvestigated in polycrystalline films of aluminum and gold.

Soon afterward it also became possible to observe the diffraction phenomena of atoms and molecules. According to formula, the following wavelength corresponds to atoms with mass *M* in gaseous state in a vessel at absolute temperature *T*:

$$\lambda = \frac{h}{\sqrt{3MKT}}$$

where *k* is the Boltzmann constant (since the average kinetic energy of an atom g = $3/2kT$). For light atoms and molecules(H, H_2, He) and temperatures of hundreds of degrees Kelvin the wavelength λ is also about 1 υ. For the most part, diffractingatoms or molecules do not penetrate to the depth of the crystal; therefore, it may be assumed that their diffraction occursthrough scattering from the surface of the crystal— that is, as if on a flat diffraction grating.

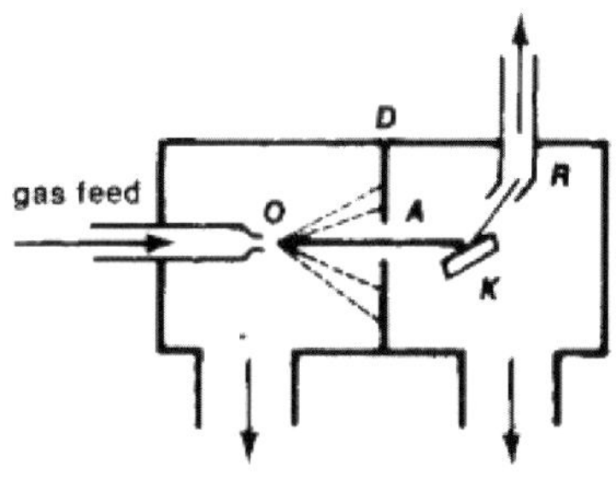

Schematic diagram of researchapparatus in the diffraction of atomic ormolecular beams. (*A*) atomic or molecularbeam, (*K*) crystal, (*O*) gas supply tube, (*D*)diaphragm, (*R*) receiver connected to amanometer. The manometer measures thepressure generated by the diffracted beam.

A molecular or atomic beam emitted from a vessel and formed by means of diaphragms is directed toward the crystal, andthe “reflected” diffraction beams are in some way fixed (Figure). In this manner, the German scientists O. Stern and I.Estermann and other researchers in the early 1930’s observed the diffraction of atomic and molecular beams (Figure).

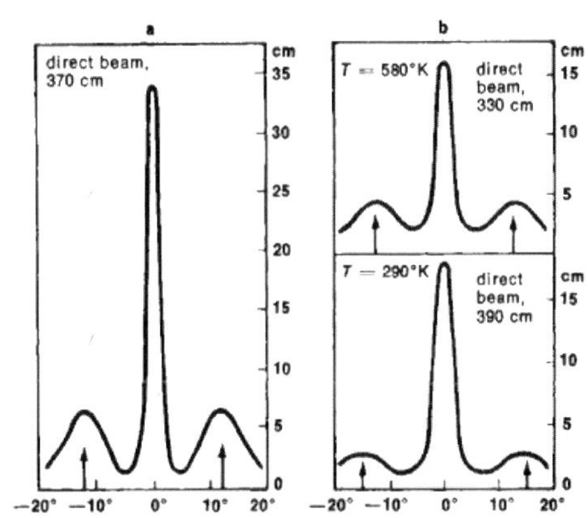

Lithium fluoride crystal diffraction of helium atoms (a) and hydrogenmolecules, at two values of absolute temperature, T (b). The angle ofdiffraction θ is plotted along the abscissa; the intensity of the diffracted beams(in cm of deflection of the needle of the measuring instrument), along theordinate. In addition to a peak at $\theta = 0°$ (from the mirror reflection of the initialbeam), two side peaks can be observed. At $T = 580°K$ the side peaks aresomewhat closer to the central peak than at $T = 290°K$. This corresponds tothe decline in wavelength λ as the temperature rises.

The diffraction of protons and neutrons was observed later and came into wide use as a method of investigating the structureof matter.

Thus it was proven experimentally that wave properties are inherent in all microparticles, without exception.

In the broad sense of the word, diffraction scattering always occurs in the elastic scattering of various elementary particles byatoms and atomic nuclei, as well as by each other. On the other hand, the concept of the particle wave dualism of matter wasstrengthened by the analysis of phenomena that had always been considered typically wave phenomena, such as thediffraction of X rays—short electromagnetic waves with a wavelength $\lambda \approx 0.5$ -5υ. At the same time, the initial and scattered X-ray beams may be considered and recorded as a flux of particles, or photons; the number of photons of X-radiation in thesebeams can be determined by using photon counters.

It should be emphasized that wave properties are inherent in each particle individually. This has been confirmed by anexperiment conducted by V. A. Fabrikant on the diffraction of the electrons that successively pass through a sample.A standard diffraction pattern gradually arose as a certain amount of time elapsed. This meant that each of the electronsconformed to all laws of wave optics, and that the diffraction effect is a result of the interaction between the de Broglie waveof each electron and the volume of the entire crystal. The initial wave ψ_0 [see formula], which describes the motion of theinitial electron, is converted to the scattered wave ψ on passing through the crystal.

The formation of a diffraction pattern upon the scattering of particles is interpreted in the following manner in quantummechanics. An electron that passes through a crystal is deflected from its original motion as a result of interaction with thecrystal lattice of the sample and strikes some point of a photographic plate mounted beyond the crystal in order to recordelectrons. Upon entering the photographic emulsion the electron is observed as a particle and induces a photochemicalreaction. At first glance the incidence of the electron at a given point of the plate would seem to be totally arbitrary. But onprolonged exposure there gradually arises an ordered pattern of diffraction maxima and minima in the distribution of theelectrons that have passed through the crystal.

It is impossible to predict precisely at what point of the photographic plate a given electron will strike, but the probability of itsincidence at some point of the plate after scattering can be indicated. This probability is determined by the wave function ofthe electron, ψ, or, more accurately, by the square of its modulus (since ψ is a complex function), $|\Psi|^2$. However, whenthere are large numbers of trials this probability is realized as a certainty, and when an electron passes through a crystalrepeatedly (or, as is the case in actual diffraction experiments, when an electron beam containing a tremendous number ofparticles passes through the sample) the quantity $|\Psi|^2$ now determines the intensity distribution in the diffracted beams.Thus, the resultant wave function of the electron, ψ, which can be calculated if we know and the potential energy ofinteraction between the electron and the crystal, gives a complete description of the diffraction experiment in a statistical sense.

Because of the common character of the geometric principles of diffraction, the theory of the diffraction of particles borroweda great deal from the previously developed theory of X-ray diffraction. However, the interaction of various types of particles(for example, electrons, neutrons, and atoms) with matter has a varying physical nature. Therefore, in considering thediffraction of particles in crystals, liquids, and elsewhere, it is essential to know how different particles are scattered by anisolated atom of a substance. It is precisely in the scattering of particles by individual atoms that the specifics of thediffraction of various particles become apparent. For example, the scattering of electrons is determined by the interaction ofthe electric charge with the electrostatic potential of the atoms $\varphi(r)$ (where r is the distance from the atom), which is the sumof the potential of the positively charged nucleus and the potential of the atom's electron shell; the potential energy of thisinteraction is $U = e\varphi(r)$. Neutron scattering is determined by the potential of their strong interaction with the atomic nucleus,as well as by the interaction of the magnetic moment of the neutron with the magnetic moment of the atom (the magneticscattering of neutrons).

Quantitatively, the scattering power of an atom is characterized by a quantity called the atomic scattering amplitude, $f(\theta)$,where θ is the scattering angle and is determined by the potential energy of the interaction of particles of a given type withatoms of the scattering substance. The scattering intensity of the particles is proportional to $f^2(\theta)$.

If the atomic amplitude is known and if we know the relative position of the scattering centers—the atoms of the substance inthe sample (that is, if we know the structure of the scattering sample)—then we can draw up a general diffraction pattern(which is formed as a result of the interference of secondary waves originating from the scattering centers).

Theoretical calculations, confirmed by experimental measurements, show that the atomic scattering amplitude of electrons $f_e$is maximal when θ = 0, and declines as θ increases. The quantity f_e also depends on the nuclear charge (atomic number) $Z$and on the structure of the electron shells of the atom, increasing with Z (on the average) approximately as $z^{1/3}$ for low vaiuesof θ and as $z^{2/3}$ for large values of θ, but exhibiting oscillations associated with the periodic character of the filling of theelectron shells.

The atomic scattering amplitude of neutrons f_n for thermal neutrons (neutrons with an energy of hundredths of an eV) doesnot depend on the scattering angle—that is, the scattering of such neutrons by the nucleus is identical in all directions (it hasspherical symmetry). This is due to the fact that the atomic nucleus, with a radius on the order of 10^{-13} cm, is a "point" for (itssize is infinitesimal in comparison to) thermal neutrons, whose wavelength is 10^{-8} cm. Moreover, there is no cleardependence in neutron scattering on the nuclear charge Z. Because of the presence in some nuclei of so-called resonancelevels, with an energy close to the energy of thermal neutrons, the f_n of such nuclei are negative.

An atom scatters electrons much more strongly than X rays and neutrons: the absolute values of the electron scatteringamplitude f_e are on the order 10^{-8} cm; for X rays, $f_x \sim 10^{-11}\ cm$; and for neutrons, $f_n \sim 10^{-12}$ cm. Since the scatteringintensity is proportional to the square of the scattering amplitude, electrons interact with matter (are scattered) approximatelya million times more strongly than X rays (and especially than neutrons). Therefore, fine films, 10^{-6}–10^{-5} cm thick, usuallyserve as samples for the observation of electron diffraction, while samples several mm thick must be used to observe thediffraction of X rays and neutrons.

The diffraction in any atomic system, such as a molecule or crystal, can be calculated if the coordinates of their centers $r_1$and the atomic amplitude f_i are known for the given type of particle.

The effects of the diffraction of particles show up most clearly in crystal diffraction. However, the thermal motion of atoms in acrystal somewhat changes the conditions of diffraction, and the intensity of the diffracted beams declines as the angle θ. In the liquid diffraction of particles, or in diffraction by amorphous bodies or gas molecules, where thedegree of order is far less than in crystals, several broad diffraction maxima are usually observed.

The diffraction of particles, which at one time played such an important role in establishing the dual nature of matter (particle-wave dualism) and which thereby served as the experimental substantiation of quantum mechanics, has long been one ofthe principal working methods in studying the structure of matter. Two important modern methods of analyzing the atomicstructure of matter—electronography and neutronography—are based on the diffraction of particles.

Bragg Law and Diffraction

Bragg's Law was introduced by Sir W.H. Bragg and his son Sir W.L. Bragg. The law states that when the x-ray is incident onto a crystal surface, its angle of incidence, θ, will reflect back with a same angle of scattering, θ. And, when the path difference, dd is equal to a whole number, n, of wavelength, a constructive interference will occur.

Consider a single crystal with aligned planes of lattice points separated by a distance *d*. Monochromatic X-rays A, B, and C are incident upon the crystal at an angle θ. They reflect off atoms X, Y, or Z.

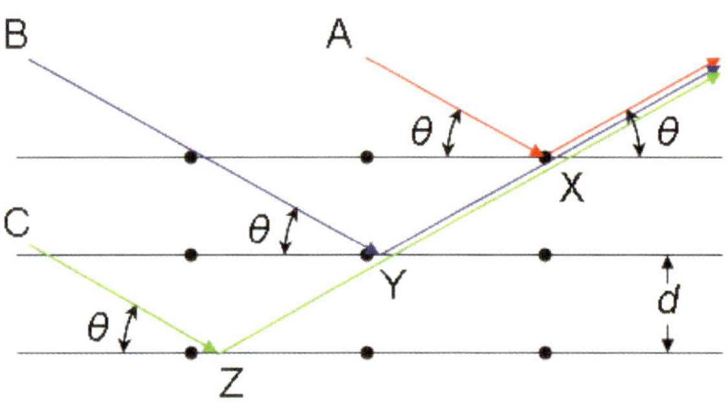

The path difference between the ray reflected at atom X and the ray reflected at atom Y can be seen to be 2YX. From the Law of Sines we can express this distance YX in terms of the lattice distance and the X-ray incident angle:

If the path difference is equal to an integer multiple of the wavelength, then X-rays A and B (and by extension C) will arrive at atom X in the same phase. In other words, given the following conditions:

then the scattered radiation will undergo constructive interference and thus the crystal will appear to have reflected the X-radiation. If, however, this condition is not satisfied, then destructive interference will occur.

Bragg's Law

$$n\lambda = 2d \sin \theta$$

where:

- λ is the wavelength of the x-ray,
- d is the spacing of the crystal layers (path difference),
- θ is the incident angle (the angle between incident ray and the scatter plane), and
- n is an integer

The principle of Bragg's law is applied in the construction of instruments such as Bragg spectrometer, which is often used to study the structure of crystals and molecules.

Deriving Bragg's Law

Bragg's Law can easily be derived by considering the conditions necessary to make the phases of the beams coincide when the incident angle equals and reflecting angle. The rays of the incident beam are always in phase and parallel up to the point at which the top beam strikes the top layer at atom z (Fig.). The second beam continues to the next layer where it is scattered by atom B. The second beam must travel the extra distance AB + BC if the two beams are to continue traveling adjacent and parallel. This extra distance must be an integral (n) multiple of the wavelength (λ) for the phases of the two beams to be the same:

(eq.) $n\lambda = AB + BC$.

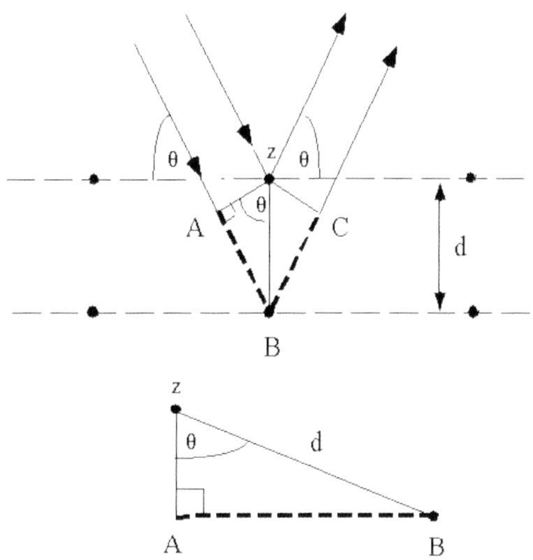

Figure: Deriving Bragg's Law using the reflection geometry and applying trigonometry. The lower beam must travel the extra distance $(AB + BC)$ to continue traveling parallel and adjacent to the top beam.

Recognizing d as the hypotenuse of the right triangle Abz, we can use trigonometry to relate d and θ to the distance $(AB + BC)$. The distance AB is opposite θ so,

$$AB = d \sin\theta .$$

Because AB = BC eq. $n\lambda = AB + BC$ becomes,

$$n\lambda = 2AB$$

Substituting eq. $AB = d \sin\theta$ in eq. $n\lambda = 2AB$ we have,

$$n\lambda = 2d \sin\theta$$

and Bragg's Law has been derived. The location of the surface does not change the derivation of Bragg's Law.

Heuristic Derivation

Suppose that a single monochromatic wave (of any type) is incident on aligned planes of lattice points, with separation d, at angle θ. Points A and C are on one plane, and B is on the plane below. Points ABCC' form a quadrilateral.

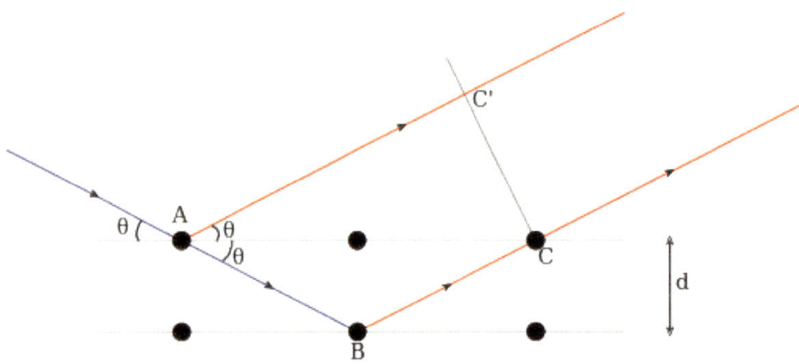

There will be a path difference between the ray that gets reflected along AC' and the ray that gets transmitted, then reflected, along AB and BC respectively. This path difference is

$$(AB + BC) - (AC').$$

The two separate waves will arrive at a point with the same phase, and hence undergo constructive interference, if and only if this path difference is equal to any integer value of the wavelength, i.e.

$$(AB + BC) - (AC') = n\lambda,$$

where the same definition of n and λ apply as above.

Therefore,

$$AB = BC = \frac{d}{\sin\theta} \text{ and } AC = \frac{2d}{\tan\theta},$$

from which it follows that

$$AC' = AC \cdot \cos\theta = \frac{2d}{\tan\theta}\cos\theta = \left(\frac{2d}{\sin\theta}\cos\theta\right)\cos\theta = \frac{2d}{\sin\theta}\cos^2\theta.$$

Putting everything together,

$$n\lambda = \frac{2d}{\sin\theta} - \frac{2d}{\tan\theta}\cos\theta = \frac{2d}{\sin\theta}(1-\cos^2\theta) = \frac{2d}{\sin\theta}\sin^2\theta$$

which simplifies to $n\lambda = 2d\sin\theta$, which is Bragg's law.

If only two planes of atoms were diffracting, as shown in the pictures, then the transition from constructive to destructive interference would be gradual as a function of angle, with gentle maxima at the Bragg angles. However, since many atomic planes are interfering in real materials, very sharp peaks surrounded by mostly destructive interference result.

A rigorous derivation from the more general Laue equations is available.

Bragg Scattering of Visible Light by Colloids

A colloidal crystal is a highly ordered array of particles that forms over a long range (from a few millimeters to one centimeter in length); colloidal crystals have appearance and properties roughly analogous to their atomic or molecular counterparts. It has been known for many years that, due to repulsive Coulombic interactions, electrically charged macromolecules in an aqueous environment can exhibit long-range crystal-like correlations, with interparticle separation distances often being considerably greater than the individual particle diameter. Periodic arrays of spherical particles give rise to interstitial voids (the spaces between the particles), which act as a natural diffraction grating for visible light waves, when the interstitial spacing is of the same order of magnitude as the incident light wave. In these cases in nature, brilliant iridescence (or play of colours) is attributed to the diffraction and constructive interference of visible light waves according to Bragg's law, in a matter analogous to the scattering of X-rays in crystalline solid. The effects occur at visible wavelengths because the separation parameter *d* is much larger than for true crystals.

Applications of Bragg's Law

In X-ray diffraction (XRD) the interplanar spacing (d-spacing) of a crystal is used for identification and characterization purposes. In this case, the wavelength of the incident X-ray is known and measurement is made of the incident angle (Θ) at which constructive interference occurs. Solving Bragg's Equation gives the d-spacing between the crystal lattice planes of atoms that produce the constructive interference. A given unknown crystal is expected to have many rational planes of atoms in its structure; therefore, the collection of "reflections" of all the planes can be used to uniquely identify an unknown crystal. In general, crystals with high symmetry (e.g. isometric system) tend to have relatively few atomic planes, whereas crystals with low symmetry (in the triclinic or monoclinic systems) tend to have a large number of possible atomic planes in their structures.

In the case of wavelength dispersive spectrometry (WDS) or X-ray fluorescence spectroscopy (XRF), crystals of known d-spacings are used as analyzing crystals in the spectrometer. Because the position of the sample and the detector is fixed in these applications, the angular position of the reflecting crystal is changed in accordance with Bragg's Law so that a particular wavelength of interest can be directed to a detector for quantitative analysis. Every element in the Periodic Table has a discrete energy difference between the orbital "shells" (e.g. K, L, M), such that every element will produce X-rays of a fixed wavelength. Therefore, by using a spectrometer crystal (with fixed d-spacing of the crystal) and positioning the crystal at a unique and fixed angle (Θ), it is possible to detect and quantify elements of interest based on the characteristic X-ray wavelengths produced by each element.

References

- Halliday, David; Resnick, Robert; Walker, Jerl (2005), Fundamental of Physics (7th ed.), USA: John Wiley and Sons, Inc., ISBN 978-0-471-23231-5
- Interference and diffraction, introchem: courses.lumenlearning.com, Retrieved 11 March 2018
- Thomas Young (1804-01-01). "The Bakerian Lecture: Experiments and calculations relative to physical optics". Philosophical Transactions of the Royal Society of London. 94: 1–16. doi:10.1098/rstl.1804.0001.. (Note: This lecture was presented before the Royal Society on 24 November 1803)
- Diffraction-grating, Theory of operation: newworldencyclopedia.org, Retrieved 11 July 2018
- Arumugam, Nadia. "Food Explainer: Why Is Some Deli Meat Iridescent?". Slate. The Slate Group. Archived from the original on 10 September 2013. Retrieved 9 September2013
- Grating-physics: newport.com, Retrieved 09 May 2018
- Hiltner, PA; IM Krieger (1969). "Diffraction of Light by Ordered Suspensions". Journal of Physical Chemistry. 73: 2306. doi:10.1021/j100727a049
- Diffraction-grating, technology: britannica.com, Retrieved 30 May 2018
- Ayahiko Ichimiya; Philip I. Cohen (13 December 2004). Reflection High-Energy Electron Diffraction. Cambridge University Press. ISBN 978-0-521-45373-8. Archived from the original on 16 July 2017

5

Polarization

Polarization is a property characteristic of transverse waves that helps to specify the orientation of oscillations. It is a complex topic that requires understanding of Jones calculus, polarizers, Brewster's angle and birefringence, which have been extensively dealt within this chapter.

Polarization is a property that is common to all types of vector waves. In classical physics, light is modeled as a sinusoidal electromagnetic wave in which an oscillating electric field and an oscillating magnetic field propagate through space. Since the magnetic field is always perpendicular to the electric field, we usually sketch just the electric field when visualizing the optical wave's oscillations. Polarization is defined in terms of the pattern traced out in the transverse plane by the electric field vector as a function of time.

Light is called natural or unpolarized if its plane of polarization fluctuates randomly around the direction of light beam propagation, so that, on average, no direction is favored. For example, most naturally produced light (sunlight, firelight) is unpolarized. In any other case, the light beam can be considered to consist of partially polarized or fully polarized light.

The polarization of a light beam can be represented by its electric field vector. Its optical power is a scalar quantity that is proportional to the mean square of the electric field amplitude.

Let z be the direction of propagation. Then, the polarization vector is in the x-y plane. By superposition of the x- and y-vector components at time t at any location, the polarization vector can be expressed:

$$\vec{E}(t) = \vec{E_x}(t) + \vec{E}_y(t)$$
$$= E_x \cos\left(\overline{\omega} t + \delta_x\right)\vec{e}_x + E_y \cos\left(\overline{\omega} t + \delta_y\right)\vec{e}_y$$

where E_x and E_y are the amplitudes of the x and y components of the electric field; e x, e y are the unit vectors of the x-y orthogonal reference system; $\overline{\omega}$ is the angular frequency; and δ_x and δ_y are the phases of the electric field in the x and y directions, respectively. The phase difference is $\delta = \delta_x - \delta_y$.

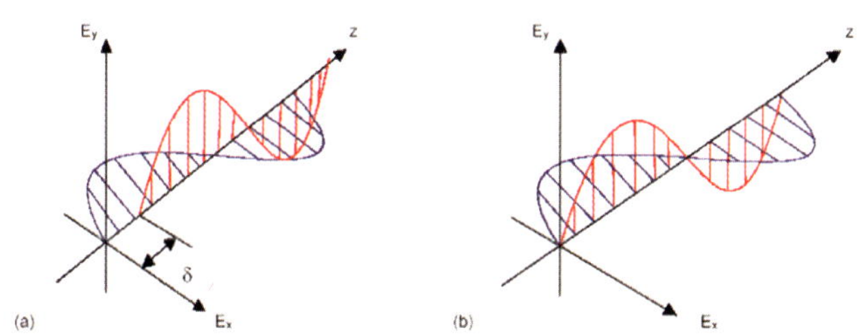

Figure: Spatial representation of x- and y- electric field components over time.
(a) with a phase difference, (b) no phase difference.

Figure shows electric field vectors in the x-y plane as a function of time. Figure shows the two orthogonal electric field components with a relative phase difference of δ. In Figure there is no phase difference between the two orthogonal components.

Polarized light may be classified into three groups by the values of δ, Ex , and Ey : linearly polarized, circularly polarized and elliptically polarized light. Linearly polarized light has no phase difference between the x and y electric field components (δ=0). Circularly polarized light has a fixed phase difference of 90 degrees, and the amplitudes of the two electric field components are the same (ie. δ = 90° and Ex = Ey). The polarization state is elliptical for all other values of δ, Ex, and Ey.

Poincaré Sphere and Stokes Vectors

The Poincaré sphere is used to describe the polarization and changes in polarization of a propagating electromagnetic wave. It provides a convenient way of representing polarized light, and of predicting how any given retarder will change the polarization form. Any given polarization state corresponds to a unique point on the sphere. The two poles of the sphere represent left and right-hand circularly polarized light. Points on the equator indicate linear polarizations. All other points on the sphere represent elliptical polarization states. An arbitrarily chosen point H on the equator designates horizontal linear polarization, and the diametrically opposite point V designates vertical linear polarization.

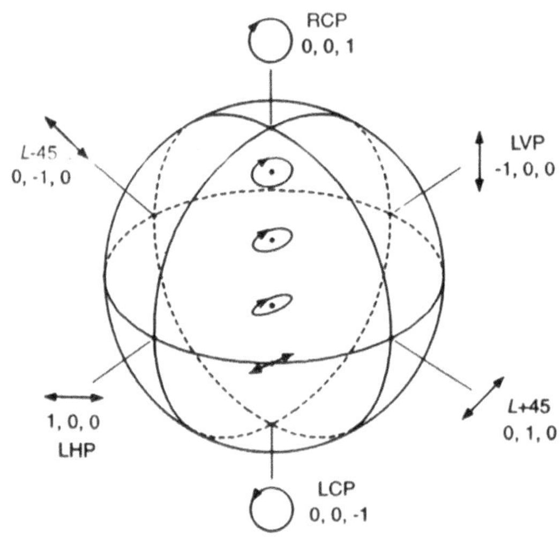

Figure: Poincaré sphere representation of polarization states

The Stokes parameters have a simple physical interpretation related to Poincaré sphere representation. They also have a physical interpretation related to intensity measurement.

In a given Oxyz reference system (Oz being the direction of propagation of light), the Stokes vector can be expressed as:

$$S = \begin{bmatrix} S_0 \\ S_1 \\ S_2 \\ S_3 \end{bmatrix} = \begin{bmatrix} I_0 \\ I_x - I_y \\ I_{+45} - I_{-45} \\ I_L - I_R \end{bmatrix}$$

where I_0 is the total power of the light, I_x, I_y are the x- and y-linearly polarized light intensities, I_{+45}, I_{-45} are the 45° and -45° linearly polarized light intensities, and I_L, I_R are the left and right hand circularly polarized light intensities.

If we normalize [S_1 S_2 S_3] by S_0, we will get the SOP [s1 s2 s3], which are the coordinates of one point on the Poincaré sphere.

The degree of polarization (DOP) is defined by:

$$DOP = \sqrt{s_1^2 + s_2^2 + s_3^2}$$

with a value between 0 and 1.

Whenever DOP=0, light is said to be unpolarized, and whenever DOP=1, it is totally polarized. Intermediate cases correspond to partially polarized light.

DOP can also be expressed as:

$$DOP = \frac{I_{pol}}{I_{pol} + I_{unp}}$$

where I_{pol}, I_{unp} are the intensity of polarized light and unpolarized light, respectively.

What is Birefringence?

In a homogenous or isotropic medium, light propagates at the same speed regardless of polarization state. This speed, v, can be expressed as:

$$\mathbf{v} = c/n$$

$$\vec{E}(z,t) = \begin{bmatrix} e_x \\ e_y \\ 0 \end{bmatrix} e^{i2\pi(z/(\lambda/n)-t/T)} = \begin{bmatrix} e_x \\ e_y \\ 0 \end{bmatrix} e^{i(kz-\omega t)}$$

$$\vec{H}(z,t) = \begin{bmatrix} h_x \\ h_y \\ 0 \end{bmatrix} e^{i2\pi(z/(\lambda/n)-t/T)} = \begin{bmatrix} h_x \\ h_y \\ 0 \end{bmatrix} e^{i(kz-\omega t)}$$

where c is the speed of light in a vacuum, and n is the index of refraction of the medium.

If the light is propagating in an anisotropic medium, the transmission speeds of light polarized along different directions are different. One may use two special orthogonal axes called principal axes to express the transmission properties of the medium.

Usually, this can be expressed through different indices of refraction, such as the ordinary index (n_o) and extraordinary index (n_e). The difference between no and ne is called birefringence. Birefringence is also called double refraction.

Therefore, the index of refraction is a measure of how fast light transmits through a medium. If a medium is birefringent, plane-polarized light launched into one principal axis travels through the medium faster than plane-polarized light in the other principal axis. As a result, the two polarization components of a light signal projected along the two axes will experience a relative phase difference as they propagate through the material.

If an optical pulse is launched into a fiber made of a birefringent material with its polarization aligned with either of the fiber's principal axes, then the pulse transmits through the fiber and emerges in the same state of polarization in which it was launched. However, if we launch the pulse with randomly oriented polarization, then part of the pulse transmits along one of the principal axes, and part of the pulse transmits along the other principal axis. Because of the medium's birefringence, the two parts of the pulse travel at different speeds. Consequently, the pulse becomes broadened because the two parts exit the fiber at slightly different times, and the state of polarization changes.

PMD

PMD stands for polarization mode dispersion. It occurs when different polarization components of light inside a fiber travel at slightly different speeds, as shown in figure. PMD is caused directly by the birefringence of the optical fiber. The major causes of the birefringence are the asymmetry of the fiber-optic strand, where the fiber core is slightly oval or out-of-round, and mechanical stresses on the fiber. The asymmetry may be inherent in the fiber from the manufacturing process, or it may be a result of mechanical stress on the deployed fiber. The inherent asymmetries of the fiber are fairly constant over time, while mechanical stress due to movement or temperature change of the fiber can vary, resulting in a dynamic aspect to PMD. 1st order PMD is called differential group delay (DGD).

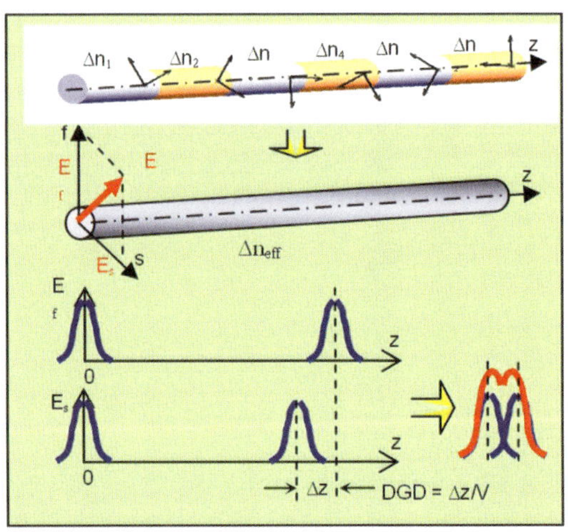

Figure: A real fiber is like many retardation plates with different orientations and birefringences in series. It is equivalent to a single retardation plate having a slow and fast axis, and an effective birefringence. An optical pulse broadens because the two polarization components travel with different speeds.

Because of its dynamic properties, PMD does not have a single, fixed value for a given section of fiber. Rather, it is described in terms of average DGD, and a fiber has a distribution of DGD values over time. The probability of a DGD with a certain value at any particular time follows the Maxwellian distribution shown in figure. As an approximation, the maximum instantaneous DGD is about 3.2 times the average DGD of a fiber.

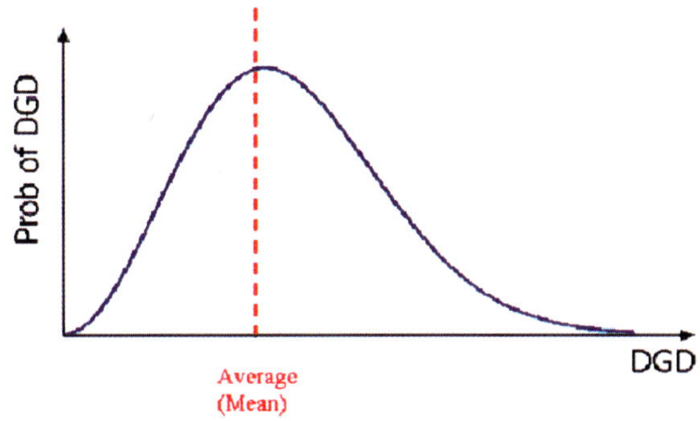

Figure: Maxwellian distribution of DGDS

Wave Propagation and Polarization

Most sources of light are classified as incoherent and unpolarized (or only "partially polarized") because they consist of a random mixture of waves having different spatial characteristics, frequencies (wavelengths), phases, and polarization states. However, for understanding electromagnetic waves and polarization in particular, it is easiest to just consider coherent plane waves; these are sinusoidal waves of one particular direction (or wavevector), frequency, phase, and polarization state. Characterizing an optical system in relation to a plane wave with those given parameters can then be used to predict its response to a more general case, since a wave with any specified spatial structure can be decomposed into a combination of plane waves (its so-called angular spectrum). And incoherent states can be modeled stochastically as a weighted combination of such uncorrelated waves with some distribution of frequencies (its *spectrum*), phases, and polarizations.

Transverse Electromagnetic Waves

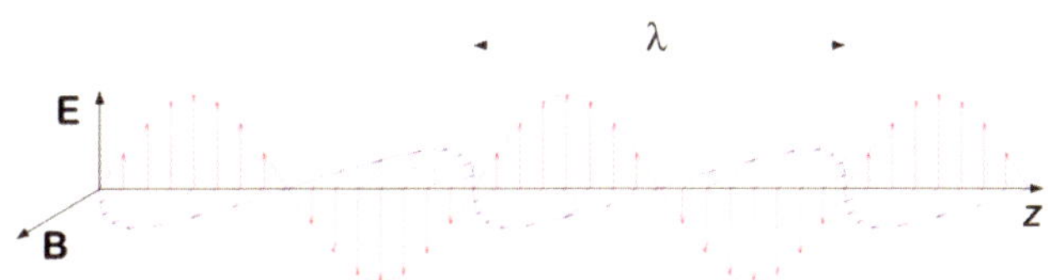

A "vertically polarized" electromagnetic wave of wavelength λ has its electric field vector E (red) oscillating in the vertical direction. The magnetic field B (or H) is always at right angles to it (blue), and both are perpendicular to the direction of propagation (z).

Electromagnetic waves (such as light), traveling in free space or another homogeneous isotropic non-attenuating medium, are properly described as transverse waves, meaning that a plane wave's electric field vector E and magnetic field H are in directions perpendicular to (or "transverse" to) the direction of wave propagation; E and H are also perpendicular to each other. Considering a monochromatic plane wave of optical frequency f (light of vacuum wavelength λ has a frequency of $f = c/\lambda$ where c is the speed of light), let us take the direction of propagation as the z axis. Being a transverse wave the E and H fields must then contain components only in the x and y directions whereas $E_z=H_z=0$. Using complex (or phasor) notation, we understand the instantaneous physical electric and magnetic fields to be given by the real partsof the complex quantities occurring in the following equations. As a function of time t and spatial position z (since for a plane wave in the $+z$ direction the fields have no dependence on x or y) these complex fields can be written as:

$$\vec{E}(z,t) = \begin{bmatrix} e_x \\ e_y \\ 0 \end{bmatrix} e^{i2\pi(z/(\lambda/n)-t/T)} = \begin{bmatrix} e_x \\ e_y \\ 0 \end{bmatrix} e^{i(kz-\omega t)}$$

and

$$\vec{H}(z,t) = \begin{bmatrix} h_x \\ h_y \\ 0 \end{bmatrix} e^{i2\pi(z/(\lambda/n)-t/T)} = \begin{bmatrix} h_x \\ h_y \\ 0 \end{bmatrix} e^{i(kz-\omega t)}$$

where λ/n is the wavelength *in the medium* (whose refractive index is n) and $T = 1/f$ is the period of the wave. Here e_x, e_y, h_x, and h_y are complex numbers. In the second more compact form, as these equations are customarily expressed, these factors are described using the wavenumber $k = 2\pi n / \lambda$ and angular frequency (or "radian frequency") $\omega = 2\pi f$. In a more general formulation with propagation *not* restricted to the +z direction, then the spatial dependence kz is replaced by $\vec{k} \cdot \vec{r}$ where $\vec{k}$ is called the wave vector, the magnitude of which is the wavenumber.

Thus the leading vectors e and h each contain up to two nonzero (complex) components describing the amplitude and phase of the wave's x and y polarization components (again, there can be no z polarization component for a transverse wave in the +z direction). For a given medium with a characteristic impedance η, h is related to e by:

$$h_y = e_x / \eta$$

And

$$h_x = -e_y / \eta .$$

In a dielectric, η is real and has the value η_0/n, where n is the refractive index and η_0 is the impedance of free space. The impedance will be complex in a conducting medium. Note that given that relationship, the dot product of E and H must be zero:

$$\vec{E}(\vec{r},t) \cdot \vec{H}(\vec{r},t) = e_x h_x + e_y h_y + e_z h_z = e_x(-e_y / \eta) + e_y(e_x / \eta) + 0 \cdot 0 = 0$$

indicating that these vectors are orthogonal (at right angles to each other), as expected.

So knowing the propagation direction (+z in this case) and η, one can just as well specify the wave in terms of just e_x and e_y describing the electric field. The vector containing e_x and e_y (but without the z component which is necessarily zero for a transverse wave) is known as a Jones vector. In addition to specifying the polarization state of the wave, a general Jones vector also specifies the overall magnitude and phase of that wave. Specifically, the intensity of the light wave is proportional to the sum of the squared magnitudes of the two electric field components:

$$I = \left(|e_x|^2 + |e_y|^2\right)\frac{1}{2\eta}$$

however the wave's *state of polarization* is only dependent on the (complex) *ratio* of e_y to e_x. So let us just consider waves whose $|e_x|^2 + |e_y|^2 = 1$; this happens to correspond to an intensity of about .00133 watts per square meter in free space (where $\eta = \eta_0$). And since the absolute phase of a wave is unimportant in discussing its polarization state, let us stipulate that the phase of e_x is zero, in other words e_x is a real number while e_y may be complex. Under these restrictions, e_x and e_y can be represented as follows:

$$e_x = \sqrt{\frac{1+Q}{2}}$$

$$e_y = \sqrt{\frac{1-Q}{2}}\, e^{i\phi}$$

where the polarization state is now fully parameterized by the value of Q (such that $-1 < Q < 1$) and the relative phase ϕ. By convention when one speaks of a wave's "polarization," if not otherwise specified, reference is being made to the polarization of the electric field. The polarization of the magnetic field always follows that of the electric field but with a 90 degree rotation, as detailed above.

Non-transverse Waves

In addition to transverse waves, there are many wave motions where the oscillation is not limited to directions perpendicular to the direction of propagation. These cases are far beyond the scope of the current article which concentrates on transverse waves (such as most electromagnetic waves in bulk media), however one should be aware of cases where the polarization of a coherent wave cannot be described simply using a Jones vector, as we have just done.

Just considering electromagnetic waves, we note that the preceding discussion strictly applies to plane waves in a homogeneous isotropic non-attenuating medium, whereas in an anisotropic medium (such as birefringent crystals as discussed below) the electric or magnetic field may have longitudinal as well as transverse components. In those cases the electric displacement *D* and magnetic flux density *B* still obey the above geometry but due to anisotropy in the electric susceptibility (or in the magnetic permeability), now given by a tensor, the direction of *E* (or *H*) may differ from that of *D* (or *B*). Even in isotropic media, so-called inhomogeneous waves can be launched into a medium whose refractive index has a significant imaginary part (or "extinction coefficient") such as metals; these fields are also not strictly transverse. Surface waves or waves propagating in a waveguide (such as an optical fiber) are generally *not* transverse waves, but might be described as an electric or magnetic transverse mode, or a hybrid mode.

Even in free space, longitudinal field components can be generated in focal regions, where the plane wave approximation breaks down. An extreme example is radially or tangentially polarized light, at the focus of which the electric or magnetic field respectively is *entirely* longitudinal (along the direction of propagation).

For longitudinal waves such as sound waves in fluids, the direction of oscillation is by definition along the direction of travel, so the issue of polarization is not normally even mentioned. On the other hand, sound waves in a bulk solid can be transverse as well as longitudinal, for a total of three polarization components. In this case, the transverse polarization is associated with the direction of the shear stress and displacement in directions perpendicular to the propagation direction, while the longitudinal polarization describes compression of the solid and vibration along the direction of propagation. The differential propagation of transverse and longitudinal polarizations is important in seismology.

Polarization State

Polarization is best understood by initially considering only pure polarization states, and only a coherent sinusoidal wave at some optical frequency. The vector in the adjacent diagram might describe the oscillation of the electric field emitted by a single-mode laser (whose oscillation frequency would be typically 10^{15} times faster). The field oscillates in the *x-y* plane, along the page, with the wave propagating in the *z* direction, perpendicular to the page. The first two diagrams below trace the electric field vector over a complete cycle for linear polarization at two different orientations; these are each considered a distinct *state of polarization* (SOP). Note that the linear polarization at 45° can also be viewed as the addition of a horizontally linearly polarized wave (as in the leftmost figure) and a vertically polarized wave of the same amplitude *in the same phase.*

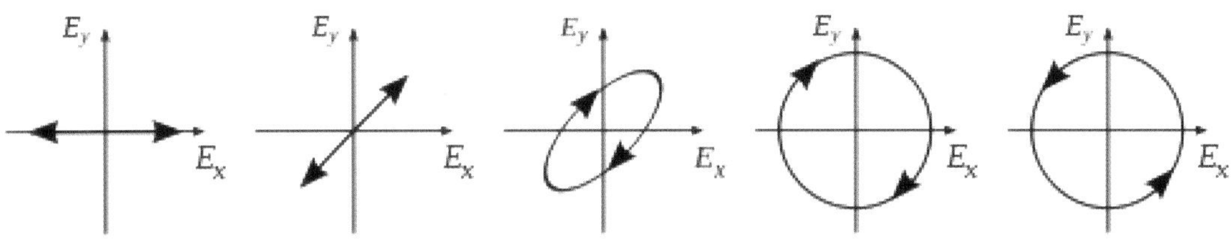

Now if one were to introduce a phase shift in between those horizontal and vertical polarization components, one would generally obtain elliptical polarization as is shown in the third figure. When the phase shift is exactly ±90°, then *circular polarization* is produced (fourth and fifth figures). Thus is circular polarization created in practice, starting with linearly polarized light and employing a quarter-wave plate to introduce such a phase shift. The result of two such phase-shifted components in causing a rotating electric field vector is depicted in the animation on the right. Note that circular or elliptical polarization can involve either a clockwise or counter clockwise rotation of the field. These correspond to distinct polarization states, such as the two circular polarizations shown above.

Of course the orientation of the *x* and *y* axes used in this description is arbitrary. The choice of such a coordinate system and viewing the polarization ellipse in terms of the *x* and *y* polarization components, corresponds to the definition of the Jones vector (below) in terms of those basis polarizations. One would typically choose axes to suit a particular problem such as *x* being in the plane of incidence. Since there are separate reflection coefficients for the linear polarizations in and orthogonal to the plane of incidence, that choice greatly simplifies the calculation of a wave's reflection from a surface.

Moreover, one can use as basis functions *any* pair of orthogonal polarization states, not just linear polarizations. For instance, choosing right and left circular polarizations as basis functions simplifies the solution of problems involving circular birefringence (optical activity) or circular dichroism.

Polarization Ellipse

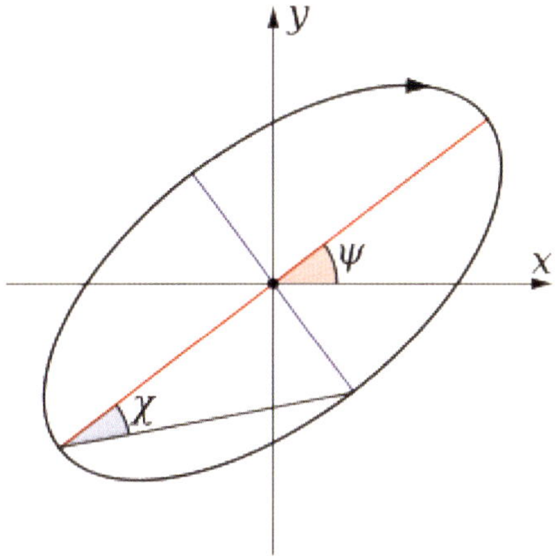

Consider a purely polarized monochromatic wave. If one were to plot the electric field vector over one cycle of oscillation, an ellipse would generally be obtained, as is shown in the figure, corresponding to a particular state of elliptical polarization. Note that linear polarization and circular polarization can be seen as special cases of elliptical polarization.

A polarization state can then be described in relation to the geometrical parameters of the ellipse, and its "handedness", that is, whether the rotation around the ellipse is clockwise or counter clockwise. One parameterization of the elliptical figure specifies the orientation angle ψ, defined as the angle between the major axis of the ellipse and the x-axis along with the ellipticity $\varepsilon = a/b$, the ratio of the ellipse's major to minor axis. (also known as the axial ratio). The ellipticity parameter is an alternative parameterization of an ellipse's eccentricity $e = \sqrt{1 - b^2 / a^2}$, or the ellipticity angle, $\chi = \arctan b/a = \arctan 1/\varepsilon$ as is shown in the figure. The angle χ is also significant in that the latitude (angle from the equator) of the polarization state as represented on the Poincaré sphere (see below) is equal to $\pm 2\chi$. The special cases of linear and circular polarization correspond to an ellipticity ε of infinity and unity (or χ of zero and 45°) respectively.

Jones Vector

Full information on a completely polarized state is also provided by the amplitude and phase of oscillations in two components of the electric field vector in the plane of polarization. This representation was used above to show how different states of polarization are possible. The amplitude and phase information can be conveniently represented as a two-dimensional complex vector:

$$\mathbf{e} = \begin{bmatrix} a_1 e^{i\theta_1} \\ a_2 e^{i\theta_2} \end{bmatrix}.$$

Here a_1 and a_2 denote the amplitude of the wave in the two components of the electric field vector, while θ_1 and represent the phases. The product of a Jones vector with a complex number of unit modulus gives a different Jones vector representing the same ellipse, and thus the same state of polarization. The physical electric field, as the real part of the Jones vector, would be altered but the polarization state itself is independent of absolute phase. The basis vectors used to represent the Jones vector need not represent linear polarization states (i.e. be real). In general any two orthogonal states can be used, where an orthogonal vector pair is formally defined as one having a zero inner product. A common choice is left and right circular polarizations, for example to model the different propagation of waves in two such components in circularly birefringent media or signal paths of coherent detectors sensitive to circular polarization.

Coordinate Frame

Regardless of whether polarization state is represented using geometric parameters or Jones vectors, implicit in the parameterization is the orientation of the coordinate frame. This permits a degree of freedom, namely rotation about the propagation direction. When considering light that is propagating parallel to the surface of the Earth, the terms "horizontal" and "vertical" polarization are often used, with the former being associated with the first component of the Jones vector, or zero azimuth angle. On the other hand, in astronomy the equatorial coordinate system is generally used instead, with the zero azimuth (or position angle, as it is more commonly called in astronomy to avoid confusion with the horizontal coordinate system) corresponding to due north.

s and p Designations

Another coordinate system frequently used relates to the *plane of incidence*. This is the plane made by the incoming propagation direction and the vector perpendicular to the plane of an interface, in other words, the plane in which the ray travels before and after reflection or refraction. The component of the electric field parallel to this plane is termed *p-like* (parallel) and the component perpendicular to this plane is termed *s-like* (from *senkrecht*, German for perpendicular). Polarized light with its electric field along the plane of incidence is thus denoted *p-polarized*, while light whose electric field is normal to the plane of incidence is called *s-polarized*. *P* polarization is commonly referred to as *transverse-magnetic* (TM), and has also been termed *pi-polarized* or *tangential plane polarized*. *S* polarization is also called *transverse-electric*(TE), as well as *sigma-polarized* or *sagittal plane polarized*.

Unpolarized and Partially Polarized Light

Definition

Natural light, and most other common sources of visible light, are incoherent: radiation is produced independently by a large number of atoms or molecules whose emissions are uncorrelated and generally of random polarizations. In this case the light is said to be *unpolarized*. This term is somewhat inexact, since at any instant of time at one location there is a definite direction to the electric and magnetic fields, however it implies that the polarization changes so quickly in time that it will not be measured or relevant to the outcome of an experiment. A so-called depolarizer acts on a polarized beam to create one which is actually *fully* polarized at every point, but in which the polarization varies so rapidly across the beam that it may be ignored in the intended applications.

Unpolarized light can be described as a mixture of two independent oppositely polarized streams, each with half the intensity. Light is said to be *partially polarized* when there is more power in one of these streams than the other. At any particular wavelength, partially polarized light can be statistically described as the superposition of a completely unpolarized component and a completely polarized one. One may then describe the light in terms of the degree of polarization and the parameters of the polarized component. That polarized component can be described in terms of a Jones vector or polarization ellipse, as is detailed above. However, in order to also describe the degree of polarization, one normally employs Stokes parameters (see below) to specify a state of partial polarization.

Motivation

The transmission of plane waves through a homogeneous medium are fully described in terms of Jones vectors and 2×2 Jones matrices. However, in practice there are cases in which all of the light cannot be viewed in such a simple manner due to spatial inhomogeneities or the presence of mutually incoherent waves. So-called depolarization, for instance, cannot be described using Jones matrices. For these cases it is usual instead to use a 4×4 matrix that acts upon the Stokes 4-vector. Such matrices were first used by Paul Soleillet in 1929, although they have come to be known as Mueller matrices. While every Jones matrix has a Mueller matrix, the reverse is not true. Mueller matrices are then used to describe the observed polarization effects of the scattering of waves from complex surfaces or ensembles of particles, as shall now be presented.

Coherency Matrix

The Jones vector perfectly describes the state of polarization *and phase* of a single monochromatic wave, representing a pure state of polarization as described above. However any mixture of waves of different polarizations (or even of different frequencies) do *not* correspond to a Jones vector. In so-called partially polarized radiation the fields are stochastic, and the variations and correlations between components of the electric field can only be described statistically. One such representation is the coherency matrix:

$$
\begin{aligned}
\emptyset &= \left\langle \mathrm{ee}^{\dagger} \right\rangle \\
&= \left\langle \begin{bmatrix} e_1 e_1^* & e_1 e_2^* \\ e_2 e_1^* & e_2 e_2^* \end{bmatrix} \right\rangle \\
&= \left\langle \begin{bmatrix} a_1^2 & a_1 a_2 e^{i(\theta_1-\theta_2)} \\ a_1 a_2 e^{-i(\theta_1-\theta_2)} & a_2^2 \end{bmatrix} \right\rangle
\end{aligned}
$$

where angular brackets denote averaging over many wave cycles. Several variants of the coherency matrix have been proposed: the Wiener coherency matrix and the spectral coherency matrix of Richard Barakat measure the coherence of a spectral decomposition of the signal, while the Wolf coherency matrix averages over all time/frequencies.

The coherency matrix contains all second order statistical information about the polarization. This matrix can be decomposed into the sum of two idempotent matrices, corresponding to the eigenvectors of the coherency matrix, each representing a polarization state that is orthogonal to the other. An alternative decomposition is into completely polarized (zero determinant) and unpolarized (scaled identity matrix) components. In either case, the operation of summing the components corresponds to the incoherent superposition of waves from the two components. The latter case gives rise to the concept of the "degree of polarization"; i.e., the fraction of the total intensity contributed by the completely polarized component.

Stokes Parameters

The coherency matrix is not easy to visualize, and it is therefore common to describe incoherent or partially polarized radiation in terms of its total intensity (*I*), (fractional) degree of polarization (*p*), and the shape parameters of the polarization ellipse. An alternative and mathematically convenient description is given by the Stokes parameters, introduced by George Gabriel Stokes in 1852. The relationship of the Stokes parameters to intensity and polarization ellipse parameters is shown in the equations and figure below.

$$S_0 = I$$
$$S_1 = Ip\cos 2\psi \cos 2\chi$$
$$S_2 = Ip\sin 2\psi \cos 2\chi$$
$$S_3 = Ip\sin 2\chi$$

Here Ip, 2ψ and 2χ are the spherical coordinates of the polarization state in the three-dimensional space of the last three Stokes parameters. Note the factors of two before ψ and χ corresponding respectively to the facts that any polarization ellipse is indistinguishable from one rotated by 180°, or one with the semi-axis lengths swapped accompanied by a 90° rotation. The Stokes parameters are sometimes denoted *I*, *Q*, *U* and *V*.

Implications for Reflection and Propagation

Polarization in Wave Propagation

In a vacuum, the components of the electric field propagate at the speed of light, so that the phase of the wave varies in space and time while the polarization state does not. That is, the electric field vector e of a plane wave in the +z direction follows:

$$\mathbf{e}(z+\Delta z, t+\Delta t) = \mathbf{e}(z,t)e^{ik(c\Delta t-\Delta z)},$$

where k is the wavenumber. As noted above, the instantaneous electric field is the real part of the product of the Jones vector times the phase factor $e^{-i\omega t}$. When an electromagnetic wave interacts with matter, its propagation is altered according to the material's (complex) index of refraction. When the real or imaginary part of that refractive index is dependent on the polarization state of a wave, properties known as birefringence and polarization dichroism (or diattenuation) respectively, then the polarization state of a wave will generally be altered.

In such media, an electromagnetic wave with any given state of polarization may be decomposed into two orthogonally polarized components that encounter different propagation constants. The effect of propagation over a given path on those two components is most easily characterized in the form of a complex 2×2 transformation matrix J known as a Jones matrix:

$$\mathbf{e}' = \mathbf{J}\mathbf{e}.$$

The Jones matrix due to passage through a transparent material is dependent on the propagation distance as well as the birefringence. The birefringence (as well as the average refractive index) will generally be dispersive, that is, it will vary as a function of optical frequency (wavelength). In the case of non-birefringent materials, however, the 2×2 Jones matrix is the identity matrix (multiplied by a scalar phase factor and attenuation factor), implying no change in polarization during propagation.

For propagation effects in two orthogonal modes, the Jones matrix can be written as

$$\mathbf{J} = \mathbf{T}\begin{bmatrix} g_1 & 0 \\ 0 & g_2 \end{bmatrix}\mathbf{T}^{-1},$$

where g_1 and g_2 are complex numbers describing the phase delay and possibly the amplitude attenuation due to propagation in each of the two polarization eigenmodes. T is a unitary matrix representing a change of basis from these propagation modes to the linear system used for the Jones vectors; in the case of linear birefringence or diattenuation the modes are themselves linear polarization states so T and T^{-1} can be omitted if the coordinate axes have been chosen appropriately.

Dichroism

Media in which transmission of one polarization mode is preferentially reduced are called *dichroic* or *diattenuating*. Like birefringence, diattenuation can be with respect to linear polarization modes (in a crystal) or circular polarization modes (usually in a liquid).

Devices that block nearly all of the radiation in one mode are known as *polarizing filters* or simply "polarizers". This corresponds to g_2=0 in the above representation of the Jones matrix. The output of an ideal polarizer is a specific polarization state (usually linear polarization) with an amplitude equal to the input wave's original amplitude in that polarization mode. Power in the other polarization mode is eliminated. Thus if unpolarized light is passed through an ideal polarizer (where g_1=1 and g_2=0) exactly half of its initial power is retained. Practical polarizers, especially inexpensive sheet polarizers, have additional loss so that $g_1 < 1$. However, in many instances the more relevant figure of merit is the polarizer's degree of polarization or extinction ratio, which involve a comparison of g_1 to g_2. Since Jones vectors refer to waves' amplitudes (rather than intensity), when illuminated by unpolarized light the remaining power in the unwanted polarization will be $(g_2/g_1)^2$ of the power in the intended polarization.

Specular Reflection

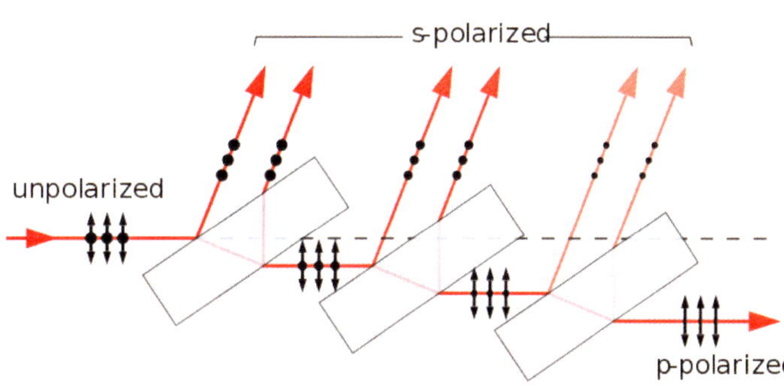

A stack of plates at Brewster's angle to a beam reflects off a fraction of the *s*-polarized light at each surface, leaving (after many such plates) a mainly *p*-polarized beam.

In addition to birefringence and dichroism in extended media, polarization effects describable using Jones matrices can also occur at (reflective) interface between two materials of different refractive index. These effects are treated by the Fresnel equations. Part of the wave is transmitted and part is reflected; for a given material those proportions (and also the phase of reflection) are dependent on the angle of incidence and are different for the *s* and *p* polarizations. Therefore, the polarization state of reflected light (even if initially unpolarized) is generally changed.

Any light striking a surface at a special angle of incidence known as Brewster's angle, where the reflection coefficient for *p* polarization is zero, will be reflected with only the *s*-polarization remaining. This principle is employed in the so-called "pile of plates polarizer" (see figure) in which part of the *s* polarization is removed by reflection at each Brewster angle surface, leaving only the *p* polarization after transmission through many such surfaces. The generally smaller reflection coefficient of the *p* polarization is also the basis of polarized sunglasses; by blocking the *s* (horizontal) polarization, most of the glare due to reflection from a wet street, for instance, is removed.

In the important special case of reflection at normal incidence (not involving anisotropic materials) there is no particular *s* or *p* polarization. Both the *x* and *y* polarization components are reflected identically, and therefore the polarization of the reflected wave is identical to that of the incident wave. However, in the case of circular (or elliptical) polarization, the handedness of the polarization state is thereby reversed, since by convention this is specified relative to the direction of propagation. The circular rotation of the electric field around the *x*-*y* axes called "right-handed" for a wave in the +*z* direction is "left-handed" for a wave in the -*z* direction. But in the general case of reflection at a nonzero angle of incidence, no such generalization can be made. For instance, right-circularly polarized light reflected from a dielectric surface at a grazing angle, will still be right-handed (but elliptically) polarized. Linear polarized light reflected from a metal at non-normal incidence will generally become elliptically polarized. These cases are handled using Jones vectors acted upon by the different Fresnel coefficients for the *s* and *p* polarization components.

Measurement Techniques Involving Polarization

Some optical measurement techniques are based on polarization. In many other optical techniques polarization is crucial or at least must be taken into account and controlled; such examples are too numerous to mention.

Measurement of Stress

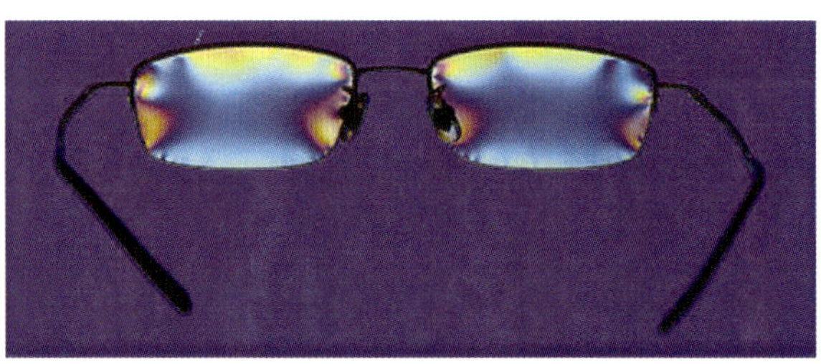

Stress in plastic glasses

In engineering, the phenomenon of stress induced birefringence allows for stresses in transparent materials to be readily observed. As noted above and seen in the accompanying photograph, the chromaticity of birefringence typically creates colored patterns when viewed in between two polarizers. As external forces are applied, internal stress induced in the material is thereby observed.

Additionally, birefringence is frequently observed due to stresses "frozen in" at the time of manufacture. This is famously observed in cellophane tape whose birefringence is due to the stretching of the material during the manufacturing process.

Ellipsometry

Ellipsometry is a powerful technique for the measurement of the optical properties of a uniform surface. It involves measuring the polarization state of light following specular reflection from such a surface. This is typically done as a function of incidence angle or wavelength (or both). Since ellipsometry relies on reflection, it is not required for the sample to be transparent to light or for its back side to be accessible.

Ellipsometry can be used to model the (complex) refractive index of a surface of a bulk material. It is also very useful in determining parameters of one or more thin film layers deposited on a substrate. Due to their reflection properties, not only are the predicted magnitude of the p and s polarization components, but their relative phase shifts upon reflection, compared to measurements using an ellipsometer. A normal ellipsometer does not measure the actual reflection coefficient (which requires careful photometric calibration of the illuminating beam) but the ratio of the p and s reflections, as well as change of polarization ellipticity (hence the name) induced upon reflection by the surface being studied. In addition to use in science and research, ellipsometers are used in situ to control production processes for instance.

Geology

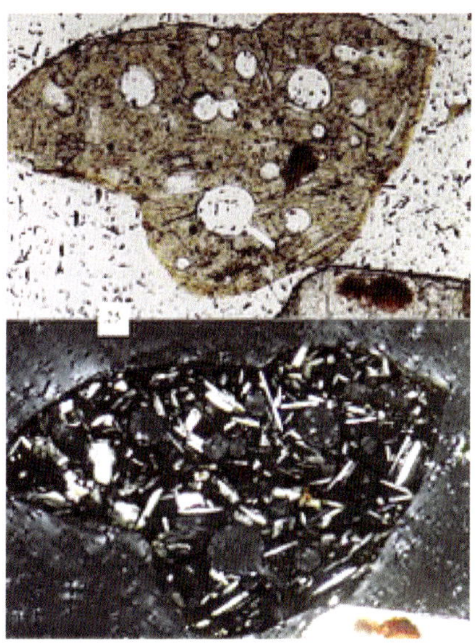

Photomicrograph of a volcanic sand grain; upper picture is plane-polarized light, bottom picture is cross-polarized light, scale box at left-center is 0.25 millimeter.

The property of (linear) birefringence is widespread in crystalline minerals, and indeed was pivotal in the initial discovery of polarization. In mineralogy, this property is frequently exploited using polarization microscopes, for the purpose of identifying minerals.

Sound waves in solid materials exhibit polarization. Differential propagation of the three polarizations through the earth is a crucial in the field of seismology. Horizontally and vertically polarized seismic waves (shear waves)are termed SH and SV, while waves with longitudinal polarization (compressional waves) are termed P-waves.

Chemistry

We have seen (above) that the birefringence of a type of crystal is useful in identifying it, and thus detection of linear birefringence is especially useful in geology and mineralogy. Linearly polarized light generally has its polarization state altered upon transmission through such a crystal, making it stand out when viewed in between two crossed polarizers, as seen in the photograph, above. Likewise, in chemistry, rotation of polarization axes in a liquid solution can be a useful measurement. In a liquid, linear birefringence is impossible, however there may be circular birefringence when a chiral molecule is in solution. When the right and left handed enantiomers of such a molecule are present in equal numbers (a so-called racemic mixture) then their effects cancel out. However, when there is only one (or a preponderance of one), as is more often the case for organic molecules, a net circular birefringence (or *optical activity*) is observed, revealing the magnitude of that imbalance (or the concentration of the molecule itself, when it can be assumed that only one enantiomer is present). This is measured using a polarimeter in which polarized light is passed through a tube of the liquid, at the end of which is another polarizer which is rotated in order to null the transmission of light through it.

Astronomy

In many areas of astronomy, the study of polarized electromagnetic radiation from outer space is of great importance. Although not usually a factor in the thermal radiation of stars, polarization is also present in radiation from coherent astronomical sources (e.g. hydroxyl or methanol masers), and incoherent sources such as the large radio lobes in active galaxies, and pulsar radio radiation (which may, it is speculated, sometimes be coherent), and is also imposed upon starlight by scattering from interstellar dust. Apart from providing information on sources of radiation and scattering, polarization also probes the interstellar magnetic field via Faraday rotation. The polarization of the cosmic microwave background is being used to study the physics of the very early universe. Synchrotron radiation is inherently polarised. It has been suggested that astronomical sources caused the chirality of biological molecules on Earth.

Applications and Examples

Polarized Sunglasses

Effect of a polarizer on reflection from mud flats. In the picture on the left, the horizontally oriented polarizer preferentially transmits those reflections; rotating the polarizer by 90° (right) as one would view using polarized sunglasses blocks almost all specularly reflected sunlight.

One can test whether sunglasses are polarized by looking through two pairs, with one perpendicular to the other. If both are polarized, all light will be blocked.

Unpolarized light, after being reflected by a specular (shiny) surface, generally obtains a degree of polarization. This phenomenon was observed in 1808 by the mathematician Étienne-Louis Malus, after whom Malus's law is named. Polarizing sunglasses exploit this effect to reduce glare from reflections by horizontal surfaces, notably the road ahead viewed at a grazing angle.

Wearers of polarized sunglasses will occasionally observe inadvertent polarization effects such as color-dependent birefringent effects, for example in toughened glass (e.g., car windows) or items made from transparent plastics, in conjunction with natural polarization by reflection or scattering. The polarized light from LCD monitors is very conspicuous when these are worn.

Sky Polarization and Photography

The effects of a polarizing filter (right image) on the sky in a photograph.

Polarization is observed in the light of the sky, as this is due to sunlight scattered by aerosols as it passes through the earth's atmosphere. The scattered light produces the brightness and color in clear skies. This partial polarization of scattered light can be used to darken the sky in photographs, increasing the contrast. This effect is most strongly observed at points on the sky making a 90° angle to the sun. Polarizing filters use these effects to optimize the results of photographing scenes in which reflection or scattering by the sky is involved.

Sky polarization has been used for orientation in navigation. The Pfund sky compass was used in the 1950s when navigating near the poles of the Earth's magnetic field when neither the sun nor stars were visible (e.g., under daytime cloud or twilight). It has been suggested, controversially, that the Vikings exploited a similar device (the "sunstone") in their extensive expeditions across the North Atlantic, before the arrival of the magnetic compass from Asia to Europe. Related to the sky compass is the "polar clock", invented by Charles Wheatstone.

Display Technologies

The principle of liquid-crystal display (LCD) technology relies on the rotation of the axis of linear polarization by the liquid crystal array. Light from the backlight (or the back reflective layer, in devices not including or requiring a backlight) first passes through a linear polarizing sheet. That polarized light passes through the actual liquid crystal layer which may be organized in pixels (for a TV or computer monitor) or in another format such as a seven-segment display or one with custom symbols for a particular product. The liquid crystal layer is produced with a consistent right (or left) handed chirality, essentially consisting of tiny helices. This causes circular birefringence, and is engineered so that there is a 90 degree rotation of the linear polarization state. However, when a voltage is applied across a cell, the molecules straighten out, lessening or totally losing the circular birefringence. On the viewing side of the display is another linear polarizing sheet, usually oriented at 90 degrees from the one behind the active layer. Therefore, when the circular birefringence is removed by the application of a sufficient voltage, the polarization of the transmitted light remains at right angles to the front polarizer, and the pixel appears dark. With no voltage, however, the 90 degree rotation of the polarization causes it to exactly match the axis of the front polarizer, allowing the light through. Intermediate voltages create intermediate rotation of the polarization axis and the pixel has an intermediate intensity. Displays based on this principle are widespread, and now are used in the vast majority of televisions, computer monitors and video projectors, rendering the previous CRT technology essentially obsolete. The use of polarization in the operation of LCD displays is immediately apparent to someone wearing polarized sunglasses, often making the display unreadable.

In a totally different sense, polarization encoding has become the leading (but not sole) method for delivering separate images to the left and right eye in stereoscopic displays used for 3D movies. This involves separate images intended for each eye either projected from two different projectors with orthogonally oriented polarizing filters or, more typically, from a single projector with time multiplexed polarization (a fast alternating polarization device for successive frames). Polarized 3D glasses with suitable polarizing filters ensure that each eye receives only the intended image. Historically such systems used linear polarization encoding because it was inexpensive and offered good separation. However circular polarization makes separation of the two images insensitive to tilting of the head, and is widely used in 3-D movie exhibition today, such as the system from RealD. Projecting such images requires screens that maintain the polarization of the projected light when viewed in reflection (such as silver screens); a normal diffuse white projection screen causes depolarization of the projected images, making it unsuitable for this application.

Although now obsolete, CRT computer displays suffered from reflection by the glass envelope, causing glare from room lights and consequently poor contrast. Several anti-reflection solutions were employed to ameliorate this problem. One solution utilized the principle of reflection of circularly polarized light. A circular polarizing filter in front of the screen allows for the transmission of (say) only right circularly polarized room light. Now, right circularly polarized light (depending on the convention used) has its electric (and magnetic) field direction rotating clockwise while propagating in the +z direction. Upon reflection, the field still has the same direction of rotation, but now propagation is in the −z direction making the reflected wave *left* circularly polarized. With the right circular polarization filter placed in front of the reflecting glass, the unwanted light reflected from the glass will thus be in very polarization state that is *blocked* by that filter, eliminating the reflection problem. The reversal of circular polarization on reflection and elimination of reflections in this manner can be easily observed by looking in a mirror while wearing 3-D movie glasses which employ left- and right-handed circular polarization in the two lenses. Closing one eye, the other eye will see a reflection in which it cannot see itself; that lens appears black. However the other lens (of the closed eye) will have the correct circular polarization allowing the closed eye to be easily seen by the open one.

Radio Transmission and Reception

All radio (and microwave) antennas used for transmitting or receiving are intrinsically polarized. They transmit in (or receive signals from) a particular polarization, being totally insensitive to the opposite polarization; in certain cases that polarization is a function of direction. Most antennas are nominally linearly polarized, but elliptical and circular polarization is a possibility. As is the convention in optics, the "polarization" of a radio wave is understood to refer to the polarization of its electric field, with the magnetic field being at a 90 degree rotation with respect to it for a linearly polarized wave.

The vast majority of antennas are linearly polarized. In fact it can be shown from considerations of symmetry that an antenna that lies entirely in a plane which also includes the observer, can *only* have its polarization in the direction of that plane. This applies to many cases, allowing one to easily infer such an antenna's polarization at an intended direction of propagation. So a typical rooftop Yagi or log-periodic antenna with horizontal conductors, as viewed from a second station toward the horizon, is necessarily horizontally polarized. But a vertical "whip antenna" or AM broadcast tower used as an antenna element (again, for observers horizontally displaced from it) will transmit in the vertical polarization. A turnstile antenna with its four arms in the horizontal plane, likewise transmits horizontally polarized radiation toward the horizon. However, when that same turnstile antenna is used in the "axial mode" (upwards, for the same horizontally-oriented structure) its radiation is circularly polarized. At intermediate elevations it is elliptically polarized.

Polarization is important in radio communications because, for instance, if one attempts to use a horizontally polarized antenna to receive a vertically polarized transmission, the signal strength will be substantially reduced (or under very controlled conditions, reduced to nothing). This principle is used in satellite television in order to double the channel capacity over a fixed frequency band. The same frequency channel can be used for two signals broadcast in opposite polarizations. By adjusting the receiving antenna for one or the other polarization, either signal can be selected without interference from the other.

Especially due to the presence of the ground, there are some differences in propagation (and also in reflections responsible for TV ghosting) between horizontal and vertical polarizations. AM and FM broadcast radio usually use vertical polarization, while television uses horizontal polarization. At low frequencies especially, horizontal polarization is avoided. That is because the phase of a horizontally polarized wave is reversed upon reflection by the ground. A distant station in the horizontal direction will receive both the direct and reflected wave, which thus tend to cancel each other. This problem is avoided with vertical polarization. Polarization is also important in the transmission of radar pulses and reception of radar reflections by the same or a different antenna. For instance, back scattering of radar pulses by rain drops can be avoided by using circular polarization. Just as specular reflection of circularly polarized light reverses the handedness of the polarization, as discussed above, the same principle applies to scattering by objects much smaller than a wavelength such as rain drops. On the other hand, reflection of that wave by an irregular metal object (such as an airplane) will typically introduce a change in polarization and (partial) reception of the return wave by the same antenna.

The effect of free electrons in the ionosphere, in conjunction with the earth's magnetic field, causes Faraday rotation, a sort of circular birefringence. This is the same mechanism which can rotate the axis of linear polarization by electrons in interstellar space as mentioned below. The magnitude of Faraday rotation caused by such a plasma is greatly exaggerated at lower frequencies, so at the higher microwave frequencies used by satellites the effect is minimal. However medium or short wave transmissions received following refraction by the ionosphere are strongly affected. Since a wave's path through the ionosphere and the earth's magnetic field vector along such a path are rather unpredictable, a wave transmitted with vertical (or horizontal) polarization will generally have a resulting polarization in an arbitrary orientation at the receiver.

Polarization and Vision

Many animals are capable of perceiving some of the components of the polarization of light, e.g., linear horizontally polarized light. This is generally used for navigational purposes, since the linear polarization of sky light is always perpendicular to the direction of the sun. This ability is very common among the insects, including bees, which use this information to orient their communicative dances. Polarization sensitivity has also been observed in species of octopus, squid, cuttlefish, and mantis shrimp. In the latter case, one species measures all six orthogonal components of polarization, and is believed to have optimal polarization vision. The rapidly changing, vividly colored skin patterns of cuttlefish, used for communication, also incorporate polarization patterns, and mantis shrimp are known to have polarization selective reflective tissue. Sky polarization was thought to be perceived by pigeons, which was assumed to be one of their aids in homing, but research indicates this is a popular myth.

The naked human eye is weakly sensitive to polarization, without the need for intervening filters. Polarized light creates a very faint pattern near the center of the visual field, called Haidinger's brush. This pattern is very difficult to see, but with practice one can learn to detect polarized light with the naked eye.

Angular Momentum using Circular Polarization

It is well known that electromagnetic radiation carries a certain linear momentum in the direction of propagation. In addition, however, light carries a certain angular momentum if it is circularly polarized (or partially so). In comparison with lower frequencies such as microwaves, the amount of angular momentum in light, even of pure circular polarization, compared to the same wave's linear momentum (or radiation pressure) is very small and difficult to even measure. However it was utilized in an experiment to achieve speeds of up to 600 million revolutions per minute.

Linear Polarization

From a technical perspective, linear polarization is defined as polarization of an electromagnetic wave in which the electric vector at a fixed point in space remains pointing in a fixed direction, although varying in magnitude. There are two forms of linear polarization: vertical, where the electric field is perpendicular to the Earth's surface, and horizontal, where the electric field is parallel to the Earth's surface. Both directions can be used simultaneously on the same frequency. Some customers consider linear polarization to be superior, if only because the specific equipment costs are marginally less. Linear polarization can be found in both C-Band and Ku-Band.

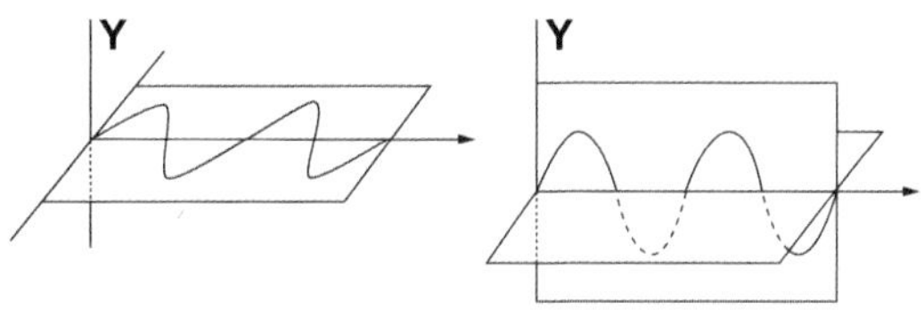

Figure: Linear Polarization

We say a plane wave is linearly polarized if there is no phase difference between E_x and E_y. We can write linear polarizations as

$$\vec{E}_0 = \left(E_x, E_y\ 0\right)$$

and choose the overall phase so that E_x and E_y are real numbers. If $E_y = 0$ but $E_x \neq 0$, we have

$$\vec{E} = E_0 \hat{x} e^{i(kz-\omega t)}$$

with $E_0 = \left|\vec{E}_0\right|$ just a number now. Then, from Eq. , since $\hat{z} \times \hat{x} = \hat{y}$,

$$\vec{B} = \frac{1}{c} E_0 \hat{y} e^{i(kz-\omega t)}$$

This configuration is said to be linear polarized in the x direction.

Similarly we can have

$$\vec{E} = E_0 \hat{y} e^{i(kz-\omega t)}$$

and using $\hat{z} \times \hat{y} = \hat{x}$:

$$\vec{B} = -\frac{1}{c} E_0 \hat{x} e^{i(kz-\omega t)}$$

This configuration is said to be linear polarized in the y direction. Note that in both cases, the magnetic field is given by rotating the electric field 90° counter clockwise in the x-y plane and dividing by c.

More generally, for any unit vector $\hat{v}$ we can have

$$\vec{E} = E_0 \hat{v} e^{i(\vec{k}.\vec{x}-\omega t)}$$

which is linearly polarized in the $\hat{v}$ direction. The magnetic field will be polarized in direction 90° behind the electric field.

Remember, we always implicitly want to have the real part of these fields. So, for linearly polarized light in the x direction, the fields are actually

$$\mathrm{Re}\left[\vec{E}\right]=\left(E_0 \cos\left(k z-\omega t\right),0,0\right),\quad \mathrm{Re}\left[\vec{B}\right]=\left(0,\frac{E_0}{c}\cos\left(k z-\omega t\right),0\right)$$

Note that there is no x or y dependence in these solutions: the fields are completely uniform in x and y – they are plane waves. At each point on the plane, the electric field points in the $\hat{x}$ direction with the same magnitude. This magnitude varies as we move in z and in t but is always uniform in the plane.

Here's a picture of how the fields vary as they move along for a wave moving in the x direction.

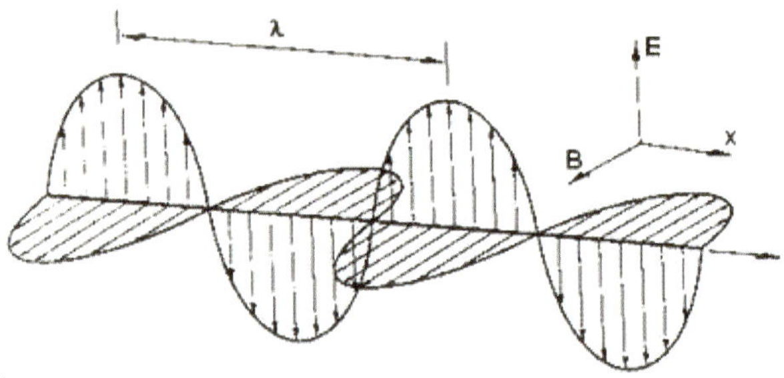

Elliptical Polarization

In electrodynamics, elliptical polarization is the polarization of electromagnetic radiation such that the tip of the electric field vector describes an ellipse in any fixed plane intersecting, and normal to, the direction of propagation. An elliptically polarized wave may be resolved into two linearly polarized waves in phase quadrature, with their polarization planes at right angles to each other. Since the electric field can rotate clockwise or counterclockwise as it propagates, elliptically polarized waves exhibit chirality.

Other forms of polarization, such as circular and linear polarization, can be considered to be special cases of elliptical polarization.

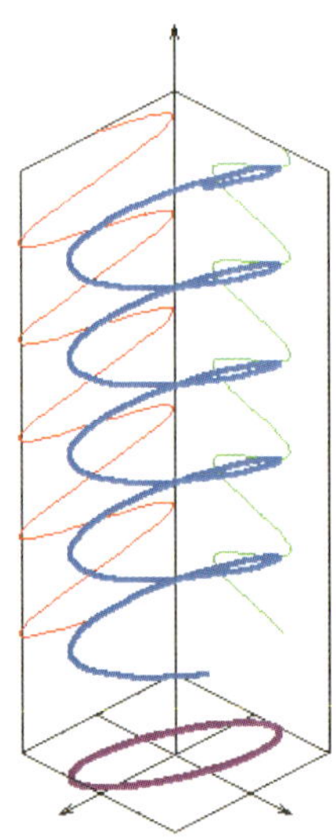

Mathematical Description of Elliptical Polarization

The classical sinusoidal plane wave solution of the electromagnetic wave equation for the electric and magnetic fields is (Gaussian units)

$$\mathbf{E}(\mathrm{r,t}) = |\mathbf{E}|\,\mathrm{Re}\{|\psi\rangle exp[i(kz-\omega t)]\}$$

$$\mathbf{B}(\mathrm{r,t}) = \hat{\mathbf{z}} \times \mathbf{E}(\mathrm{r,t})$$

for the magnetic field, where k is the wavenumber,

$$\omega = ck$$

is the angular frequency of the wave propagating in the +z direction, and c is the speed of light.

Here $|\mathbf{E}|$ is the amplitude of the field and

$$|\psi\rangle \stackrel{\mathrm{def}}{=} \begin{pmatrix} \psi_x \\ \psi_y \end{pmatrix} = \begin{pmatrix} cos\,\theta\, exp(i\alpha_x) \\ sin\,\theta\, exp(i\alpha_y) \end{pmatrix}$$

is the normalized Jones vector. This is the most complete representation of polarized electromagnetic radiation and corresponds in general to elliptical polarization.

Polarization Ellipse

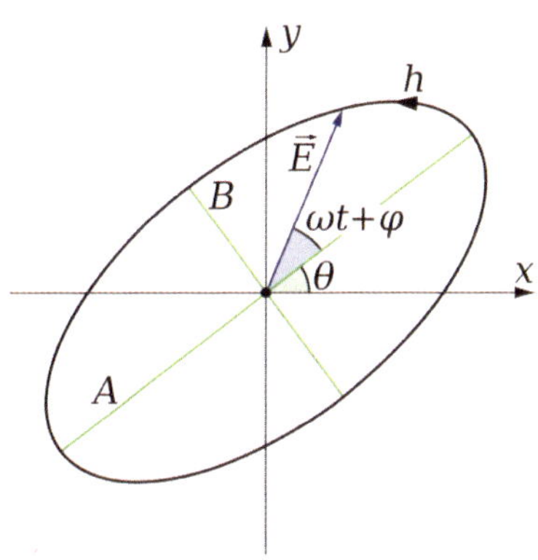

At a fixed point in space (or for fixed z), the electric vector **E** traces out an ellipse in the x-y plane. The semi-major and semi-minor axes of the ellipse have lengths A and B, respectively, that are given by

$$A = |\mathbf{E}|\sqrt{\frac{1+\sqrt{1-sin^2(2\theta)sin^2\beta}}{2}}$$

and

$$B = |\mathbf{E}|\sqrt{\frac{1-\sqrt{1-sin^2(2\theta)sin^2\beta}}{2}},$$

where $\beta = \alpha_y - \alpha_x$. The orientation of the ellipse is given by the angle ϕ the semi-major axis makes with the x-axis. This angle can be calculated from

$$tan\,2\phi = tan\,2\theta\,cos\,\beta.$$

If $\beta = 0$, the wave is linearly polarized. The ellipse collapses to a straight line $(A = |\mathbf{E}|, B = 0)$ oriented at an angle $\phi = \theta$. This is the case of superposition of two simple harmonic motions (in phase), one in the x direction with an amplitude $|\mathbf{E}|\,cos\,\theta$, and the other in the y direction with an amplitude $|\mathbf{E}|\,sin\,\theta$. When β increases from zero, i.e., assumes positive values, the line evolves into an ellipse that is being traced out in the counterclockwise direction (looking in the direction of the propagating wave); this then corresponds to *left-handed elliptical polarization*; the semi-major axis is now oriented at an angle $\phi \neq \theta$. Similarly, if β becomes negative from zero, the line evolves into an ellipse that is being traced out in the clockwise direction; this corresponds to *right-handed elliptical polarization*.

If $\beta = \pm\pi/2$ and $\theta = \pi/4, A = B = |\mathbf{E}|/\sqrt{2}$, i.e., the wave is circularly polarized. When $\beta = \pi/2$, the wave is left-circularly polarized, and when $\beta = -\pi/2$, the wave is right-circularly polarized.

In Nature

The reflected light from some beetles (e.g. Cetonia aurata) is elliptical polarized.

Circular Polarization

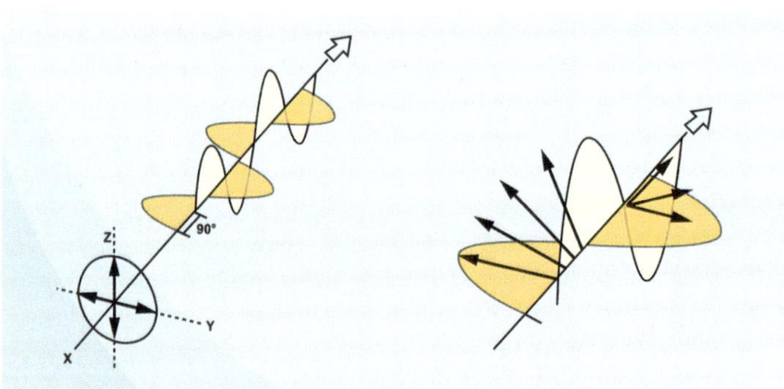

Figure: Circular Polarization

Some customers feel that circular polarization is not as desirable as linear polarization. The reasons for this view are not clear; one possible reason is due to the customer not being fully educated on the benefits or the price sensitivity towards equipment. Some customers may also feel that since Intelsat is the only satellite provider offering C-Band, the availability of antennas could become an issue. However, it is important to realize that the increased reliability in signal strength, resistance to weather conditions, and ease of installation outweigh the expense of the feed horn. Technically speaking, circular polarization involves the plane of polarization rotating in a corkscrew pattern, making one complete revolution during each wavelength. The circularly polarized wave will radiate energy in the horizontal and vertical plane, as well as every plane in between. There are two directions of propagation that come with circular polarization: Right-Hand-Circular (RHC) which follows a clockwise pattern, and Left Hand-Circular (LHC) which follows a counter clockwise pattern. As with linear polarization, both directions can be used simultaneously on the same frequency, allowing higher revenue generation through the doubling of capacity on the satellites. Circular polarization can be found on both C-Band and Ku-Band. It is important to note that Intelsat does not have any Ku-Band fleet with circular polarization.

Feed Horns

Switching from linear polarization to circular polarization is easier than many think; it simply requires changing the feed horn that is mounted at the antenna's focal point. The price difference can be almost negligible. For example, a low cost VSAT antenna with two port feeds is a mere $200 more than a linear feed horn. However, in some cases the feed horn can be significantly more expensive for circular polarization, an investment which will be at least partially offset by its advantages.

Operationally, the feed horn conveys radio waves between the transceiver and the reflector. The feed horn also separates the two polarities (vertical from horizontal or left from right hand) of the signal being received. This removes unwanted signals, (for example, left hand circular), from the desired signal, (for example, right hand circular) of a given frequency. For a linear feed horn to properly separate the polarizations, it must be precisely rotated to the exact alignment with the satellite's signals. A circular feed horn does not have this requirement.

Reversal of Handedness by Phase Shift

To convert a given handedness of polarized light to the other handedness one can use a half-waveplate. A half-waveplate shifts a given component of light one half of a wavelength relative to the component to which it is orthogonal.

Reversal of Handedness by Reflection

The handedness of polarized light is also reversed when it is reflected off a surface at normal incidence. Upon such reflection, the rotation of the plane of polarization of the reflected light is identical to that of the incident field. However, with propagation now in the *opposite* direction, the same rotation direction that would be described as "right handed" for the incident beam, is "left-handed" for propagation in the reverse direction, and vice versa. Aside from the reversal of handedness, the ellipticity of polarization is also preserved (except in cases of reflection by a birefringent surface).

Note that this principle only holds strictly for light reflected at normal incidence. For instance, right circularly polarized light reflected from a dielectric surface at grazing incidence (an angle beyond the Brewster angle) will still emerge as right handed, but elliptically, polarized. Light reflected by a metal at non-normal incidence will generally have its ellipticity changed as well. Such situations may be solved by decomposing the incident circular (or other) polarization into components of linear polarization parallel and perpendicular to the plane of incidence, commonly denoted p and s respectively. The reflected components in the p and s linear polarizations are found by applying the Fresnel coefficients of reflection, which are generally different for those two linear polarizations. Only in the special case of normal incidence, where there is no distinction between p and s, are the Fresnel coefficients for the two components identical, leading to the above property.

A 3-slide series of pictures taken with and without a pair of masterImage 3D circularly polarized movie glasses of some dead European rose chafers (Cetonia aurata) whose shiny green color comes from left-polarized light. Note that without glasses both the beetles and their images have shiny color. The right-polarizer removes the color of the beetles but leaves the color of the images. The left-polarizer does the opposite showing reversal of handedness of the reflected light.

Conversion to and from Linear Polarization

Circularly polarized light can be converted into linearly polarized light by passing it through a quarter-waveplate. Passing linearly polarized light through a quarter-waveplate with its axes at 45° to its polarization axis will convert it to circular polarization. In fact, this is the most common way of producing circular polarization in practice. Note that passing linearly polarized light through a quarter-waveplate at an angle *other* than 45° will generally produce elliptical polarization.

Left-/right-handedness Conventions

Circular polarization may be referred to as right-handed or left-handed, and clockwise or anti-clockwise, depending on the direction in which the electric field vector rotates. Unfortunately, two opposing historical conventions exist.

From the Point of View of the Source

Using this convention, polarization is defined from the point of view of the source. When using this convention, left or right handedness is determined by pointing one's left or right thumb away from the source, in the same direction that the wave is propagating, and matching the curling of one's fingers to the direction of the spatial rotation of the field at a given point in space. When determining if the wave is clockwise or anti-clockwise circularly polarized, one again takes the point of view of the source, and while looking away from the source and in the same direction of the wave's propagation, one observes the direction of the field's spatial rotation.

Using this convention, the electric field vector of a right handed circularly polarized wave is as follows: $\left(E_x, E_y, E_z\right) \propto \left(cos\frac{2\pi}{\lambda}(ct-z), sin\frac{2\pi}{\lambda}(ct-z), 0 \right)$.

As a specific example, refer to the circularly polarized wave in the first animation. Using this convention that wave is defined as right-handed because when one points one's right thumb in the same direction of the wave's propagation, the fingers of that hand curl in the same direction of the field's temporal rotation. It is considered clockwise circularly polarized because from the point of view of the source, looking in the same direction of the wave's propagation, the field rotates in the clockwise direction. The second animation is that of left-handed or anti-clockwise light using this same convention.

This convention is in conformity with the Institute of Electrical and Electronics Engineers (IEEE) standard and as a result it is generally used in the engineering community.

Quantum physicists also use this convention of handedness because it is consistent with their convention of handedness for a particle's spin.

Radio astronomers also use this convention in accordance with an International Astronomical Union (IAU) resolution made in 1973.

From the Point of view of the Receiver

In this alternative convention, polarization is defined from the point of view of the receiver. Using this convention, left or right handedness is determined by pointing one's left or right thumb toward the source, against the direction of propagation, and then matching the curling of one's fingers to the temporal rotation of the field.

When using this convention, in contrast to the other convention, the defined handedness of the wave matches the handedness of the screw type nature of the field in space. Specifically, if one freezes a right-handed wave in time, when one curls the fingers of one's right hand around the helix, the thumb will point in the direction which the helix progresses given that sense of rotation. Note that it is the nature of all screws and helices that it does not matter in which direction you point your thumb when determining its handedness.

When determining if the wave is clockwise or anti-clockwise circularly polarized, one again takes the point of view of the receiver and, while looking toward the source, against the direction of propagation, one observes the direction of the field's temporal rotation.

Just as in the other convention, right-handedness corresponds to a clockwise rotation and left-handedness corresponds to an anti-clockwise rotation.

Many optics textbooks use this second convention. It is also used by SPIE.

Uses of the Two Conventions

As stated earlier, there is significant confusion with regards to these two conventions. As a general rule the engineering, quantum physics, and radio astronomy communities use the first convention where the wave is observed from the point of view of the source. In many physics textbooks dealing with optics the second convention is used where the light is observed from the point of view of the receiver.

To avoid confusion, it is good practice to specify "as defined from the point of view of the source" or "as defined from the point of view of the receiver" when discussing polarization matters.

The archive of the US Federal Standard 1037C proposes two contradictory conventions of handedness.

FM Radio

The term "circular polarization" is often used erroneously to describe mixed polarity signals used mostly in FM radio (87.5 to 108.0 MHz in the USA), where a vertical and a horizontal component are propagated simultaneously by a single or a combined array. This has the effect of producing greater penetration into buildings and difficult reception areas than a signal with just one plane of polarization. This would be an instance where the polarization would more appropriately be called random polarization because the polarization at a receiver, although constant, will vary depending on the direction from the transmitter and other factors in the transmitting antenna design. The term "FM radio" above refers to broadcast radio, not 2-way radio (more properly called land mobile radio), which uses vertical polarization almost exclusively.

Circular Dichroism

Circular dichroism (CD) is the differential absorption of left- and right-handed circularly polarized light. Circular dichroism is the basis of a form of spectroscopy that can be used to determine the optical isomerism and secondary structure of molecules.

In general, this phenomenon will be exhibited in absorption bands of any optically active molecule. As a consequence, circular dichroism is exhibited by most biological molecules, because of the dextrorotary (e.g. some sugars) and levorotary (e.g. some amino acids) molecules they contain. Noteworthy as well is that a secondary structure will also impart a distinct CD to its respective molecules. Therefore, the alpha helix, beta sheet and random coil regions of proteins and the double helix of nucleic acids have CD spectral signatures representative of their structures.

Also, under the right conditions, even non-chiral molecules will exhibit magnetic circular dichroism, that is, circular dichroism induced by a magnetic field.

Circularly Polarized Luminescence

Circularly polarized luminescence (CPL) can occur when either a luminophore or an ensemble of luminophores is chiral. The extent to which emissions are polarized is quantified in the same way it is for circular dichroism, in terms of the *dissymmetry factor*, also sometimes referred to as the *anisotropy factor*. This value is given by

$$g_{em} = 2\left(\frac{\theta_{\text{left}} - \theta_{\text{right}}}{\theta_{\text{left}} + \theta_{\text{right}}}\right)$$

where θ_{left} corresponds to the quantum yield of left-handed circularly polarized light, and θ_{right} to that of right-handed light. The maximum absolute value of g_{em}, corresponding to purely left- or right-handed circular polarization, is therefore 2. Meanwhile, the smallest absolute value that g_{em} can achieve, corresponding to linearly polarized or unpolarized light, is zero.

Mathematical Description

The classical sinusoidal plane wave solution of the electromagnetic wave equation for the electric and magnetic fields is

$$\mathbf{E}(\mathbf{r},t) = |\mathbf{E}|\mathrm{Re}\left\{\mathbf{Q}|\psi\rangle exp\left[i(kz-\omega t)\right]\right\}$$
$$\mathbf{B}(\mathbf{r},t) = \hat{\mathbf{z}}\times\mathbf{E}(\mathbf{r},t)$$

where k is the wavenumber,

$$\omega = ck$$

is the angular frequency of the wave, $\mathbf{Q} = \left[\hat{\mathbf{x}},\hat{\mathbf{y}}\right]$ is an orthogonal 2×2 matrix whose columns span the transverse x-y plane and c is the speed of light.

Here

$$|\mathbf{E}|$$

is the amplitude of the field and

$$|\psi\rangle \stackrel{\text{def}}{=} \begin{pmatrix}\psi_x \\ \psi_y\end{pmatrix} = \begin{pmatrix} cos\theta exp(i\alpha_x) \\ sin\theta exp(i\alpha_y)\end{pmatrix}$$

is the normalized Jones vector in the x-y plane.

If α_y is rotated by $\pi/2$ radians with respect to α_x and the x amplitude equals the y amplitude the wave is circularly polarized. The Jones vector is

$$|\psi\rangle = \frac{1}{\sqrt{2}}\begin{pmatrix}1 \\ \pm i\end{pmatrix} exp(i\alpha_x)$$

where the plus sign indicates left circular polarization and the minus sign indicates right circular polarization. In the case of circular polarization, the electric field vector of constant magnitude rotates in the x-y plane.

If basis vectors are defined such that

$$|R\rangle \stackrel{\text{def}}{=} \frac{1}{\sqrt{2}}\begin{pmatrix}1 \\ -i\end{pmatrix}$$

and

$$|L\rangle \stackrel{\text{def}}{=} \frac{1}{\sqrt{2}}\begin{pmatrix}1 \\ i\end{pmatrix}$$

then the polarization state can be written in the “R-L basis” as

$$|\psi\rangle = \psi_R |R\rangle + \psi_L |L\rangle$$

where

$$\psi_R \stackrel{\text{def}}{=} \frac{1}{\sqrt{2}}(cos\,\theta + i\,sin\,\theta\,exp(i\delta))exp(i\alpha_x)$$

$$\psi_L \stackrel{\text{def}}{=} \frac{1}{\sqrt{2}}(cos\,\theta - i\,sin\,\theta\,exp(i\delta))exp(i\alpha_x)$$

and

$$\delta = \alpha_y - \alpha_x.$$

Antennas

A number of different types of antenna elements can be utilized to produce circularly polarized (or nearly so) radiation; following Balanis, one can use *dipole elements*:

> "two crossed dipoles provide the two orthogonal field components. If the two dipoles are identical, the field intensity of each along zenith. would be of the same intensity. Also, if the two dipoles were fed with a 90° degree time-phase difference (phase quadrature), the polarization along zenith would be circular. One way to obtain the 90° time-phase difference between the two orthogonal field components, radiated respectively by the two dipoles, is by feeding one of the two dipoles with a transmission line which is 1/4 wavelength longer or shorter than that of the other".

or *helical elements*:

> "To achieve circular polarization the circumference *C* of the helix must be with *C*/wavelength = 1 near optimum, and the spacing about *S* = wavelength/4.

or *patch elements*:

> "circular and elliptical polarizations can be obtained using various feed arrangements or slight modifications made to the elements... Circular polarization can be obtained if two orthogonal modes are excited with a 90° time-phase difference between them. This can be accomplished by adjusting the physical dimensions of the patch. For a square patch element, the easiest way to excite ideally circular polarization is to feed the element at two adjacent edges ... The quadrature phase difference is obtained by feeding the element with a 90° power divider".

Quantum Mechanics

In the quantum mechanical view, light is composed of photons. Polarization is a manifestation of the intrinsic angular momentum (the spin) of the photon. More specifically, in quantum mechanics the direction of spin of a photon is tied to the handedness of the circularly polarized light and the spin of a beam of photons is similar to the spin of a beam of particles, such as electrons.

In Nature

The rose chafer's external surface reflects almost exclusively left-circularly polarized light.

Only a few mechanisms in nature are known to systematically produce circularly polarized light. In 1911, Albert Abraham Michelson discovered that light reflected from the golden scarab beetle *Chrysina resplendens* is preferentially left-polarized. Since then, circular polarization has been measured in several other scarab beetles like *Chrysina gloriosa*, as well as some crustaceans such as the mantis shrimp. In these cases, the underlying mechanism is the molecular-level helicity of the chitinous cuticle.

The bioluminescence of the larvae of fireflies is also circularly polarized, as reported in 1980 for the species *Photuris lucicrescens* and *Photuris versicolor*. For fireflies, it is more difficult to find a microscopic explanation for the polarization, because the left and right lanterns of the larvae were found to emit polarized light of opposite senses. The authors suggest that the light begins with a linear polarization due to inhomogeneties inside aligned photocytes, and it picks up circular polarization while passing through linearly birefringent tissue.

Water-air interfaces provide another source of circular polarization. Sunlight that gets scattered back up towards the surface is linearly polarized. If this light is then totally internally reflected back down, its vertical component undergoes a phase shift. To an underwater observer looking up, the faint light outside Snell's window therefore is (partially) circularly polarized.

Weaker sources of circular polarization in nature include multiple scattering by linear polarizers, as in the circular polarization of starlight, and selective absorption by circularly dichroic media.

Two species of mantis shrimp have been reported to be able to detect circular polarized light.

Starlight

The circular polarization of starlight has been observed to be a function of the linear polarization of starlight.

Starlight becomes partially linearly polarized by scattering from elongated interstellar dust grains whose long axes tend to be oriented perpendicular to the galactic magnetic field. According to the Davis–Greenstein mechanism, the grains spin rapidly with their rotation axis along the magnetic field. Light polarized along the direction of the magnetic field perpendicular to the line of sight is transmitted, while light polarized in the plane defined by the rotating grain is blocked. Thus the polarization direction can be used to map the galactic magnetic field. The degree of polarization is on the order of 1.5% for stars at 1,000 parsecs' distance.

Normally, a much smaller fraction of circular polarization is found in starlight. Serkowski, Mathewson and Ford measured the polarization of 180 stars in UBVR filters. They found a maximum fractional circular polarization of $q = 6 \times 10^{-4}$, in the R filter.

The explanation is that the interstellar medium is optically thin. Starlight traveling through a kiloparsec column undergoes about a magnitude of extinction, so that the optical depth ~ 1. An optical depth of 1 corresponds to a mean free path, which is the distance, on average that a photon travels before scattering from a dust grain. So on average, a starlight photon is scattered from a single interstellar grain; multiple scattering (which produces circular polarization) is much less likely. Observationally, the linear polarization fraction p ~ 0.015 from a single scattering; circular polarization from multiple scattering goes as p^2, so we expect a circularly polarized fraction of $q \sim 2 \times 10^{-4}$.

Light from early-type stars has very little intrinsic polarization. Kemp et al. measured the optical polarization of the Sun at sensitivity of 3×10^{-7}; they found upper limits of 10^{-6} for both p (fraction of linear polarization) and q (fraction of circular polarization).

The interstellar medium can produce circularly polarized (CP) light from unpolarized light by sequential scattering from elongated interstellar grains aligned in different directions. One possibility is twisted grain alignment along the line of sight due to variation in the galactic magnetic field; another is the line of sight passes through multiple clouds. For these mechanisms the maximum expected CP fraction is $q \sim p^2$, where p is the fraction of linearly polarized (LP) light. Kemp & Wolstencroft found CP in six early-type stars (no intrinsic polarization), which they were able to attribute to the first mechanism mentioned above. In all cases, $q \sim 10^{-4}$ in blue light.

Martin showed that the interstellar medium can convert LP light to CP by scattering from partially aligned interstellar grains having a complex index of refraction. This effect was observed for light from the Crab Nebula by Martin, Illing and Angel.

An optically thick circumstellar environment can potentially produce much larger CP than the interstellar medium. Martin suggested that LP light can become CP near a star by multiple scattering in an optically thick asymmetric circumstellar dust cloud. This mechanism was invoked by Bastien, Robert and Nadeau, to explain the CP measured in 6 T-Tauri stars at a wavelength of 768 nm. They found a maximum CP of $q \sim 7 \times 10^{-4}$. Serkowski measured CP of $q = 7 \times 10^{-3}$ for the red supergiant NML Cygni and $q = 2 \times 10^{-3}$ in the long-period variable M star VY Canis Majoris in the H band, ascribing the CP to multiple scattering in circumstellar envelopes. Chrysostomou et al. found CP with q of up to 0.17 in the Orion OMC-1 star-forming region, and explained it by reflection of starlight from aligned oblate grains in the dusty nebula.

Circular polarization of zodiacal light and Milky Way diffuse galactic light was measured at wavelength of 550 nm by Wolstencroft and Kemp. They found values of $q \sim 5 \times 10^{-3}$, which is higher than for ordinary stars, presumably because of multiple scattering from dust grains.

Advantages of Circular Polarization

There are several key advantages for circular polarization over linear polarization, which make it more appealing:

The Faraday Effect

The Faraday effect deals with the interaction between light and magnetic fields. It affects linear, but not circular, polarized signals, and the effects are more severe at lower frequencies, such as C-Band, and not noticeable at higher ones, such as Ku-Band. As signals pass through the atmosphere they become de-polarized, causing undesirable reception of the opposing polarity. Linear polarized feeds are aligned in such a way to compensate for the Faraday effect, usually with the help of a tracking device; corrections can be made either by rotating the feed system or using adjustable polarizer's within the feed system. This can be very time consuming because the alignment must be exact. One result of incorrect alignment is increased interference. Fortunately, this is not a concern for circular polarization since there is no need for exact signal alignment. Ku-Band is at a high enough frequency that Faraday's effect is not a factor.

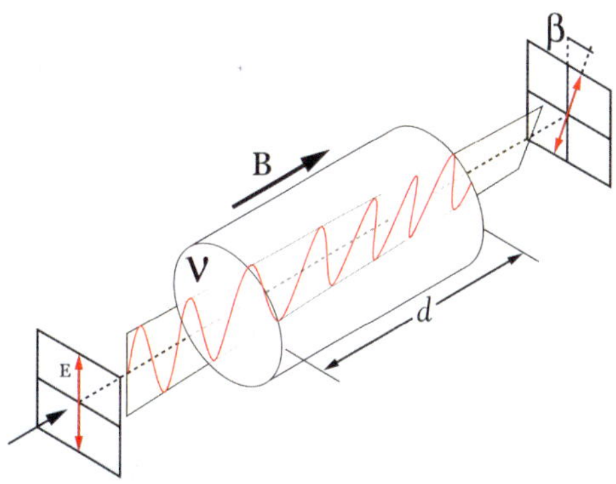

Figure: Faraday's Effect: The Linear Polarized Signal Twists

Atmospheric Conditions

Circular polarization is more resistant to signal degradation due to atmospheric conditions. These conditions can cause changes in the rotation of the signal, and will more adversely affect linear polarization than circular polarization. The effect of a high frequency signal passing through rain can cause signal attenuation and accounts for the majority of the problems with rain fade. Moisture laden clouds are also a factor; by the time a signal passes through a cloud system it can be attenuated by as much as 1dB. Water droplets on the feed horn may also cause detrimental effects. However, the most important aspect to note is that higher frequencies (like Ku-Band) degrade faster, harder, and longer than their frequency counterparts (C-Band).

Easier Installation

The only requirement is ensuring that the antenna is aimed in the correct direction on the satellite; simply point and transmit. This allows for circular feeds to be set up quicker, and there is less of a risk of being misaligned.

Higher Link Reliability

There is higher link reliability since there is a low risk of misalignment, and encountering interference. Faraday's effect will not affect transmission with circular C-band, so there will be no need to readjust the alignment. Finally, because transmission is sent and/or received at different frequencies, interference (cross polarization) is less of a concern.

Availability on the Intelsat Fleet

Circular polarized C-band is available on key roles on the Intelsat fleet.

Table: Breakdown of Polarization on Intelsat satellites.

Circular Polarization:	
C-Band	IS-601, IS-602, IS-603, IS-605, IS-701, IS-702, IS-704, IS-705, IS-706, IS-707, IS-709, IS-801, IS-802, IS-901, IS-902, IS-903, IS-904, IS-905, IS-906, IS-907, IS-10-02
Ku-Band	None
Linear Polarization:	
C-Band	G-3C, G-4R, G-9, G-10R, G-11, G-12, G-13, G-14, G-15, G-16, G-23, G-25, G-26, G-27, G-28, Horizons 1, APR-1, IS-1R, IS-2, IS-3R, IS-5, IS-6B, IS-7, IS-8, IS-805, IS-9, IS-10 and IS-12
Ku-Band	All Satellites

Jones Calculus

The Jones matrix calculus is a matrix formulation of polarized light that consists of 2 × 1 Jones vectors to describe the field components and 2 × 2 Jones matrices to describe polarizing components. While a 2 × 2 formulation is "simpler" than the Mueller matrix formulation the Jones formulation is limited to treating only completely polarized light; it cannot describe unpolarized or partially polarized light. The Jones formulation is used when treating interference phenomena or in problems where field amplitudes must be superposed. A polarized beam propagating through a polarizing element is shown below.

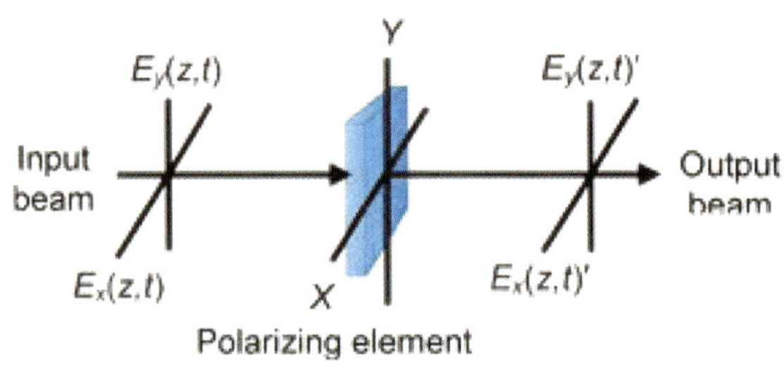

This relationship can be expressed in equations as:

$$E_x' = j_{xx}E_x + j_{xy}E_y$$

$$E_y' = j_{yx}E_x + j_{yy}E_y$$

where E_x', E_y' are the components of the emerging light beam, and E_x, E_y are the components of the incident light beam.

The quantities J_{xx}, J_{xy}, J_{yx}, J_{yy} are the transforming factors of this polarizing element.

We can rewrite in a matrix form:

$$\begin{bmatrix} E_x' \\ E_y' \end{bmatrix} = \begin{bmatrix} J_{xx} & J_{xy} \\ J_{yx} & J_{yy} \end{bmatrix} \begin{bmatrix} E_x \\ E_y \end{bmatrix}$$

Or

$$E' = JE$$

Where

$$J = \begin{bmatrix} J_{xx} & J_{xy} \\ J_{yx} & J_{yy} \end{bmatrix}$$

Here J is a 2x2 matrix, and is called the Jones matrix for the polarizing element.

Jones Matrix for a Polarizer

A linear polarizer doesn't affect the vibration direction of either E_x (vibration along x axis) or E_y (vibration along y axis).

What this means is that E_y contributes nothing to E_x 'since their vibration directions are perpendicular to each other, and a linear polarizer doesn't affect their vibration directions in any way (unlike a wave plate).

This also means that E_x contributes nothing to E_y' for the same reason.

So based on this fact, we can get the following relations:

$$\begin{aligned} E_x' &= p_x E_x \\ E_y' &= p_y E_y \end{aligned} \qquad 0 \le p_x, p_y \le 1$$

$$p_{x,y} = 1 \quad \textit{for } 100\% \text{ transmission}$$

$$p_{x,y} = 0 \quad \textit{for } 0\% \text{ transmission}$$

We can then write in Jones matrix form as:

$$\begin{bmatrix} E_x' \\ E_y' \end{bmatrix} = \begin{bmatrix} p_x & 0 \\ 0 & p_y \end{bmatrix} \begin{bmatrix} E_x \\ E_y \end{bmatrix}$$

So the Jones matrix for a polarizer is

$$J_p = \begin{bmatrix} p_x & 0 \\ 0 & p_y \end{bmatrix} \quad 0 \le p_{x,y} \le 1$$

1. Linear Horizontal Polarizer

For a linear horizontal polarizer, E_x passes through completely, and E_y is blocked completely. so (7) becomes:

$$J_{PH} = \begin{bmatrix} 1 & 0 \\ 0 & 0 \end{bmatrix}$$

2. Linear Vertical Polarizer

For a linear vertical polarizer, E_x is blocked completely, and E_y passes through completely. so (7) becomes:

$$J_{PV} = \begin{bmatrix} 0 & 0 \\ 0 & 1 \end{bmatrix}$$

3. Linear Polarizer Rotated Through an Angle θ

It is more useful to know the Jones matrix for a linear polarizer rotated through an angle θ.

In this sense, linear horizontal polarizer (θ = 0 or π) and vertical polarizer (θ = π/2 or 3π/2) are just two special cases.

We can use rotation transformation to achieve the new J' as:

$$J' = J(-\theta) J \, J(\theta)$$

where J(θ) is the rotation matrix:

$$J(\theta) = \begin{bmatrix} \cos\theta & \sin\theta \\ -\sin\theta & \cos\theta \end{bmatrix}$$

So for a rotated linear polarizer represented by and rotated by angle θ we can get:

$$J' = \begin{bmatrix} \cos\theta & -\sin\theta \\ \sin\theta & \cos\theta \end{bmatrix} \begin{bmatrix} p_x & 0 \\ 0 & p_y \end{bmatrix} \begin{bmatrix} \cos\theta & \sin\theta \\ -\sin\theta & \cos\theta \end{bmatrix}$$

By doing the matrix multiplication, we find that the Jones matrix for a rotated linear polarizer is:

$$J_p(\theta) = \begin{bmatrix} p_x \cos^2\theta + p_y \sin^2\theta & (p_x - p_y)\sin\theta\cos\theta \\ (p_x - p_y)\sin\theta\cos\theta & p_x \sin^2\theta + p_y \cos^2\theta \end{bmatrix}$$

For an ideal linear polarizer we can set $p_x = 1$ and $p_y = 0$, so that Jones matrix for a rotated linear horizontal polarizer is:

$$J_P(\theta) = \begin{bmatrix} cos^2\theta & sin\theta cos\theta \\ sin\theta cos\theta & sin^2\theta \end{bmatrix}$$

We can see that for $\theta = 0°$ (horizontal) and $\theta = 90°$ (vertical), gives the Jones matrices for a linear horizontal polarizer and vertical polarizer as shown by and respectively.

Based on the Jones matrix for a linear polarizer rotated +45° can be calculated as:

$$J_P(45^\circ) = \frac{1}{2}\begin{bmatrix} 1 & 1 \\ 1 & 1 \end{bmatrix}$$

4. Non-ideal Linear Polarizer

If the linear polarizer is not ideal, then the Jones matrix for a polarizer at θ = +45°:

$$J_P(45^\circ) = \frac{1}{2}\begin{bmatrix} p_x + p_y & p_x - p_y \\ p_x - p_y & p_x + p_y \end{bmatrix}$$

Jones Matrix for a Retarder (Phase Shifter, Waveplate)

A waveplate or retarder is an optical device that alters the polarization state of a light wave travelling through it.

Two common types of waveplates are

- Half-wave plate, which shifts the polarization direction of linearly polarized light.
- Quarter-wave plate, which converts linearly polarized light into circularly polarized light and vice versa. A quarter wave plate can be used to produce elliptical polarization as well.

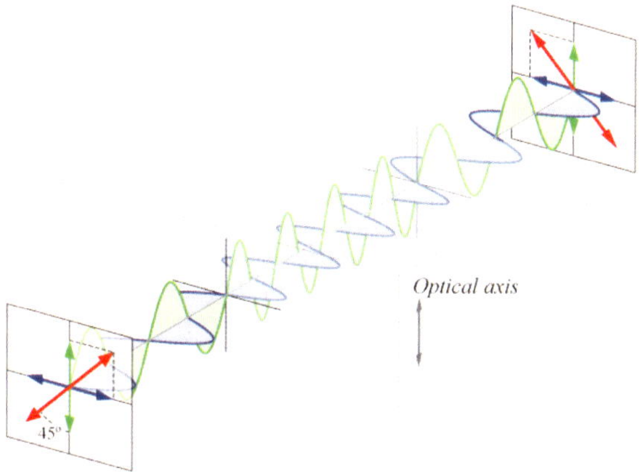

Waveplates are constructed out of a birefringent material (such as quartz or mica), for which the index of refraction is different for different orientations of light passing through it.

The behavior of a waveplate (that is, whether it is a half-wave plate, a quarter-wave plate, etc.) depends on the thickness of the crystal, the wavelength of light, and the variation of the index of refraction.

By appropriate choice of the relationship between these parameters, it is possible to introduce a controlled phase shift between the two polarization components of a light wave, thereby altering its polarization.

The retarder (waveplate) increases the phase by +φ/2 along the fast (x) axis and retards the phase by -φ/2 along the slow (y) axis.

The relationship between the incident light beam components Ex, Ey and the emerging light beam components Ex', Ey' is:

$$E'_x = e^{i\left(+\frac{\varphi}{2}\right)} E_x$$

$$E'_y = e^{i\left(+\frac{\varphi}{2}\right)} E_y$$

We can write in Jones formalism as:

$$J' = \begin{bmatrix} E'_x \\ E'_y \end{bmatrix} = \begin{bmatrix} e^{i\left(+\frac{\varphi}{2}\right)} & 0 \\ 0 & e^{i\left(-\frac{\varphi}{2}\right)} \end{bmatrix} \begin{bmatrix} E_x \\ E_y \end{bmatrix}$$

From we can get the Jones matrix for a retarder (phase shifter) as:

$$J_R(\varphi) = \begin{bmatrix} e^{i\left(+\frac{\varphi}{2}\right)} & 0 \\ 0 & e^{i\left(-\frac{\varphi}{2}\right)} \end{bmatrix}$$

where φ is the total phase shift between Ex' and Ey'.

1. Jones matrix for quarter-waveplate

For a quarter-waveplate, the phase different φ = 90°. So we get:

$$J_R\left(\frac{ë}{4}\right) = \begin{bmatrix} e^{i\left(+\frac{\pi}{2}\right)} & 0 \\ 0 & e^{i\left(-\frac{\pi}{2}\right)} \end{bmatrix} = e^{i\frac{\pi}{4}} \begin{bmatrix} 1 & 0 \\ 0 & e^{-i\frac{\pi}{4}} \end{bmatrix}$$

2. Jones matrix for half-waveplate

For a half-waveplate, the phase difference φ = 180°. So we get:

$$J_R\left(\frac{ë}{2}\right) = \begin{bmatrix} e^{i\left(+\frac{\pi}{2}\right)} & 0 \\ 0 & e^{i\left(-\frac{\pi}{2}\right)} \end{bmatrix} = \begin{bmatrix} \cos\left(\frac{\pi}{2}\right) + i\sin\left(\frac{\pi}{2}\right) & 0 \\ 0 & \cos\left(\frac{-\pi}{2}\right) + i\sin\left(\frac{-\pi}{2}\right) \end{bmatrix}$$

$$= \begin{bmatrix} i & 0 \\ 0 & -i \end{bmatrix} i \begin{bmatrix} 1 & 0 \\ 0 & -1 \end{bmatrix}$$

In , we used Euler's formula as shown below. Learn more about Euler's formula for complex analysis here.

$$e^{i\theta} = \cos\theta + i\sin\theta$$

3. Jones matrix for a rotated waveplate (retarder)

Just as the case for linear polarizer rotated through an angle θ as shown in (10), we can also conclude the Jones matrix for a rotated waveplate by angle θ.

The conclusion process involves using (19), (10), and (11) and the resultant Jones matrix is:

$$J_R(\varphi,\theta) = \begin{bmatrix} e^{i\frac{\varphi}{0}}\cos^2\theta + e^{i\frac{-\varphi}{2}}\sin^2\theta & \left(e^{i\frac{\varphi}{2}} - e^{i\frac{-\varphi}{2}}\right)\sin\theta\cos\theta \\ \left(e^{i\frac{\varphi}{2}} - e^{i\frac{-\varphi}{2}}\right)\sin\theta\cos\theta & e^{i\frac{\varphi}{2}}\sin^2\theta + e^{i\frac{-\varphi}{2}}\cos^2\theta \end{bmatrix}$$

With the half-angle formulas, can be written as:

$$J_R(\varphi,\theta) = \begin{bmatrix} \cos\frac{\varphi}{2} + i\sin\frac{\varphi}{2}\cos 2\theta & i\sin\frac{\varphi}{2}\sin 2\theta \\ i\sin\frac{\varphi}{2}\sin 2\theta & \cos\frac{\varphi}{2} - i\sin\frac{\varphi}{2}\cos 2\theta \end{bmatrix}$$

From we get the Jones matrix for a quarter-waveplate as:

$$J_R\left(\frac{\lambda}{4},\theta\right) = \begin{bmatrix} \frac{1}{\sqrt{(2)}} + \frac{i}{\sqrt{(2)}}\cos 2\theta & \frac{i}{\sqrt{(2)}}\sin 2\theta \\ \frac{i}{\sqrt{(2)}}\sin 2\theta & \frac{1}{\sqrt{(2)}} - \frac{i}{\sqrt{(2)}}\cos 2\theta \end{bmatrix}$$

Similarly, we get the Jones matrix for a half-waveplate as:

$$J_R\left(\frac{\lambda}{2},\theta\right) = i\begin{bmatrix} \cos 2\theta & \sin 2\theta \\ \sin 2\theta & -\cos 2\theta \end{bmatrix}$$

The factor i in is unimodular and can be suppressed. It is common, therefore, to write simply as:

$$J_R\left(\frac{\lambda}{2},\theta\right) = \begin{bmatrix} \cos 2\theta & \sin 2\theta \\ \sin 2\theta & -\cos 2\theta \end{bmatrix}$$

Jones Matrix for a Rotator (I.e. Faraday Rotator)

A polarization rotator rotates the polarization state (polarization ellipse) of an incident beam by an angle θ as shown below.

We define angle θ to be positive for counterclockwise rotation, and negative for clockwise rotation.

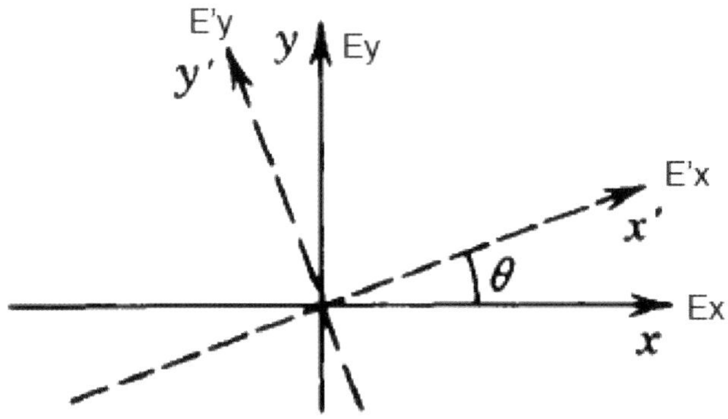

So from amplitudes projection, we see that:

- Ex contributes (cosθ•Ex) to E'x, Ey contributes (sinθ•Ey) to E'x
- Ex contributes (-sinθ•Ex) to E'y, Ey contributes (cosθ•Ey) to E'y

Thus we write them into an equation:

$$E'_x = \cos\theta\, E_x + \sin\theta\, E_y$$
$$E'_y = -\sin\theta\, E_x + \cos\theta\, E_y$$

We put them into the matrix form as:

$$E' = \begin{bmatrix} E'_x \\ E'_y \end{bmatrix} = \begin{bmatrix} \cos\theta & \sin\theta \\ -\sin\theta & \cos\theta \end{bmatrix} \begin{bmatrix} E_x \\ E_y \end{bmatrix}$$

So the Jones matrix for a polarization rotator is:

$$J_{ROT} = \begin{bmatrix} \cos\theta & \sin\theta \\ -\sin\theta & \cos\theta \end{bmatrix}$$

Jones matrix for a rotated (angle φ) rotator

fact:

The incident beam's polarization ellipse (state) can only be rotated by an amount θ intrinsic to a rotator. The mechanical rotation of a rotator does not affect the rotation angle θ at all!

In other words, no matter how you rotate a rotator by any angle φ, the emerging beam's polarization ellipse is only rotated by an angle θ in respective to the incident beam's polarization ellipse.

Proof:

In reference to $J' = J(-\theta) J\, J(\theta)$, $J(\theta) = \begin{bmatrix} \cos\theta & \sin\theta \\ -\sin\theta & \cos\theta \end{bmatrix}$ and $J_{ROT} = \begin{bmatrix} \cos\theta & \sin\theta \\ -\sin\theta & \cos\theta \end{bmatrix}$, the Jones matrix for a rotated rotator (by an angle φ) is:

$$J_{ROT}(\varphi) = J(-\varphi) J_{ROT} J(\varphi)$$

we then get:

$$J_{ROT}(\varphi)=\begin{bmatrix}\cos\varphi & -\sin\varphi\\ \sin\varphi & \cos\varphi\end{bmatrix}\begin{bmatrix}\cos\theta & \sin\theta\\ -\sin\theta & \cos\theta\end{bmatrix}\begin{bmatrix}\cos\theta & \sin\theta\\ -\sin\theta & \cos\theta\end{bmatrix}$$

After carrying out the matrix multiplication in, we then get:

$$J_{ROT}(\varphi)=\begin{bmatrix}\cos\theta & \sin\theta\\ -\sin\theta & \cos\theta\end{bmatrix}=J_{ROT}$$

Isn't that amazing?

Jones Matrix Calculus Application Examples

An application of the Jones matrix calculus is to determine the intensity of an output beam when a rotating polarizer is placed between two crossed polarizers as shown below.

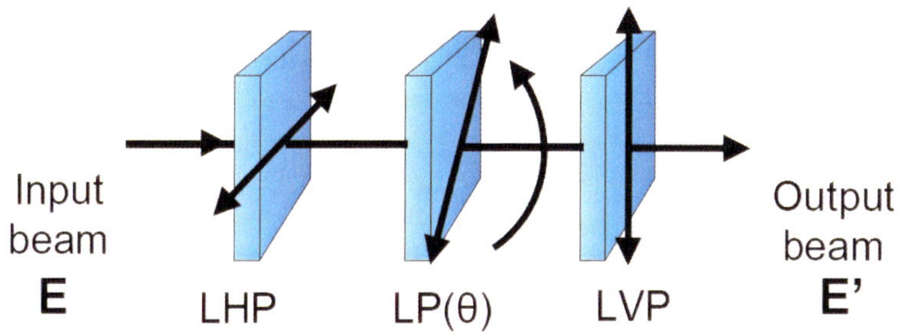

LHP = Linear Horizontal Polarizer
LP = Linear Polarizer
LVP = Linear Vertical Polarizer

So first we need to find the Jones matrix for the whole system. The output beam E' is

$$E' = J \bullet E$$

In reference we then get the Jones matrix for the whole system J as:

$$\begin{aligned}J &= J_{LVP}\bullet J_{LP}(\theta)\bullet J_{LHP}\\ &=\begin{bmatrix}0 & 0\\ 0 & 1\end{bmatrix}\begin{bmatrix}\cos^2\theta & \sin\theta\cos\theta\\ \sin\theta\cos\theta & \sin^2\theta\end{bmatrix}\begin{bmatrix}1 & 0\\ 0 & 0\end{bmatrix}\\ &=\begin{bmatrix}0 & 0\\ \sin\theta\cos\theta & 0\end{bmatrix}\end{aligned}$$

Since the first LHP is a linear polarizer on x axis, as seen equation the normalized Jones vector for linear horizontally polarized light is

$$E = \hat{x} = \begin{bmatrix}1\\ 0\end{bmatrix}$$

Thus since $E' = J \bullet E$, we then get

$$E' = J \cdot E = \begin{bmatrix} 0 & 0 \\ \sin\theta\cos\theta & 0 \end{bmatrix} \begin{bmatrix} 1 \\ 0 \end{bmatrix} = \begin{bmatrix} 0 \\ \sin\theta\cos\theta \end{bmatrix}$$

From the same reference article, we find the complex transpose $E^{\dagger}$ of output beam E' as

$$E^{\dagger} = \begin{bmatrix} 0 & \sin\theta\cos\theta \end{bmatrix}$$

And intensity I' of the output beam is

$$I' = E^{\dagger} \cdot E$$

We then get I'

$$I' = E^{\dagger} \cdot E \begin{bmatrix} 0 & \sin\theta\cos\theta \end{bmatrix} \begin{bmatrix} 0 \\ \sin\theta\cos\theta \end{bmatrix} = \sin^2\theta\cos^2\theta$$

By repeatedly using the following trigonometry formulas:

$$\sin^2\alpha = \frac{1}{2}(1 - \cos 2\alpha)$$

$$\cos^2\alpha = \frac{1}{2}(1 + \cos 2\alpha)$$

We can then get:

$$I' = \frac{1 - \cos 4\theta}{8}$$

We can plot I' in respective to θ in a graph.

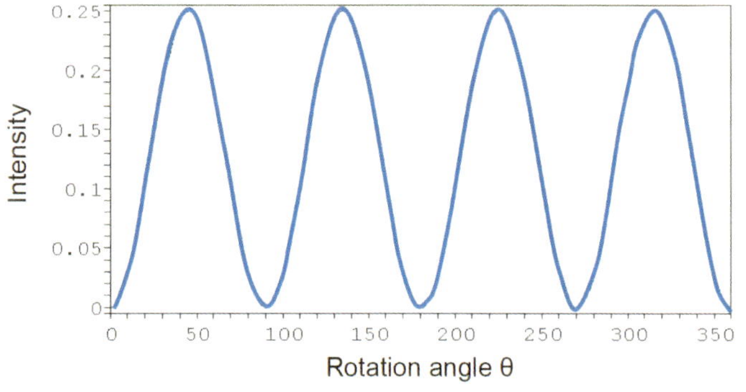

Polarizer

Polarizer Selection Guide

Polarization is an important characteristic of light. Polarizers are key optical elements for controlling your polarization, transmitting a desired polarization state while reflecting, absorbing or deviating the rest. There is a wide variety of polarizer designs, each with its own advantages and disadvantages. To help you select the best polarizer for your application, we will discuss polarizer specifications as well as the different classes of polarizer designs to help you select the best polarizer for your application.

Polarizer Characteristics

Polarizers are defined by a few key parameters, some of which are specific to polarization optics. The most important characteristics are:

Extinction Ratio and Degree of Polarization: The polarizing properties of a linear polarizer are typically defined by the degree of polarization or polarization efficiency, P, and its extinction ratio, ρp. Following the formalism given in the Handbook of Optics, the principal transmittances of the polarizer are T_1 and T_2. T_1 is the maximum transmission of the polarizer and occurs when the axis of the polarizer is parallel to the plane of polarization of the incident polarized beam. T_2 is the minimum transmission of the polarizer and occurs when the axis of the polarizer is perpendicular to the plane of polarization of the incident polarized beam.

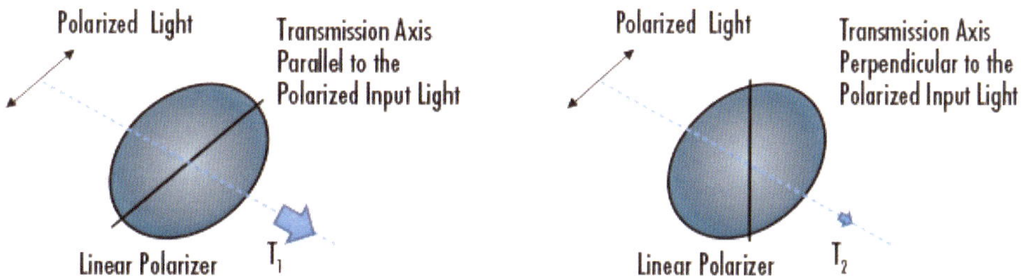

$$P = (T_1 - T_2) / (T_1 + T_2) \text{ and, } \rho_p = T_2 / T_1$$

The extinction performance of a linear polarizer is often expressed as $1 / \rho_p : 1$ This parameter ranges from less than 100:1 for economical sheet polarizers to 10^6:1 for high quality birefringent crystalline polarizers. The extinction ratio typically varies with wavelength and incident angle and must be evaluated along with other factors like cost, size, and polarized transmission for a given application.

Transmission: This value either refers to the transmission of light polarized linearly in the direction of polarization axis, or to the transmission of unpolarized light through the polarizer. Parallel transmission is the transmission of unpolarized light through two polarizers with their polarization axes aligned in parallel, while crossed transmission is the transmission of unpolarized light through two polarizers with their polarization axes crossed. For ideal polarizers transmission of linearly polarized light parallel to the polarization axis is 100%, parallel transmission is 50% and crossed transmission is 0%. This can be calculated with Malus' law.

Acceptance angle: The acceptance angle is the largest deviation from design incidence angle at which the polarizer will still perform within specifications. Most polarizers are designed to work at an incidence angle of 0° or 45°, or at Brewster's angle. The acceptance angle is important for alignment but has particular importance when working with non-collimated beams. Wire grid and dichroic polarizers have the largest acceptance angles, up to a full acceptance angle of almost 90°.

Construction: Polarizers come in many forms and designs. Thin film polarizers are thin films similar to optical filters. Polarizing plate beamsplitters are thin, flat plates placed at an angle to the beam. Polarizing cube beamsplitters consist of two right angle prisms mounted together at the hypothenus. Birefringent polarizers consist of two crystalline prisms mounted together, where the angle of the prisms is determined by the specific polarizer design.

Clear aperture: the clear aperture is typically most restrictive for birefringent polarizers as the availability of optically pure crystals limits the size of these polarizers. Dichroic polarizers have the largest available clear apertures as their fabrication lends itself to larger sizes.

Optical path length: the length light must travel through the polarizer. Important for dispersion, damage thresholds and space constraints, optical path lengths can be significant in birefringent polarizers but are usually short in dichroic polarizers.

Damage threshold: the laser damage threshold is determined by the material used as well as the polarizer design, with birefringent polarizers typically having the highest damage threshold. Cement is often the most susceptible element to laser damage, which is why optically contacted beamsplitters or air spaced birefringent polarizers have higher damage thresholds.

Cost: Some polarizers require large, very pure crystals, which are expensive, while others are made of stretched plastic, which make them more economical.

Polarizers can be categorized into two groups:

- Absorptive Polarizers
- Beam-Splitting Polarizers

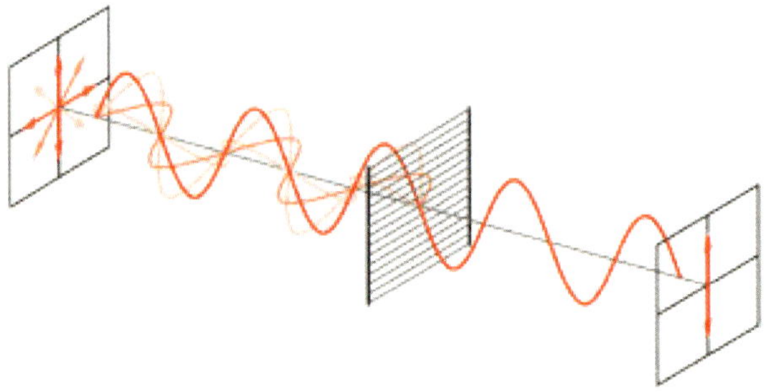

Polarizers only allow light to pass if it has a particular angle of polarization.

Absorptive Polarizers

Wire-grid polarizer is the simplest polarizer in concept. Wire-grid polarizer is composed of a regular array of fine metallic wires that is parallel to one another. It is then placed in a plane at 90 degrees to the incident beam that is unpolarized.

If the waves in the unpolarized beam that are at 90 degrees to the fine metallic wire, the unpolarized beam is able to travel through the grid with very little energy being reflected or lost. However if the waves in the unpolarized beam is parallel to the wires, then it will be absorbed or reflected. In simple terms, it is only when light travels in a certain direction will pass through the polarizer, and the rest of the light beam will be absorbed or reflected.

The distance separating the wires must be less than the radiation wavelength for any practical uses of the wire-grid polarizer. Wire-grid polarizers are generally applied only for microwaves and for infrared light in the far- and mid-range.

Beam-Splitting Polarizers

The incident beam is divided into two differing polarization beams in beam-splitting polarizers. Only one of the two differing polarization beam is fully polarized for many common beam-splitting polarizers.

Since beam splitting polarizers does not absorb or reflect the energy of the rejected polarization state, beam-splitting polarizers are more suited for applications such as laser light. True polarizing beamsplitters are also practical where the two polarization components are to be studied or simultaneously used.

Reflective polarizers transmit the desired polarization and reflect the rest. They either use a wire grid, Brewster's angle or interference effects. Brewster's angle is the angle at which, based on the Fresnel equations, only s-polarized light is reflected. Because the p-polarized light is not reflected while the s-polarized light is partially reflected, the transmitted light is enriched in p-polarization.

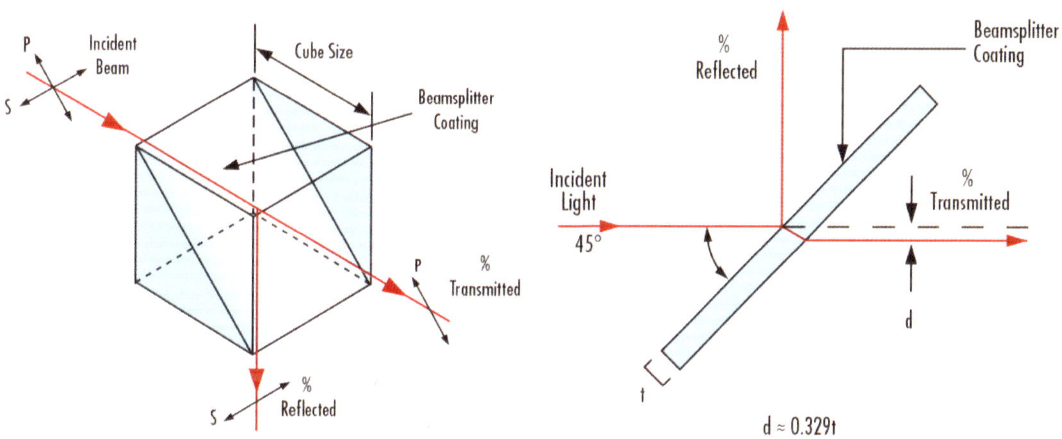

Figure: Reflective polarizers, available as cube beamsplitters, plate beamsplitters or thin films, reflect the unwanted polarization state

Type	Applications	Wavelength Range (nm)	Laser Damage Threshold	Cost
Brewster Windows	Laser Cavities	200 - 2200	High	$
Polarizing Plate Beamsplitters	Space/Weight Constrained Applications, Low Cost & Low Extinction Ratio Applications, Femtosecond Laser Applications	250 - 1550	Low to High	$
Polarizing Cube Beamsplitters	Beam Combining	400 - 1100	Low to High	$$
Wire Grid Polarizers	Demanding Environments, Broadband Applications, Infrared, Uncollimated Light	300 - 15000	Low to High	$$ - $$$
Wire Grid Polarizing Cube Beamsplitters	Broadband Applications, Uncollimated Light	400 - 700	Low to High	$$

Polarization by Fresnel Reflection

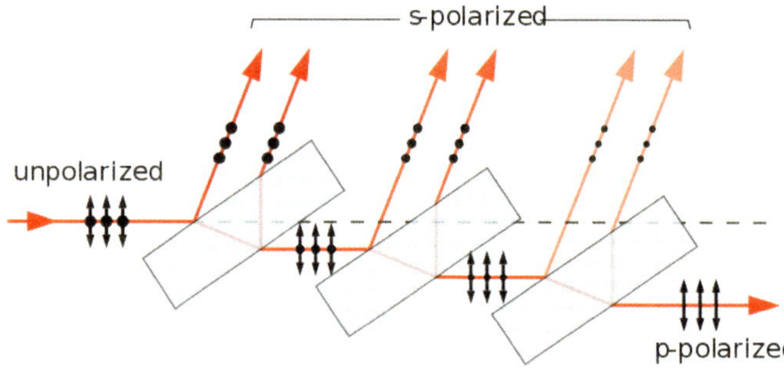

A stack of plates at Brewster's angle to a beam reflects off a fraction of the *s*-polarized light at each surface, leaving a *p*-polarized beam. Full polarization at Brewster's angle requires many more plates than shown. The arrows indicate the direction of the electrical field, not the magnetic field, which is perpendicular to the electric field

When light reflects (by Fresnel reflection) at an angle from an interface between two transparent materials, the reflectivity is different for light polarized in the plane of incidence and light polarized perpendicular to it. Light polarized in the plane is said to be *p*-polarized, while that polarized perpendicular to it is *s*-polarized. At a special angle known as Brewster's angle, no *p*-polarized light is reflected from the surface, thus all reflected light must be *s*-polarized, with an electric field perpendicular to the plane of incidence.

A simple linear polarizer can be made by tilting a stack of glass plates at Brewster's angle to the beam. Some of the *s*-polarized light is reflected from each surface of each plate. For a stack of plates, each reflection depletes the incident beam of *s*-polarized light, leaving a greater fraction of *p*-polarized light in the transmitted beam at each stage. For visible light in air and typical glass, Brewster's angle is about 57°, and about 16% of the *s*-polarized light present in the beam is reflected for each air-to-glass or glass-to-air transition. It takes many plates to achieve even mediocre polarization of the transmitted beam with this approach. For a stack of 10 plates (20 reflections), about 3% (= $(1-0.16)^{20}$) of the *s*-polarized light is transmitted. The reflected beam, while fully polarized, is spread out and may not be very useful.

A more useful polarized beam can be obtained by tilting the pile of plates at a steeper angle to the incident beam. Counterintuitively, using incident angles greater than Brewster's angle yields a higher degree of polarization of the *transmitted* beam, at the expense of decreased overall transmission. For angles of incidence steeper than 80° the polarization of the transmitted beam can approach 100% with as few as four plates, although the transmitted intensity is very low in this case. Adding more plates and reducing the angle allows a better compromise between transmission and polarization to be achieved.

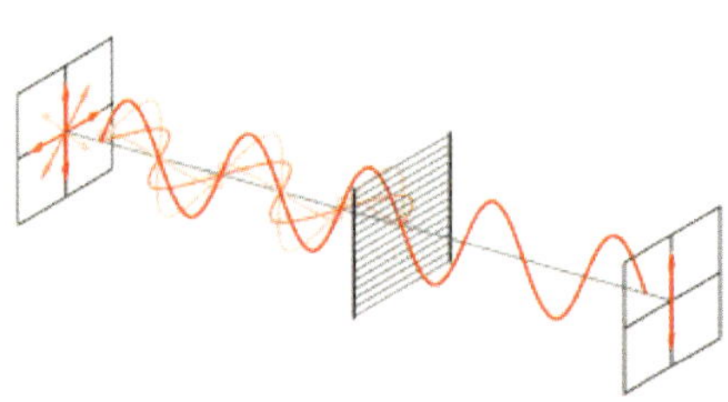

A wire-grid polarizer converts an unpolarized beam into one with a single linear polarization. Coloured arrows depict the electric field vector. The diagonally polarized waves also contribute to the transmitted polarization. Their vertical components are transmitted (shown), while the horizontal components are absorbed and reflected (not shown).

Because their polarization vectors depend on incidence angle, polarizers based on Fresnel reflection inherently tend to produce *s–p* polarization rather than Cartesian polarization, which limits their use in some applications.

Birefringent Polarizers

Other linear polarizers exploit the birefringent properties of crystals such as quartz and calcite. In these crystals, a beam of unpolarized light incident on their surface is split by refraction into two rays. Snell's law holds for both of these rays, the *ordinary* or *o*-ray, and the *extraordinary* or *e*-ray, with each ray experiencing a different index of refraction (this is called double refraction). In general the two rays will be in different polarization states, though not in linear polarization states except for certain propagation directions relative to the crystal axis.

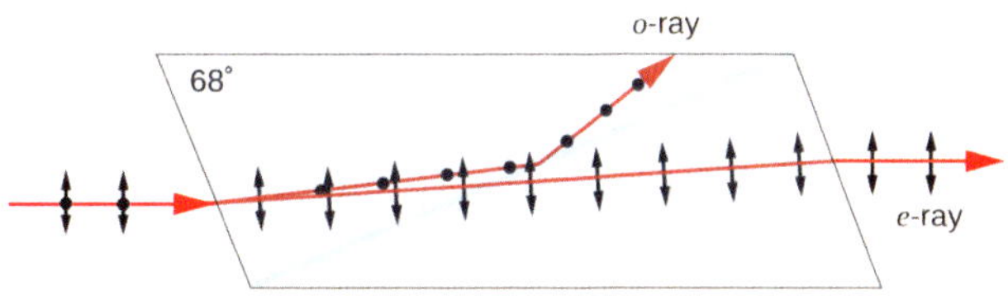

A Nicol prism

A Nicol prism was an early type of birefringent polarizer, that consists of a crystal of calcite which has been split and rejoined with Canada balsam. The crystal is cut such that the *o*- and *e*-rays are in orthogonal linear polarization states. Total internal reflection of the *o*-ray occurs at the balsam interface, since it experiences a larger refractive index in calcite than in the balsam, and the ray is deflected to the side of the crystal. The *e*-ray, which sees a smaller refractive index in the calcite, is transmitted through the interface without deflection. Nicol prisms produce a very high purity of polarized light, and were extensively used in microscopy, though in modern use they have been mostly replaced with alternatives such as the Glan–Thompson prism, Glan–Foucault prism, and Glan–Taylor prism. These prisms are not true polarizing beamsplitters since only the transmitted beam is fully polarized.

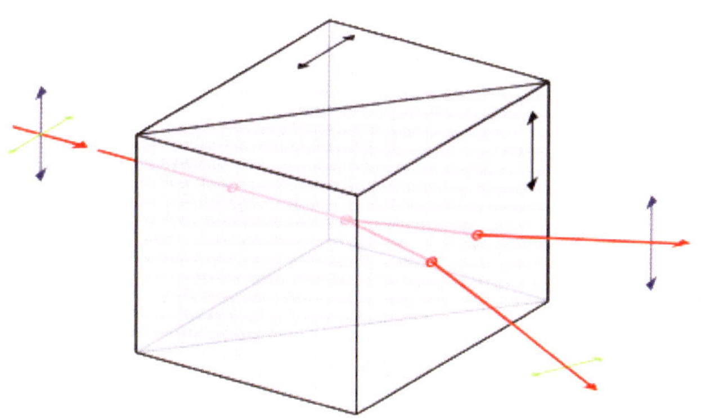

A Wollaston prism

A Wollaston prism is another birefringent polarizer consisting of two triangular calcite prisms with orthogonal crystal axes that are cemented together. At the internal interface, an unpolarized beam splits into two linearly polarized rays which leave the prism at a divergence angle of 15°–45°. The Rochon and Sénarmont prisms are similar, but use different optical axis orientations in the two prisms. The Sénarmont prism is air spaced, unlike the Wollaston and Rochon prisms.

These prisms truly split the beam into two fully polarized beams with perpendicular polarizations. The Nomarski prism is a variant of the Wollaston prism, which is widely used in differential interference contrast microscopy.

Thin Film Polarizers

Thin-film linear polarizers are glass substrates on which a special optical coating is applied. Either Brewster's angle reflections or interferenceeffects in the film cause them to act as beam-splitting polarizers. The substrate for the film can either be a plate, which is inserted into the beam at a particular angle, or a wedge of glass that is cemented to a second wedge to form a cube with the film cutting diagonally across the center (one form of this is the very common MacNeille cube). Thin-film polarizers generally do not perform as well as Glan-type polarizers, but they are inexpensive and provide two beams that are about equally well polarized. The cube-type polarizers generally perform better than the plate polarizers. The former are easily confused with Glan-type birefringent polarizers.

Wire-grid Polarizers

One of the simplest linear polarizers is the *wire-grid polarizer* (WGP), which consists of many fine parallel metallic wires that are placed in a plane. WGPs mostly reflect the non-transmitted polarization and can thus be used as polarizing beam splitters. The parasitic absorption is relatively high compared to most of the dielectric polarizers though much lower than in absorptive polarizers.

Electromagnetic waves which have a component of their electric fields aligned parallel to the wires will induce the movement of electrons along the length of the wires. Since the electrons are free to move in this direction, the polarizer behaves in a similar manner to the surface of a metal when reflecting light, and the wave is reflected backwards along the incident beam (minus a small amount of energy lost to Joule heating of the wire).

For waves with electric fields perpendicular to the wires, the electrons cannot move very far across the width of each wire. Therefore, little energy is reflected and the incident wave is able to pass through the grid. In this case the grid behaves like a dielectric material.

Overall, this causes the transmitted wave to be linearly polarized with an electric field that is completely perpendicular to the wires. The hypothesis that the waves "slip through" the gaps between the wires is incorrect.

For practical purposes, the separation between wires must be less than the wavelength of the incident radiation. In addition, the width of each wires should be small compared to the spacing between wires. Therefore, it is relatively easy to construct wire-grid polarizers for microwaves, far-infrared, and mid-infrared radiation. In addition, advanced lithographictechniques can also build very tight pitch metallic grids, allowing for the polarization of visible light to a useful degree. Since the degree of polarization depends little on wavelength and angle of incidence, they are used for broad-band applications such as projection.

Analytical solutions using rigorous coupled-wave analysis for wire grid polarizers have shown that for electric field components perpendicular to the wires, the medium behaves like a dielectric, and for electric field components parallel to the wires, the medium behaves like a metal (reflective).

Dichroic Polarizers

Dichroic polarizers transmit the desired polarization and absorb the rest. This is achieved via anisotropy in the polarizer; common examples are oriented polymer molecules and stretched nanoparticles. This is a broad class of polarizers, going from low cost laminated plastic polarizers to precision high cost glass nanoparticle polarizers. Most dichroic polarizers have good extinction ratios relative to their cost. Their damage thresholds and environmental stability are often limited, although glass dichroic polarizers outperform plastic dichroic polarizers in this aspect. Dichroic polarizers are well suited for microscopy, imaging and display applications, and are often the only choice when very large apertures are necessary.

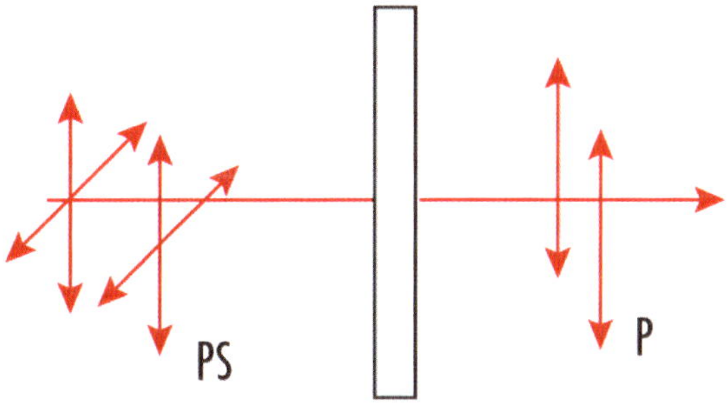

Figure : Dichroic polarizers absorb the unwanted polarization state

Type	Applications	Wavelength Range (nm)	Laser Damage Threshold	Cost
Dichroic	Microscopy, Display, Imaging, Intensity Adjustment	300 - 2700	Low to Medium	$ - $$$

Specifications	
Bandwidth (nm)	100 to > 1000
Extinction Ratio	10^2 - 10^6
Transmission	> 50% to > 90%
Acceptance Angle (°)	40 - 80
Clear Aperture	10 - 900
Optical Path Length (mm)	0.2 - 2
Transmitted Wavefront Quality	λ/5 - λ/2

Malus's Law and other Properties

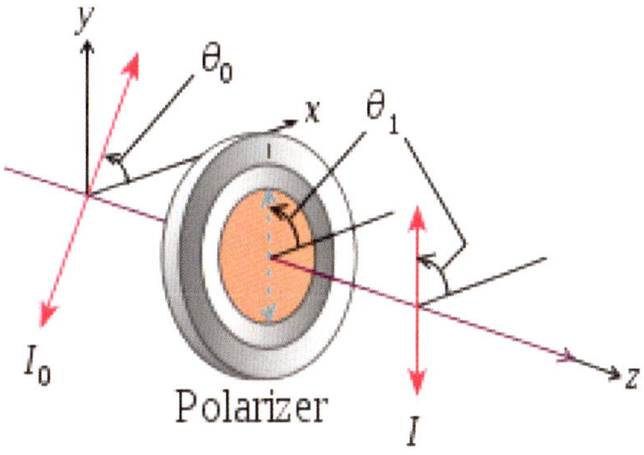

Polarization of light. In this picture, $\theta_1 - \theta_o = \theta_i$.

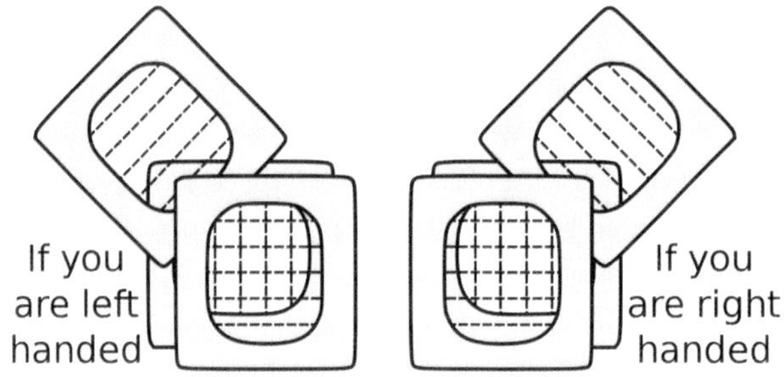

If you have 3 linear filters, hold two filters crossed to block the light with your clumsy hand and use your clever hand to insert third at 45°.

Malus's law, which is named after Étienne-Louis Malus, says that when a perfect polarizer is placed in a polarized beam of light, the irradiance, *I*, of the light that passes through is given by

$$I = I_0 \cos^2 \theta_i ,$$

where I_0 is the initial intensity and θ_i is the angle between the light's initial polarization direction and the axis of the polarizer.

A beam of unpolarized light can be thought of as containing a uniform mixture of linear polarizations at all possible angles. Since the average value of $\cos^2 \theta$ is 1/2, the transmission coefficient becomes

$$\frac{I}{I_0} = \frac{1}{2}.$$

In practice, some light is lost in the polarizer and the actual transmission will be somewhat lower than this, around 38% for Polaroid-type polarizers but considerably higher (>49.9%) for some birefringent prism types.

If two polarizers are placed one after another (the second polarizer is generally called an *analyzer*), the mutual angle between their polarizing axes gives the value of θ in Malus's law. If the two axes are orthogonal, the polarizers are *crossed* and in theory no light is transmitted, though again practically speaking no polarizer is perfect and the transmission is not exactly zero (for example, crossed Polaroid sheets appear slightly blue in colour). If a transparent object is placed between the crossed polarizers, any polarization effects present in the sample (such as birefringence) will be shown as an increase in transmission. This effect is used in polarimetry to measure the optical activity of a sample.

Real polarizers are also not perfect blockers of the polarization orthogonal to their polarization axis; the ratio of the transmission of the unwanted component to the wanted component is called the *extinction ratio*, and varies from around 1:500 for Polaroid to about $1{:}10^6$ for Glan–Taylor prism polarizers.

In X-ray the Malus's law (relativistic form):

$$I = I_0 \frac{f}{f_0} \left[1 + \frac{\lambda(f_0 - f)}{2c} \right] \cos^2 \theta_i$$

where f_0 – frequency of the polarized radiation falling on the polarizer, f – frequency of the radiation passes through polarizer, λ – Compton wavelength of electron, c – speed of light in vacuum.

Circular Polarizers

Circular polarizers, also referred to as *circular polarizing filters*, can be used to create circularly polarized light or alternatively to selectively absorb or pass clockwise and counter-clockwise circularly polarized light. They are used as polarizing filters in photography to reduce oblique reflections from non-metallic surfaces, and are the lenses of the 3D glassesworn for viewing some stereoscopic movies, where the polarization of light is used to differentiate which image should be seen by the left and right eye.

Creating Circularly Polarized Light

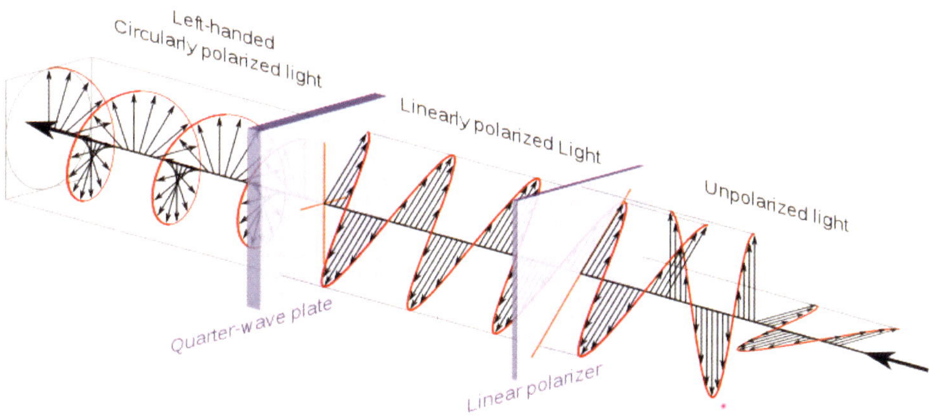

Circular polarizer creating left-handed circularly polarized light. It is considered left-handed as viewed from the receiver and right-handed as viewed from the source.

There are several ways to create circularly polarized light, the cheapest and most common involves placing a quarter-wave plate after a linear polarizer and directing unpolarized lightthrough the linear polarizer. The linearly polarized light leaving the linear polarizer is transformed into circularly polarized light by the quarter wave plate. The transmission axis of the linear polarizer needs to be half way (45°) between the fast and slow axes of the quarter-wave plate.

In the arrangement above, the transmission axis of the linear polarizer is at a positive 45° angle relative to the right horizontal and is represented with an orange line. The quarter-wave plate has a horizontal slow axis and a vertical fast axis and they are also represented using orange lines. In this instance the unpolarized light entering the linear polarizer is displayed as a single wave whose amplitude and angle of linear polarization are suddenly changing.

When one attempts to pass unpolarized light through the linear polarizer, only light that has its electric field at the positive 45° angle leaves the linear polarizer and enters the quarter-wave plate. In the illustration, the three wavelengths of unpolarized light represented would be transformed into the three wavelengths of linearly polarized light on the other side of the linear polarizer.

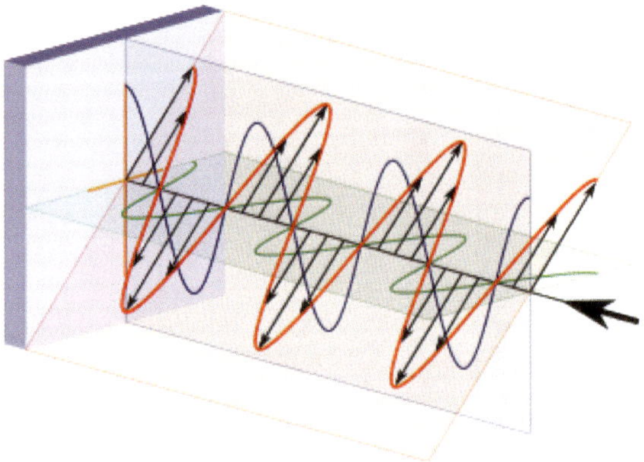

Linearly polarized light, represented using components, entering a quarter-wave plate.
The blue and green curves are projections of the red line on the vertical and horizontal planes respectively.

In the illustration toward the right is the electric field of the linearly polarized light just before it enters the quarter-wave plate. The red line and associated field vectors represent how the magnitude and direction of the electric field varies along the direction of travel. For this plane electromagnetic wave, each vector represents the magnitude and direction of the electric field for an entire plane that is perpendicular to the direction of travel. Refer to these two images in the plane wave article to better appreciate this.

Light and all other electromagnetic waves have a magnetic field which is in phase with, and perpendicular to, the electric field being displayed in these illustrations.

To understand the effect the quarter-wave plate has on the linearly polarized light it is useful to think of the light as being divided into two components which are at right angles (orthogonal) to each other. Towards this end, the blue and green lines are projections of the red line onto the vertical and horizontal planes respectively and represent how the electric field changes in the direction of those two planes. The two components have the same amplitude and are in phase.

Because the quarter-wave plate is made of a birefringent material, when in the wave plate, the light travels at different speeds depending on the direction of its electric field. This means that the horizontal component which is along the slow axis of the wave plate will travel at a slower speed than the component that is directed along the vertical fast axis. Initially the two components are in phase, but as the two components travel through the wave plate the horizontal component of the light drifts farther behind that of the vertical. By adjusting the thickness of the wave plate one can control how much the horizontal component is delayed relative to vertical component before the light leaves the wave plate and they begin again to travel at the same speed. When the light leaves the quarter-wave plate the rightward horizontal component will be exactly one quarter of a wavelength behind the vertical component making the light left-hand circularly polarized when viewed from the receiver.

At the top of the illustration toward the right, is the circularly polarized light after it leaves the wave plate, and again directly below it, for comparison purposes, the linearly polarized light that entered the quarter-wave plate. In the upper image, because this is a plane wave, each vector leading from the axis to the helix represents the magnitude and direction of the electric field for an entire plane that is perpendicular to the direction of travel. All the electric field vectors have the same magnitude indicating that the strength of the electric field does not change. The direction of the electric field however steadily rotates.

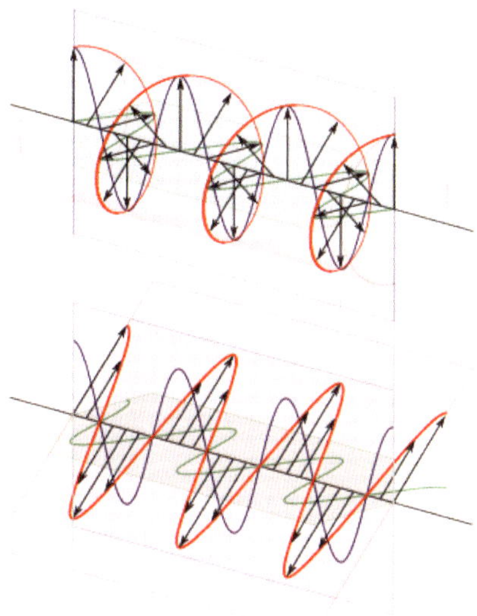

The top image is left-handed/counter-clockwise circularly polarized, as viewed from the receiver. The bottom image is that of linearly polarized light. The blue and green curves are projections of the red lines on the vertical and horizontal planes respectively.

The blue and green lines are projections of the helix onto the vertical and horizontal planes respectively and represent how the electric field changes in the direction of those two planes. Notice how the rightward horizontal component is now one quarter of a wavelength behind the vertical component. It is this quarter of a wavelength phase shift that results in the rotational nature of the electric field. It is significant to note that when the magnitude of one component is at a maximum the magnitude of the other component is always zero. This is the reason that there are helix vectors which exactly correspond to the maxima of the two components.

In the instance just cited, using the handedness convention used in many optics textbooks, the light is considered left-handed/counter-clockwise circularly polarized. Referring to the accompanying animation, it is considered left-handed because if one points one's left thumb *against* the direction of travel, ones fingers curl in the direction the electric field rotates as the wave passes a given point in space. The helix also forms a left-handed helix in space. Similarly this light is considered counter-clockwise circularly polarized because if a stationary observer faces *against* the direction of travel, the person will observe its electric field rotate in the counter-clockwise direction as the wave passes a given point in space.

To create right-handed, clockwise circularly polarized light one simply rotates the axis of the quarter-wave plate 90° relative to the linear polarizer. This reverses the fast and slow axes of the wave plate relative to the transmission axis of the linear polarizer reversing which component leads and which component lags.

In trying to appreciate how the quarter-wave plate transforms the linearly polarized light, it is important to realize that the two components discussed are not entities in and of themselves but are merely mental constructs one uses to help appreciate what is happening. In the case of linearly and circularly polarized light, at each point in space, there is always a single electric field with a distinct vector direction, the quarter-wave plate merely has the effect of transforming this single electric field.

Absorbing and Passing Circularly Polarized Light

Circular polarizers can also be used to selectively absorb or pass right-handed or left-handed circularly polarized light. It is this feature which is utilized by the 3D glasses in stereoscopic cinemas such as RealD Cinema. A given polarizer which creates one of the two polarizations of light will pass that same polarization of light when that light is sent through it in the other direction. In contrast it will block light of the opposite polarization.

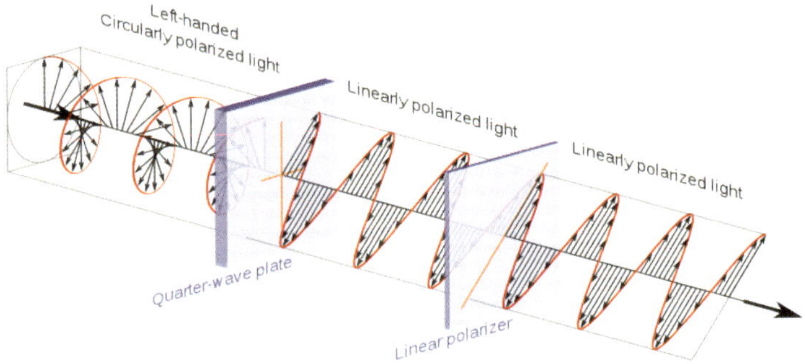

Circular polarizer passing left-handed, counter-clockwise circularly polarized light.
(Left-handed as viewed from the receiver.)

The illustration above is identical to the previous similar one with the exception that the left-handed circularly polarized light is now approaching the polarizer from the opposite direction and linearly polarized light is exiting the polarizer toward the right.

First note that a quarter-wave plate always transforms circularly polarized light into linearly polarized light. It is only the resulting angle of polarization of the linearly polarized light that is determined by the orientation of the fast and slow axes of the quarter-wave plate and the handedness of the circularly polarized light. In the illustration, the left-handed circularly polarized light entering the polarizer is transformed into linearly polarized light which has its direction of polarization along the transmission axis of the linear polarizer and it therefore passes. In contrast right-handed circularly polarized light would have been transformed into linearly polarized light that had its direction of polarization along the absorbing axis of the linear polarizer, which is at right angles to the transmission axis, and it would have therefore been blocked.

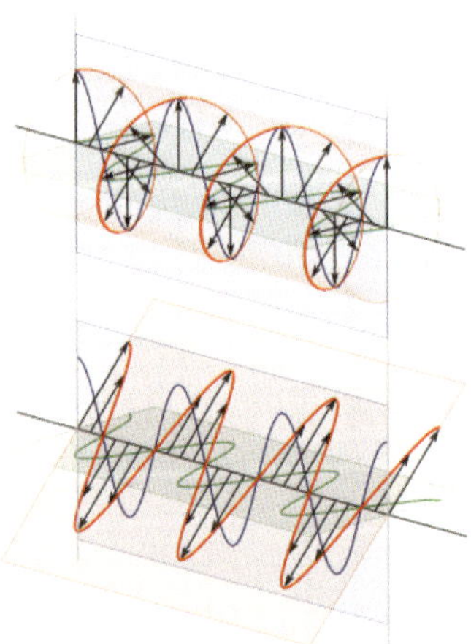

Left-handed/Counter-Clockwise circularly polarized light displayed above linearly polarized light.
The blue and green curves are projections of the helix on the vertical and horizontal planes respectively.

To understand this process, refer to the illustration on the right. It is absolutely identical to the earlier illustration even though the circularly polarized light at the top is now considered to be approaching the polarizer from the left. One can observe from the illustration that the leftward horizontal (as observed looking along the direction of travel) component is leading the vertical component and that when the horizontal component is retarded by one quarter of a wavelength it will be transformed into the linearly polarized light illustrated at the bottom and it will pass through the linear polarizer.

There is a relatively straightforward way to appreciate why a polarizer which creates a given handedness of circularly polarized light also passes that same handedness of polarized light. First, given the dual usefulness of this image, begin by imagining the circularly polarized light displayed at the top as still leaving the quarter-wave plate and traveling toward the left. Observe that had the horizontal component of the linearly polarized light been retarded by a quarter of wavelength twice, which would amount to a full half wavelength, the result would have been linearly polarized light that was at a right angle to the light that entered. If such orthogonally polarized light were rotated on the horizontal plane and directed back through the linear polarizer section of the circular polarizer it would clearly pass through given its orientation. Now imagine the circularly polarized light which has already passed through the quarter-wave plate once, turned around and directed back toward the circular polarizer again. Let the circularly polarized light illustrated at the top now represent that light. Such light is going to travel through the quarter-wave plate a second time before reaching the linear polarizer and in the process, its horizontal component is going to be retarded a second time by one quarter of a wavelength. Whether that horizontal component is retarded by one quarter of a wavelength in two distinct steps or retarded a full half wavelength all at once, the orientation of the resulting linearly polarized light will be such that it passes through the linear polarizer.

Had it been right-handed, clockwise circularly polarized light approaching the circular polarizer from the left, its horizontal component would have also been retarded, however the resulting linearly polarized light would have been polarized along the absorbing axis of the linear polarizer and it would not have passed.

To create a circular polarizer that instead passes right-handed polarized light and absorbs left-handed light, one again rotates the wave plate and linear polarizer 90° relative to each another. It is easy to appreciate that by reversing the positions of the transmitting and absorbing axes of the linear polarizer relative to the quarter-wave plate, one changes which handedness of polarized light gets transmitted and which gets absorbed.

Homogeneous Circular Polarizer

A homogeneous circular polarizer passes one handedness of circular polarization unaltered and blocks the other handedness. This is similar to the way that a linear polarizer would fully pass one angle of linearly polarized light unaltered, but would fully block any linearly polarized light that was orthogonal to it.

A homogeneous circular polarizer can be created by sandwiching a linear polarizer between two quarter-wave plates. Specifically we take the circular polarizer described previously, which transforms circularly polarized light into linear polarized light, and add to it a second quarter-wave plate rotated 90° relative to the first one.

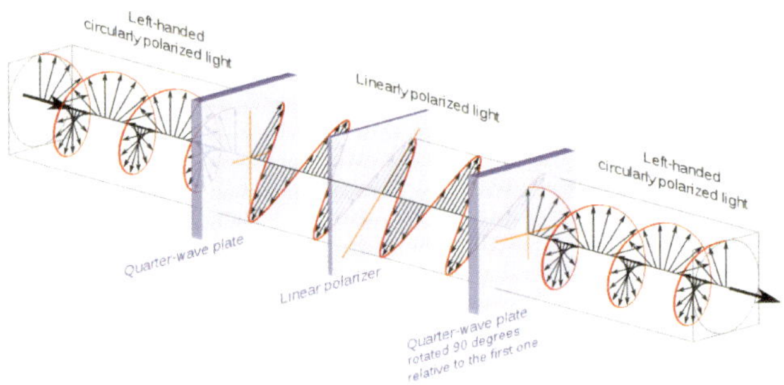

Homogeneous circular polarizer passing left-handed, counter-clockwise circularly polarized light. (Left-handed as viewed from the receiver.)

Generally speaking, and not making direct reference to the above illustration, when either of the two polarizations of circularly polarized light enters the first quarter-wave plate, one of a pair of orthogonal components is retarded by one quarter of a wavelength relative to the other. This creates one of two linear polarizations depending on the handedness the circularly polarized light. The linear polarizer sandwiched between the quarter wave plates is oriented so that it will pass one linear polarization and block the other. The second quarter-wave plate then takes the linearly polarized light that passes and retards the orthogonal component that was not retarded by the previous quarter-wave plate. This brings the two components back into their initial phase relationship, reestablishing the selected circular polarization.

Note that it does not matter in which direction one passes the circularly polarized light.

Circular and Linear Polarizing Filters for Photography

Linear polarizing filters were the first types to be used in photography and can still be used for non-reflex and older SLR cameras. However, cameras with through-the-lens metering and autofocusing systems – that is, all modern SLR and DSLR – rely on optical elements that pass linearly polarized light. If light entering the camera is already linearly polarized, it can upset the exposure or autofocus systems. Circular polarizing filters cut out linearly polarized light and so can be used to darken skies or remove reflections, but the circular polarized light it passes does not impair through-the-lens systems.

Applications of Optical Polarizers

Polarizers have many different applications. Some examples:

- Polarizing sunglasses and photographic filters can help to suppress disturbing reflections of sunlight (glare) on water surfaces, for example, because that reflected light is partially polarized.
- A polarizer within a laser resonator can force the laser to emit linearly polarized light. Here, a high polarization extinction ratio is usually not required; a simple Brewster plate is normally sufficient despite its weak extinction.
- When a laser beam is already linearly polarized, but not with a high degree of polarization, that degree can be improved (and the exact polarization direction stabilized) by sending the beam through a polarizer with a high polarization extinction ratio.

- The combination of a $\lambda/2$ waveplate and a polarizer can act as a variable optical attenuator: rotating the waveplate rotates the direction of polarization and thus the transmission at the polarizer.
- Faraday isolators usually contain two polarizers around a Faraday rotator.
- Two polarized laser beams can be polarization-combined (→ polarization beam combining) at a polarizer. Coherent polarization combining is a technique for power scaling.
- Interferometers often contain polarizing beam splitters.

Brewster's Angle

When light impinges on a flat boundary between two different transparent media, generally at least some part of its optical power is reflected. However, for a particular angle of incidence, which is called Brewster's angle (after Sir David Brewster) or sometimes polarizing angle, that reflection does not occur provided that the light is p-polarized. The latter condition means that the polarization direction (i.e., the direction in which the electric field vector oscillates) lies within the plane of incidence. For s-polarized light, the reflectivity is even higher than for light with normal incidence on the boundary.

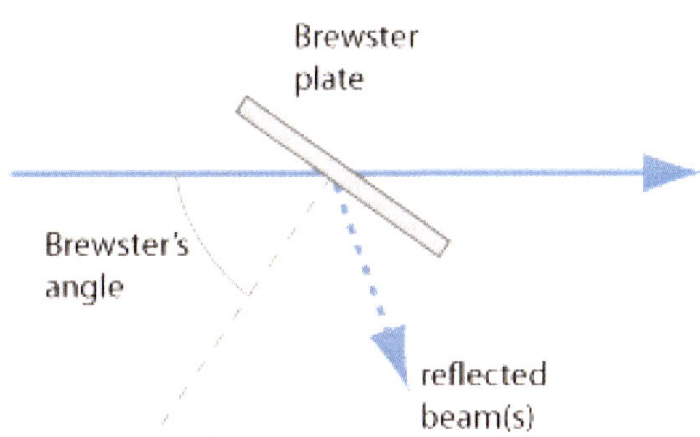

Figure: For obtaining vanishing reflection losses at a Brewster plate, the angle of incidence has to be close to Brewster's angle, and the light must be p-polarized, i.e., the polarization direction is in the plane of incidence.

The magnitude of Brewster's angle depends on the refractive indices of the involved optical media and can be calculated with Brewster's law:

$$\theta_B = \arctan \frac{n_2}{n_1}$$

Here, n_1 and n_2 are the refractive indices of the medium of the incoming beam and the other medium, respectively. One can show that the sum of the angles in both media (relative to the direction for normal incidence) is 90°.

As a numerical example, one can consider light coming from air ($n_1 \approx 1$) to a glass with $n_2 = 1.5$. For that situation, one can calculate that Brewster's angle is ≈56.3°. Figure shows Brewster's angle as a function of the ratio of refractive indices – or the refractive index of the second medium, if the first medium has $n_1 = 1$.

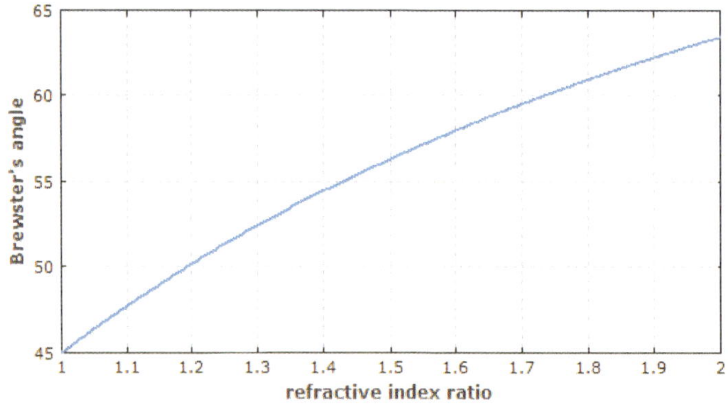

Figure: Dependence of Brewster's angle on the ratio of refractive indices of the media.

For interfaces between media with similar refractive indices, Brewster's angle is close to 45°. Reflections are anyway weak in such cases.

For unpolarized incident light at Brewster's angle, the reflected light is fully s-polarized, as there is no reflection for p-polarized light. The term "polarizing angle" comes from that fact.

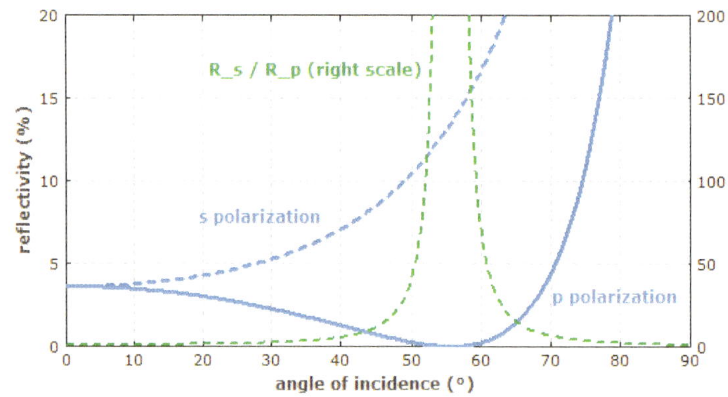

Figure: Dependence of the power reflectivity on the angle of incidence, if the light at 1064 nm wavelength impinges from air on a silica surface.

Generally, reflectivities at such surfaces can be calculated using Fresnel equations. With that, one can for example calculate the effective reflectivity for a laser beam with finite divergence as a weighted average of reflectivities for different angular components. Generally, such effective reflectivities are rather small, because most laser beams have a small divergence.

For a qualitative understanding of the vanishing reflectivity, one can consider the oscillating electric polarization in the second medium, which is perpendicular to the direction of the vanishing reflected beam. It is well known that dipoles do not emit radiation exactly in the direction of their oscillation. However, that picture is not providing a real explanation; for example, it is not clear why the direction of electric polarization in the second medium should be relevant, although the reflected beam would propagate in the first medium.

Technical Exploitation of Brewster's Angle

Optical elements in laser resonators or other optical setups are often oriented such that a laser beam propagating through them is at Brewster's angle. That way, reflection losses are avoided for p-polarized light without requiring any anti-reflection coatings. The advantage is not only that such coatings are not required, but also that the realistically achieved effective reflectivities are typically lower with that technique. However, various disadvantages can result from that approach – for example, astigmatism of focused beams, which may require additional components or design details in order to avoid detrimental effects.

It is quite common in laser technology to use optical elements at Brewster's angle.

Examples for optical elements which are often used at Brewster's angle are laser crystals, birefringent tuners, prism pairs for wavelength tuning or dispersion compensation, and Brewster windows in gas lasers. In case of prisms, this technique can of course work at the input and output face only if the angle between the two surfaces is correctly chosen. For example, for a prism made of fused silica (SiO_2) and the wavelength of 1064 nm, where the refractive index is 1.486, the ideal angle between the prism surfaces would be 67.9°. When an angle of 65° is used instead, the reflection losses for a symmetric path through the crystal are ≈0.7% in total. For 60°, that loss would already rise to 3.8%. For a glass like SF10, having a higher refractive index of ≈1.7 at 1064 nm, the ideal prism angle would be at 60.8°, and when choosing the common value of 60°, the reflection losses are still quite small (0.2% total).

In principle, one can obtain a strong and adjustable attenuation of a polarized laser beam by reflecting it e.g. on a glass surface with an angle of incidence close to Brewster's angle. The problem of that technique, however, is that the reflectivity for s-polarized light is then much higher; even if the incoming beam is quite well polarized, a small s-polarized component of it could play a dominant role in the reflected beam. As explained in the article on depolarization loss, some s-polarized components can easily be generated under certain circumstances.

Disturbing optical reflections from water surfaces can be substantially attenuated with polarizing glasses, transmitting only vertically polarized light. Although the angle of incidence of sunlight may often deviate substantially from Brewster's angle, the reflectivity for p-polarized is generally much lower than for s-polarized light.

Applications

Polarized sunglasses use the principle of Brewster's angle to reduce glare from the sun reflecting off horizontal surfaces such as water or road. In a large range of angles around Brewster's angle, the reflection of *p*-polarized light is lower than *s*-polarized light. Thus, if the sun is low in the sky, reflected light is mostly *s*-polarized. Polarizing sunglasses use a polarizing material such as Polaroid sheets to block horizontally-polarized light, preferentially blocking reflections from horizontal surfaces. The effect is strongest with smooth surfaces such as water, but reflections from roads and the ground are also reduced.

Photographers use the same principle to remove reflections from water so that they can photograph objects beneath the surface. In this case, the polarizing filter camera attachment can be rotated to be at the correct angle.

When recording a hologram, light is typically incident at Brewster's angle. Because the incident light is p-polarized, it is not back reflected from the transparent back-plane of the holographic film. This avoids unwanted interference effects in the hologram.

Photographs taken of a window with a camera polarizer filter rotated to two different angles. In the picture at left, the polarizer is aligned with the polarization angle of the window reflection. In the picture at right, the polarizer has been rotated 90° eliminating the heavily polarized reflected sunlight.

Brewster angle prisms are used in laser physics. The polarized laser light enters the prism at Brewster's angle without any reflective losses.

In surface science, Brewster angle microscopes are used in imaging layers of particles or molecules at air-liquid interfaces. By using a laser aimed at Brewster's angle to the interface, the pure liquid appears black in the image whereas molecule layers give out a reflection that can be detected and presented with a camera.

Brewster Windows

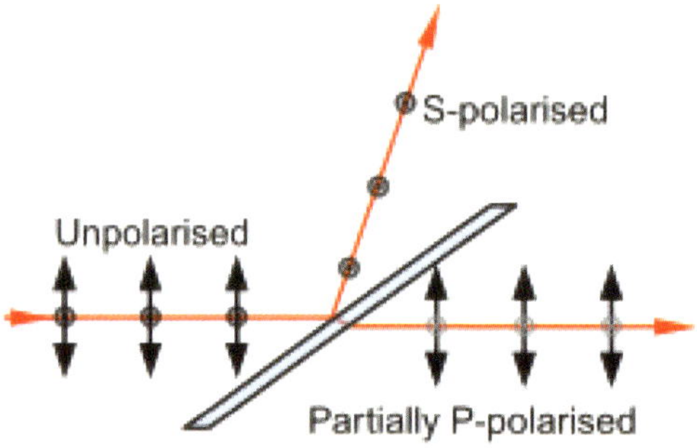

A Brewster window

Gas lasers typically use a window tilted at Brewster's angle to allow the beam to leave the laser tube. Since the window reflects some *s*-polarized light but no *p*-polarized light, the round trip loss for the *s* polarization is higher than that of the *p* polarization. This causes the laser's output to be *p* polarized due to competition between the two modes.

Pseudo-Brewster's Angle

When the reflecting surface is absorbing, reflectivity at parallel polarization (*p*) goes through a non-zero minimum at the so-called pseudo-Brewster's angle.

Birefringence

Birefringence is the property of optically non-isotropic transparent materials that the refractive index depends on the polarization direction (direction of the electric field). For example, it is observed for crystalline quartz, calcite, sapphire and ruby, also in nonlinear crystal materials like $LiNbO_3$, LBO and KTP. Figure shows that for α-quartz.

Often, birefringence results from non-cubic crystal structures. In other cases, originally isotropic materials (e.g. crystals with cubic structure and glasses) can become anisotropic due to the application of mechanical stress, or sometimes by application of a strong electric field; both can break their original symmetry. In optical fibers, birefringence can result from an elliptical shape of the fiber core, from other asymmetries of the fiber design (particularly for photonic crystal fibers), or from mechanical stress (e.g. caused by by bending). In case of polymers (plastics), birefringence can result from the ordering of molecules which is caused by an extrusion processes, for example.

The term birefringence is sometimes also used as a quantity, usually defined as the difference between extraordinary and ordinary refractive index at some optical wavelength.

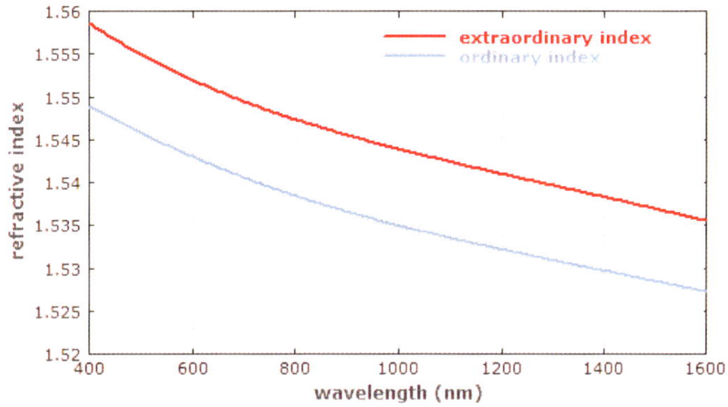

Figure: Refractive indices of α-quartz vs. wavelength. F
or this positive uniaxial material, the extraordinary index is higher.

Uniaxial and Biaxial Optical Materials

Depending on the symmetry of the crystal structure, a crystalline optical material can be uniaxial or biaxial.

The simpler case is that of uniaxial crystals (examples: calcite, quartz, sapphire, $LiNbO_3$). Those have a so-called optical axis, and the refractive index for given wavelength depends on the relative orientation of electric field director and optical axis:

- If the electric field has the direction of the optical axis, one obtains the extraordinary index n_e. This is possible only if the propagation direction (more precisely, the direction of the k vector) is perpendicular to the optical axis. For the other polarization direction, one then obtains the ordinary index n_o.
- For propagation along the optical axis, the electric field can only be perpendicular to that axis, so that one obtains the ordinary index for any polarization direction. In that situation, no birefringence is experienced.
- For an arbitrary angle θ between propagation direction and optical axis, one can find two linear polarization directions exhibiting different refractive indices. The first one is perpendicular to the k vector and the optical axis; here, we have the ordinary index n_o,

and such a wave is called an ordinary wave. The other polarization direction is perpendicular to that and to the *k* vector. The latter has a refractive index which is generally not the extraordinary index n_e, but a rather a mixture of n_e and n_o. This can be calculated with the following equation:

$$n = \frac{1}{\sqrt{\frac{\cos^2\theta}{n_o^2} + \frac{\sin^2\theta}{n_e^2}}}$$

The equation shows that for θ→o (i.e., propagation along the optical axis) we obtain n→no, and the observed birefringence vanishes: one obtains no for *any* polarization direction.

One distinguishes positive and negative uniaxial crystals; in the former case, the extraordinary index is higher than the ordinary index.

For biaxial crystals (examples: mica, $CaTiO_3$ = perovskite, LiB_3O_5 = LBO, β-BaB_2O_4 = BBO), such calculations are substantially more complicated, at least for arbitrary propagation directions. There are three mutually orthogonal principal axes associated with different refractive indices. Frequently, however, one deals with cases where the propagation direction is in one of the planes spanned by the principal axes of index ellipsoid, and in such cases the calculation is again reasonably simple. This is usually the case in calculations for phase matching of nonlinear frequency conversion processes.

For optical fibers and other waveguides, the distinction between uniaxial and biaxial does not apply, since the propagation direction is essentially determined by the waveguide.

Consequences of a Polarization-dependent Refractive Index

The polarization dependence of the refractive index can have a variety of effects, some of which are highly important in nonlinear optics and laser technology:

- When a beam is refracted at the surface of a birefringent crystal, the refraction angle depends on the polarization direction. An unpolarized light beam can then be split into two linearly polarized beams when hitting surfaces of the material with non-normal incidence (double refraction). When some object, which is illuminated with unpolarized light, is viewed through a birefringent crystal (e.g. made of calcite), two images occur which are slightly displaced.
- If a linearly polarized laser beam propagates through a birefringent medium, there are generally two polarization components with different wavenumbers. Therefore, the polarization state changes during propagation due to the change in relative phase of the two linearly polarized components. This effect can be applied, for example, in birefringent tuners, because it is wavelength-dependent (even if the difference in refractive indices is not wavelength-dependent). It can also be power-dependent (→ nonlinear polarization rotation) through self- and cross-phase modulation, e.g. in an optical fiber, and this effect is sometimes used for passive mode locking of fiber lasers.
- Similarly, the polarization state of a laser beam in a laser crystal with thermally induced

birefringence is distorted. The kind of distortion depends on the position, since the birefringent axis has a varying (e.g. always radial) orientation. This effect (combined with a polarizing optical element in the laser resonator) is the origin of depolarization loss.

- The birefringence of nonlinear crystal materials allows for birefringent phase matching of nonlinear interactions. Essentially, this means that birefringence compensates the wavelength dependence of the refractive index.
- For extraordinary waves, where the refractive index depends on the angular orientation, there is a spatial walk-off: the direction of power propagation is slightly tilted against that of the *k* vector.

Quantifying Birefringence

The magnitude of birefringence can be specified in different ways:

- For bulk optical materials, it is natural to consider the difference of refractive indices for the two polarization directions.
- For optical fibers and other waveguides, it is more appropriate to consider the difference of effective refractive indices. This is directly related to the difference in imaginary values of the propagation constants.
- Alternatively, one may specify the polarization beat length, which is 2π divided by the difference of the propagation constants. If waves with different polarization directions propagate together in the waveguide, their phase relation is restored after integer multiples of the propagation beat length.

Terminology

Much of the work involving polarization preceded the understanding of light as a transverse electromagnetic wave, and this has affected some terminology in use. Isotropic materials have symmetry in all directions and the refractive index is the same for any polarization direction. An anisotropic material is called "birefringent" because it will generally refract a single incoming ray in two directions, which we now understand correspond to the two different polarizations. This is true of either a uniaxial or biaxial material.

In a uniaxial material, one ray behaves according to the normal law of refraction (corresponding to the ordinary refractive index), so an incoming ray at normal incidence remains normal to the refracting surface. However, as explained above, the other polarization can be deviated from normal incidence, which cannot be described using the law of refraction. This thus became known as the *extraordinary ray*. The terms "ordinary" and "extraordinary" are still applied to the polarization components perpendicular to and not perpendicular to the optic axis respectively, even in cases where no double refraction is involved.

A material is termed *uniaxial* when it has a single direction of symmetry in its optical behavior, which we term the optic axis. It also happens to be the axis of symmetry of the index ellipsoid (a spheroid in this case). The index ellipsoid could still be described according to the refractive indices, n_α, n_β and n_γ, along three coordinate axes, however in this case two are equal. So if $n_\alpha = n_\beta$

corresponding to the *x* and *y* axes, then the extraordinary index is n_γ corresponding to the *z* axis, which is also called the *optic axis* in this case.

However materials in which all three refractive indices are different are termed *biaxial* and the origin of this term is more complicated and frequently misunderstood. In a uniaxial crystal, different polarization components of a beam will travel at different phase velocities, *except* for rays in the direction of what we call the optic axis. Thus the optic axis has the particular property that rays in that direction do *not* exhibit birefringence, with all polarizations in such a beam experiencing the same index of refraction. It is very different when the three principal refractive indices are all different; then an incoming ray in any of those principle directions will still encounter two different refractive indices. But it turns out that there are two special directions (at an angle to all of the 3 axes) where the refractive indices for different polarizations are again equal. For this reason, these crystals were designated as *biaxial*, with the two "axes" in this case referring to ray directions in which propagation does not experience birefringence.

Fast and Slow Rays

In a birefringent material, a wave consists of two polarization components which generally are governed by different effective refractive indices. The so-called *slow ray* is the component for which the material has the higher effective refractive index (slower phase velocity), while the *fast ray* is the one with a lower effective refractive index. When a beam is incident on such a material from air (or any material with a lower refractive index), the slow ray is thus refracted more towards the normal than the fast ray. In the figure at the top of the page, it can be seen that refracted ray with *s* polarization in the direction of the optic axis (thus the extraordinary ray) is the slow ray in this case.

Using a thin slab of that material at normal incidence, one would implement a waveplate. In this case there is essentially no spatial separation between the polarizations, however the phase of the wave in the parallel polarization (the slow ray) will be retarded with respect to the perpendicular polarization. These directions are thus known as the slow axis and fast axis of the waveplate.

Positive or Negative

Uniaxial birefringence is classified as positive when the extraordinary index of refraction n_e is greater than the ordinary index n_o. Negative birefringence means that $\Delta n = n_e - n_o$ is less than zero. In other words, the polarization of the fast (or slow) wave is perpendicular to the optic axis when the birefringence of the crystal is positive (or negative, respectively). In the case of biaxial crystals, all three of the principal axes have different refractive indices, so this designation does not apply. But for any defined ray direction one can just as well designate the fast and slow ray polarizations.

Sources of Optical Birefringence

While birefringence is usually obtained using an anisotropic crystal, it can result from an optically isotropic material in a few ways:

- *Stress birefringence* results when isotropic materials are stressed or deformed (i.e., stretched or bent) causing a loss of physical isotropy and consequently a loss of isotropy in the material's permittivity tensor.

- *Circular birefringence* in liquids where there is an enantiomeric excess in a solution containing a molecule which has stereo isomers.
- Form birefringence, whereby structure elements such as rods, having one refractive index, are suspended in a medium with a different refractive index. When the lattice spacing is much smaller than a wavelength, such a structure is described as a metamaterial.
- By the Kerr effect, whereby an applied electric field induces birefringence at optical frequencies through the effect of nonlinear optics;
- By the Faraday effect, where a magnetic field causes some materials to become *circularly birefringent* (having slightly different indices of refraction for left- and right-handed circular polarizations), making the material *optically active* until the field is removed;
- By the self or forced alignment into thin films of amphiphilic molecules such as lipids, some surfactants or liquid crystals

Common Birefringent Materials

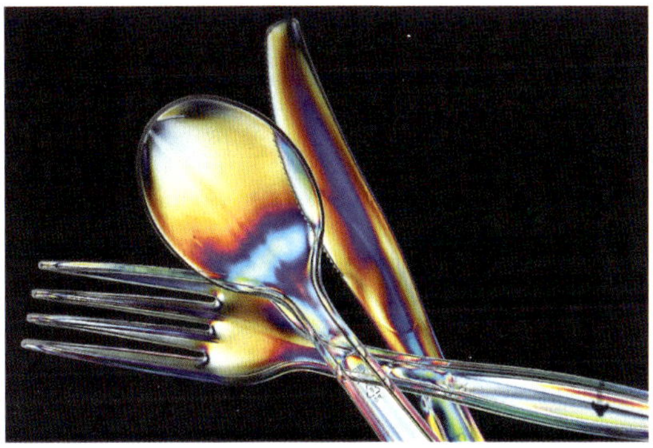

Light polarization shown on clear polystyrene cutlery between crossed polarizers

The best characterized birefringent materials are crystals. Due to their specific crystal structures their refractive indices are well defined. Depending on the symmetry of a crystal structure (as determined by one of the 32 possible crystallographic point groups), crystals in that group may be forced to be isotropic (not birefringent), to have uniaxial symmetry, or neither in which case it is a biaxial crystal. The crystal structures permitting uniaxial and biaxial birefringence are noted in the two tables, below, listing the two or three principal refractive indices (at wavelength 590 nm) of some better known crystals.

Many plastics are birefringent because their molecules are "frozen" in a stretched conformation when the plastic is molded or extruded. For example, ordinary cellophane is birefringent. Polarizers are routinely used to detect stress in plastics such as polystyrene and polycarbonate.

Cotton fiber is birefringent because of high levels of cellulosic material in the fiber's secondary cell wall.

Polarized light microscopy is commonly used in biological tissue, as many biological materials are birefringent. Collagen, found in cartilage, tendon, bone, corneas, and several other areas in the body, is birefringent and commonly studied with polarized light microscopy. Some proteins are also birefringent, exhibiting form birefringence.

Inevitable manufacturing imperfections in optical fiber leads to birefringence, which is one cause of pulse broadening in fiber-optic communications. Such imperfections can be geometrical (lack of circular symmetry), due to stress applied to the optical fiber and/or due to bending of the fiber. Birefringence is *intentionally* introduced (for instance, by making the cross-section elliptical) in order to produce polarization-maintaining optical fibers.

In addition to anisotropy in the electric polarizability (electric susceptibility), anisotropy in the magnetic polarizability (magnetic permeability) can also cause birefringence. However, at optical frequencies, values of magnetic permeability for natural materials are not measurably different from μ_0, so this is not a source of optical birefringence in practice.

Measurement

Birefringence and other polarization-based optical effects (such as optical rotation and linear or circular dichroism) can be measured by measuring the changes in the polarization of light passing through the material. These measurements are known as polarimetry. Polarized light microscopes, which contain two polarizers that are at 90° to each other on either side of the sample, are used to visualize birefringence. The addition of quarter-wave plates permit examination of circularly polarized light. Birefringence measurements have been made with phase-modulated systems for examining the transient flow behavior of fluids.

Birefringence of lipid bilayers can be measured using dual polarization interferometry. This provides a measure of the degree of order within these fluid layers and how this order is disrupted when the layer interacts with other biomolecules.

Applications

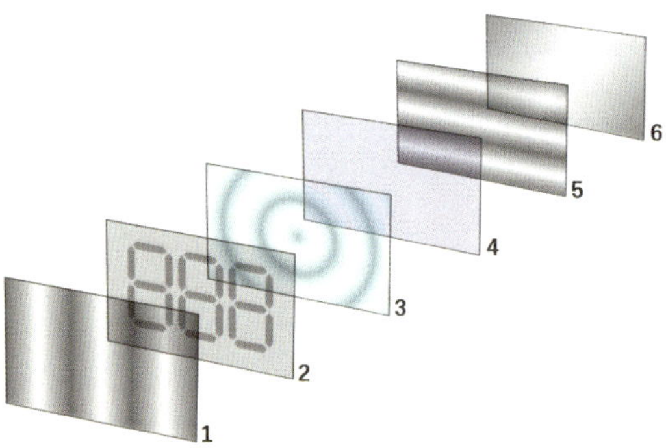

Reflective twisted-nematic liquid-crystal display. Light reflected by surface (6) (or coming from a backlight) is horizontally polarized (5) and passes through the liquid-crystal modulator (3) sandwiched in between transparent layers (2, 4) containing electrodes. Horizontally polarized light is blocked by the vertically oriented polarizer (1), except where its polarization has been rotated by the liquid crystal (3), appearing bright to the viewer.

Birefringence is used in many optical devices. Liquid-crystal displays, the most common sort of flat panel display, cause their pixels to become lighter or darker through rotation of the polarization (circular birefringence) of linearly polarized light as viewed through a sheet polarizer at the screen's surface. Similarly, light modulators modulate the intensity of light through electrically induced birefringence of polarized light followed by a polarizer. The Lyot filter is a specialized narrowband spectral filter employing the wavelength dependence of birefringence. Wave plates are thin birefringent sheets widely used in certain optical equipment for modifying the polarization state of light passing through it.

Birefringence also plays an important role in second-harmonic generation and other nonlinear optical components, as the crystals used for this purpose are almost always birefringent. By adjusting the angle of incidence, the effective refractive index of the extraordinary ray can be tuned in order to achieve phase matching, which is required for efficient operation of these devices.

Medicine

Birefringence is utilized in medical diagnostics. One powerful accessory used with optical microscopes is a pair of crossed polarizingfilters. Light from the source is polarized in the x direction after passing through the first polarizer, but above the specimen is a polarizer (a so-called *analyzer*) oriented in the y direction. Therefore, no light from the source will be accepted by the analyzer, and the field will appear dark. However areas of the sample possessing birefringence will generally couple some of the x-polarized light into the $y$polarization; these areas will then appear bright against the dark background. Modifications to this basic principle can differentiate between positive and negative birefringence.

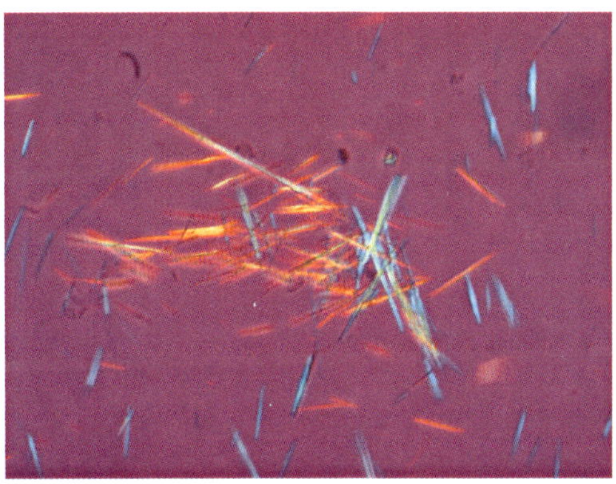

Urate crystals, with the crystals' long axis seen as horizontal in this view being parallel to that of a red compensator filter. These appear as yellow, and are thereby of negative birefringence.

For instance, needle aspiration of fluid from a gouty joint will reveal negatively birefringent monosodium urate crystals. Calcium pyrophosphate crystals, in contrast, show weak positive birefringence. Urate crystals appear yellow, and calcium pyrophosphate crystals appear blue when their long axes are aligned parallel to that of a red compensator filter, or a crystal of known birefringence is added to the sample for comparison.

Birefringence can be observed in amyloid plaques such as are found in the brains of Alzheimer's patients when stained with a dye such as Congo Red. Modified proteins such as immunoglobulin light chains abnormally accumulate between cells, forming fibrils. Multiple folds of these fibers line up and take on a beta-pleated sheet conformation. Congo red dye intercalates between the folds and, when observed under polarized light, causes birefringence.

In ophthalmology, binocular retinal birefringence screening of the *Henle fibers* (photoreceptor axons that go radially outward from the fovea) provides a reliable detection of strabismus and possibly also of anisometropic amblyopia. Furthermore, scanning laser polarimetry utilises the birefringence of the optic nerve fibre layer to indirectly quantify its thickness, which is of use in the assessment and monitoring of glaucoma.

Birefringence characteristics in sperm heads allow the selection of spermatozoa for intracytoplasmic sperm injection. Likewise, *zona imaging* uses birefringence on oocytes to select the ones with highest chances of successful pregnancy. Birefringence of particles biopsied from pulmonary nodules indicates silicosis.

Dermatologists use dermatascopes to view skin lesions. Dermatascopes use polarized light, allowing the user to view crystalline structures corresponding to dermal collagen in the skin. These structures may appear as shiny white lines or rosette shapes and are only visible under polarized dermoscopy.

Stress-induced Birefringence

Color pattern of a plastic box with "frozen in" mechanical stress placed between two crossed polarizers

Isotropic solids do not exhibit birefringence. However, when they are under mechanical stress, birefringence results. The stress can be applied externally or is "frozen in" after a birefringent plastic ware is cooled after it is manufactured using injection molding. When such a sample is placed between two crossed polarizers, colour patterns can be observed, because polarization of a light ray is rotated after passing through a birefringent material and the amount of rotation is dependent on wavelength. The experimental method called photoelasticity used for analyzing stress distribution in solids is based on the same principle. There has been recent research on using stress induced birefringence in a glass plate to generate an Optical vortex and full Poincare beams (optical beams that have every possible polarization states across its cross-section).

Other Cases of Birefringence

Birefringence is observed in anisotropic elastic materials. In these materials, the two polarizations split according to their effective refractive indices, which are also sensitive to stress.

The study of birefringence in shear waves traveling through the solid Earth (the Earth's liquid core does not support shear waves) is widely used in seismology.

Birefringence is widely used in mineralogy to identify rocks, minerals, and gemstones.

Theory

In an isotropic medium (including free space) the so-called electric displacement (D) is just proportional to the electric field (E) according to D = εE where the material's permittivity ε is just a scalar (and equal to $n^{2\varepsilon}{}_0$ where n is the index of refraction). However, in an anisotropic material exhibiting birefringence, the relationship between D and E must now be described using a tensor equation:

$$\mathbf{D} = \varepsilon \mathbf{E}$$

where ε is now a 3 × 3 permittivity tensor. We assume linearity and no magnetic permeability in the medium: $\mu = \mu_0$. The electric field of a plane wave of angular frequency ω can be written in the general form:

$$\mathbf{E} = \mathbf{E}_0 e^{i(\mathbf{k}\cdot\mathbf{r} - \omega t)}$$

where r is the position vector, t is time, and E_0 is a vector describing the electric field at r = 0, t = 0. Then we shall find the possible wave vectors k. By combining Maxwell's equations for ∇ × E and ∇ × H, we can eliminate $H \frac{1}{\mu_0}\mathbf{B}$ to obtain:

$$-\nabla \times \nabla \times \mathbf{E} = \mu_0 \frac{\partial^2}{\partial t^2}\mathbf{D}$$

With no free charges, Maxwell's equation for the divergence of D vanishes:

$$\nabla \cdot \mathbf{D} = 0$$

We can apply the vector identity $\nabla \times (\nabla \times A) = \nabla(\nabla \cdot A) - \nabla^2 A$ to the left hand side of $-\nabla \times \nabla \times \mathbf{E} = \mu_0 \frac{\partial^2}{\partial t^2}\mathbf{D}$ and use the spatial dependence in which each differentiation in x (for instance) results in multiplication by ik_x to find:

$$-\nabla \times \nabla \times \mathbf{E} = (\mathbf{k}\cdot\mathbf{E})\mathbf{k} - (\mathbf{k}\cdot\mathbf{k})\mathbf{E}$$

The right hand side of $-\nabla \times \nabla \times \mathbf{E} = \mu_0 \frac{\partial^2}{\partial t^2}\mathbf{D}$ can be expressed in terms of E through application of the permittivity tensor ε and noting that differentiation in time results in multiplication by $-i\omega$, $-\nabla \times \nabla \times \mathbf{E} = \mu_0 \frac{\partial^2}{\partial t^2}\mathbf{D}$ then becomes:

$$(-k\cdot k)E + (k\cdot E)k = -\mu_0 \omega^2 (\varepsilon E)$$

Applying the differentiation rule to eq. we find:

$$\mathbf{k}\cdot\mathbf{D} = 0$$

$\mathbf{k}\cdot\mathbf{D} = 0$ indicates that D is orthogonal to the direction of the wavevector k, even though that is no longer generally true for E as would be the case in an isotropic medium. $\mathbf{k}\cdot\mathbf{D} = 0$. will not be needed for the further steps in the following derivation.

Finding the allowed values of k for a given ω is easiest done by using Cartesian coordinates with the x, y and z axes chosen in the directions of the symmetry axes of the crystal (or simply choosing z in the direction of the optic axis of a uniaxial crystal), resulting in a diagonal matrix for the permittivity tensor ε:

$$\mathbf{å} = \varepsilon_0 \begin{bmatrix} n_x^2 & 0 & 0 \\ 0 & n_y^2 & 0 \\ 0 & 0 & n_z^2 \end{bmatrix}$$

where the diagonal values are squares of the refractive indices for polarizations along the three principal axes x, y and z. With ε in this form, and substituting in the speed of light c using $c^2 = \frac{1}{\mu_0\varepsilon_0}$, $(-k\cdot k)E+(k\cdot E)k=-\mu_0\omega^2(\varepsilon E)$ becomes

$$\left(-k_x^2-k_y^2-k_z^2\right)E_x+k_x^2E_x+k_xk_yE_y+k_xk_zE_z=-\frac{\omega^2n_x^2}{c^2}E_x$$

where E_x, E_y, E_z are the components of E (at any given position in space and time) and k_x, k_y, k_z are the components of k. Rearranging, we can write (and similarly for the y and z components of $(-k\cdot k)E+(k\cdot E)k=-\mu_0\omega^2(\varepsilon E)$.

$$\left(-k_y^2-k_z^2+\frac{\omega^2n_x^2}{c^2}\right)E_x+k_xk_yE_y+k_xk_zE_z=0$$

$$k_xk_yE_x+\left(-k_x^2-k_z^2+\frac{\omega^2n_y^2}{c^2}\right)E_y+k_yk_zE_z=0$$

$$k_xk_zE_x+k_yk_zE_y+\left(-k_x^2-k_y^2+\frac{\omega^2n_z^2}{c^2}\right)E_z=0$$

This is a set of linear equations in E_x, E_y, E_z, so it can have a nontrivial solution (that is, one other than E = 0) as long as the following determinant is zero:

$$\begin{vmatrix} \left(-k_y^2-k_z^2+\frac{\omega^2n_x^2}{c^2}\right) & k_xk_y & k_xk_z \\ k_xk_y & \left(-k_x^2-k_z^2+\frac{\omega^2n_y^2}{c^2}\right) & k_yk_z \\ k_xk_z & k_yk_z & \left(-k_x^2-k_y^2+\frac{\omega^2n_z^2}{c^2}\right) \end{vmatrix}=0$$

Evaluating the determinant of eq., and rearranging the terms, we obtain

$$\frac{\omega^4}{c^4}-\frac{\omega^2}{c^2}\left(\frac{k_x^2+k_y^2}{n_z^2}+\frac{k_x^2+k_z^2}{n_y^2}+\frac{k_y^2+k_z^2}{n_x^2}\right)+\left(\frac{k_x^2}{n_y^2n_z^2}+\frac{k_y^2}{n_x^2n_z^2}+\frac{k_z^2}{n_x^2n_y^2}\right)\left(k_x^2+k_y^2+k_z^2\right)=0$$

In the case of a uniaxial material, choosing the optic axis to be in the z direction so that $n_x = n_y = n_o$ and $n_z = n_e$, this expression can be factored into

$$\left(\frac{k_x^2}{n_o^2}+\frac{k_y^2}{n_o^2}+\frac{k_z^2}{n_o^2}-\frac{\omega^2}{c^2}\right)\left(\frac{k_x^2}{n_e^2}+\frac{k_y^2}{n_e^2}+\frac{k_z^2}{n_o^2}-\frac{\omega^2}{c^2}\right)=0$$

Setting either of the factors in $\left(\frac{k_x^2}{n_o^2}+\frac{k_y^2}{n_o^2}+\frac{k_z^2}{n_o^2}-\frac{\omega^2}{c^2}\right)\left(\frac{k_x^2}{n_e^2}+\frac{k_y^2}{n_e^2}+\frac{k_z^2}{n_o^2}-\frac{\omega^2}{c^2}\right)=0$ to zero will define an ellipsoidal surface in the space of wavevectors k that are allowed for a given ω. The first factor being zero defines a sphere; this is the solution for so-called ordinary rays, in which the effective refractive index is exactly n_o regardless of the direction of k. The second defines a spheroid symmetric about the z axis. This solution corresponds to the so-called extraordinary rays in which the effective refractive index is in between n_o and n_e, depending on the direction of k. Therefore, for any arbitrary direction of propagation (other than in the direction of the optic axis), two distinct wavevectors k are allowed corresponding to the polarizations of the ordinary and extraordinary rays.

For a biaxial material a similar but more complicated condition on the two waves can be described; the allowed k vectors in a specified direction now lie on one of two ellipsoids. By inspection one can see that

$$\begin{vmatrix} \left(-k_y^2-k_z^2+\frac{\omega^2 n_x^2}{c^2}\right) & k_x k_y & k_x k_z \\ k_x k_y & \left(-k_x^2-k_z^2+\frac{\omega^2 n_y^2}{c^2}\right) & k_y k_z \\ k_x k_z & k_y k_z & \left(-k_x^2-k_y^2+\frac{\omega^2 n_z^2}{c^2}\right) \end{vmatrix}=0$$

is generally satisfied for two positive values of ω. Or, for a specified optical frequency ω and direction normal to the wavefronts $\frac{\mathrm{k}}{|\mathrm{k}|}$, it is satisfied for two wavenumbers (or propagation constants) $|\mathrm{k}|$ (and thus effective refractive indices) corresponding to the propagation of two linear polarizations in that direction.

When those two propagation constants are equal then the effective refractive index is independent of polarization, and there is consequently no birefringence encountered by a wave traveling in that particular direction. For a uniaxial crystal, this is the optic axis, the z direction according to the above construction. But when all three refractive indices (or permittivities), n_x, n_y and n_z are distinct, it can be shown that there are exactly two such directions (where the two ellipsoids intersect); these directions are not at all obvious and do not lie along any of the three principal axes (x, y, z according to the above convention). Historically that accounts for the use of the term "biaxial" for such crystals, as the existence of exactly two such special directions (considered "axes") was discovered well before polarization and birefringence were understood physically. However these two special directions are generally not of particular interest; biaxial crystals are rather specified by their three refractive indices corresponding to the three axes of symmetry.

A general state of polarization launched into the medium can always be decomposed into two waves, one in each of those two polarizations, which will then propagate with different wavenumbers $|k|$. Applying the different phase of propagation to those two waves over a specified propagation distance will result in a generally *different* net polarization state at that point; this is the principle of the waveplate for instance. However, when you have a wave launched into a birefringent material at non-normal incidence, the problem is yet more complicated since the two polarization components will now not only have distinct wavenumbers but the k vectors will not even be in exactly the same direction. In this case the two k vectors are rather solutions of constrained by the boundary condition which requires that the components of the two transmitted waves' k vectors, and the k vector of the incident wave, as projected onto the surface of the interface, must all be identical.

Examples of Birefringent Materials

Birefringence was first described in calcite crystals by the Danish scientist Rasmus Bartholin in 1669. Since then, many birefringent crystals have been discovered.

Silicon carbide, also known as Moissanite, is strongly birefringent.

Many plastics are birefringent because their molecules are ‹frozen' in a stretched conformation when the plastic is molded or extruded. For example, cellophane is a cheap birefringent material.

Cotton (*Gossypium hirsutum*) fiber is birefringent because of high levels of cellulosic material in the fiber's secondary cell wall.

Slight imperfections in optical fibers can cause birefringence, which can lead to distortion in fiber-optic communication.

Birefringence can also arise in magnetic (not dielectric) materials, but substantial variations in magnetic permeability of materials are rare at optical frequencies.

Birefringence can be observed in amyloid plaque deposits, such as are found in the brains of Alzheimer's victims. Modified proteins such as immunoglobulin light chains abnormally accumulate between cells, forming fibrils. Multiple folds of these fibers line up and take on a beta-pleated sheet conformation. Congo red dye intercalates between the folds and, when observed under polarized light, causes birefringence.

Measuring Birefringence by Polarimetry

Birefringence and related optical effects (such as optical rotation and linear or circular dichroism) can be measured by measuring changes in the polarization of light passing through the material. These measurements are known as polarimetry.

A common feature of optical microscopes is a pair of crossed polarizing filters. Between the crossed polarizers, a birefringent sample will appear bright against a dark (isotropic) background.

References

- Shipman, James; Wilson, Jerry D.; Higgins, Charles A. (2015). An Introduction to Physical Science, 14th Ed. Cengage Learning. p. 187. ISBN 1305544676
- What-is-polarization, techpubs: generalphotonics.com, Retrieved 29 March 2018
- Le Tiec, A.; Novak, J. (July 2016). "Theory of Gravitational Waves". arXiv:1607.04202 [gr-qc]. doi:10.1142/9789813141766_0001
- Fosalba, Pablo; Lazarian, Alex; Prunet, Simon; Tauber, Jan A. (2002). "Statistical Properties of Galactic Starlight Polarization". Astrophysical Journal. 564 (2): 722. arXiv:astro-ph/0105023. Bibcode:2002ApJ...564..762F. doi:10.1086/324297
- Polarizer-selection-guide, optics, application-notes: edmundoptics.com, Retrieved 08 July 2018
- Bekefi, George; Barrett, Alan (1977). Electromagnetic Vibrations, Waves, and Radiation. USA: MIT Press. ISBN 0-262-52047-8
- Dorn, R.; Quabis, S. & Leuchs, G. (Dec 2003). "Sharper Focus for a Radially Polarized Light Beam". Physical Review Letters. 91 (23): 233901. Bibcode:2003PhRvL..91w3901D. doi:10.1103/PhysRevLett.91.233901. PMID 14683185
- Polarizer-selection-guide, optics, application-notes: edmundoptics.com, Retrieved 18 July 2018
- Serkowski, K.; Mathewson and Ford (1975). "Wavelength dependence of interstellar polarization and ratio of total to selective extinction". Astrophysical Journal. 196: 261. Bibcode:1975ApJ...196..261S. doi:10.1086/153410
- Horváth, Gábor; Dezsö Varjú (2003). Polarized Light in Animal Vision: Polarization Patterns in Nature. Springer. pp. 100–103. ISBN 3-540-40457-0
- Birefringence, calculation-of-birefringence: newworldencyclopedia.org, Retrieved 28 March 2018

Permissions

All chapters in this book are published with permission under the Creative Commons Attribution Share Alike License or equivalent. Every chapter published in this book has been scrutinized by our experts. Their significance has been extensively debated. The topics covered herein carry significant information for a comprehensive understanding. They may even be implemented as practical applications or may be referred to as a beginning point for further studies.

We would like to thank the editorial team for lending their expertise to make the book truly unique. They have played a crucial role in the development of this book. Without their invaluable contributions this book wouldn't have been possible. They have made vital efforts to compile up to date information on the varied aspects of this subject to make this book a valuable addition to the collection of many professionals and students.

This book was conceptualized with the vision of imparting up-to-date and integrated information in this field. To ensure the same, a matchless editorial board was set up. Every individual on the board went through rigorous rounds of assessment to prove their worth. After which they invested a large part of their time researching and compiling the most relevant data for our readers.

The editorial board has been involved in producing this book since its inception. They have spent rigorous hours researching and exploring the diverse topics which have resulted in the successful publishing of this book. They have passed on their knowledge of decades through this book. To expedite this challenging task, the publisher supported the team at every step. A small team of assistant editors was also appointed to further simplify the editing procedure and attain best results for the readers.

Apart from the editorial board, the designing team has also invested a significant amount of their time in understanding the subject and creating the most relevant covers. They scrutinized every image to scout for the most suitable representation of the subject and create an appropriate cover for the book.

The publishing team has been an ardent support to the editorial, designing and production team. Their endless efforts to recruit the best for this project, has resulted in the accomplishment of this book. They are a veteran in the field of academics and their pool of knowledge is as vast as their experience in printing. Their expertise and guidance has proved useful at every step. Their uncompromising quality standards have made this book an exceptional effort. Their encouragement from time to time has been an inspiration for everyone.

The publisher and the editorial board hope that this book will prove to be a valuable piece of knowledge for students, practitioners and scholars across the globe.

Index

P

R

S

T

U

X

Y